BEYOND THE NUMBERS

Student-Centered Activities for Learning Statistical Reasoning

Ninth Edition | William Rayens

Beyond the Numbers

A Student-Centered Approach for Learning Statistical Reasoning
9th Edition
William Rayens

Printed in the United States of America
10 9 8 7 6 5 4 3
ISBN: 978-1-61740-814-4

Van-Griner Publishing
Cincinnati, Ohio
www.van-griner.com

President: Dreis Van Landuyt
Project Manager: Brenda Schwieterman
Customer Care Lead: Lauren Wendel

Rayens 814-4 Su21
322596-APTF23
Copyright © 2022

Table of Contents

CHAPTER 3

Statistical Experiments and the Problem of Confounding 45

Chapter 3 Exercises 53

CHAPTER 4

Sampling: Purpose and Challenges 71

CHAPTER 7
Sampling: When Probability Isn't Enough 141

CHAPTER 8
The Language of Decision Making 157

CHAPTER 9
Hypothesis Testing: Concepts and Consumption..

CHAPTER 10
Hypothesis Testing: Computations.....223

CHAPTER 11
Hypothesis Testing: Importance of Clinical Significance 251

CHAPTER 12
More than One Variable: Association and Correlation 273

Beyond the Numbers
A Student-Centered Approach to Learning Statistical Reasoning

Overview of the Book

Dr. William Rayens has taught statistical reasoning at the University of Kentucky (U.K.) for about thirty-five years. This book addresses the needs of the statistical reasoning community, as well as the ever-growing universal need for statistically-literate citizens. There are several conceptually-focused statistical reasoning books on the market, but *Beyond the Numbers: A Student-Centered Approach to Learning Statistical Reasoning* is distinct in two noteworthy ways:

1. **This text is unambiguously focused on statistical reasoning.** Our goal is to help students competently consume statistical ideas that meet them where they live—both academically and personally. If you measure worth by the number of computations performed within this text, then it might seem light. If you measure worth based on the weight of the statistical reasoning that the chosen methods surface, then the book is heavy, indeed. Computations and exposure to different methods are still valued since they surely increase students' life skills. But we have been careful not to allow students to be distracted by unnecessary computations since it has been our experience that they often find those easier to grasp than the more important understanding of what is gained from such computations.

2. **This text is designed to shift the primary responsibility of learning to the student.** Borrowing the words of STEM (science, technology, engineering, and mathematics) educator Robert Talbert, the goal of this shift is to change the mindset of the student from a "renter" to an "owner." In a classroom of renters, fees are paid and the management is expected to deliver. In a classroom of owners, however, students realize that it is their responsibility to engage, absorb, and retain. The instructor's difficult, but important, job is to prepare an environment in which that can happen. This book, mostly through how its exercises are structured, is designed to help facilitate students' transition from renters to owners. While the content is carefully constructed to prepare students to do the accompanying exercises, they will be asked repeatedly to read, comprehend, and assimilate in order to complete them.

How This Book Has Changed and Why

Beyond the Numbers: A Student-Centered Approach to Learning Statistical Reasoning has undergone several important revisions over the years. The change in layout with the 8th edition (including a slight change in the name) is the most drastic. These changes are the result of feedback from colleagues and Dr. Rayens' own experiences after nearly ten years using the book in slightly different forms along the way. We are convinced that what we have now will be easier for both the instructors and the students. In order to understand where we are now and why we are here, it is useful to take a broad view of where we have been. While Dr. Rayens doesn't presume to speak for all his colleagues, and certainly not for others who have used this book, the comments that are made below reflect a fairly wide set of experiences presented with fully open eyes and a good sense of humor at the importance of being adaptable.

	Characterized By	Reality on the Ground
Period of Early Naiveté	■ No on-board content. ■ Flipped classroom. ■ All exercises required open response. ■ Material presented in non-traditional module format. ■ All student work was mapped to modules but appeared as a set of activities and not as exercises.	■ This was the purest teaching that most of us had ever done. ■ It was a lot of work for the instructors. ■ Students in large introductory classes struggle with the focus and independence required by this challenging format. ■ Instructors found they had to make a lot of the connections for the students.
Period of Increasing Brain Size	■ Added on-board content. ■ Flipped classroom replaced with what we really intended: a student-centered classroom. ■ All exercises still required open response ■ Material presented in non-traditional module format. ■ All student work was mapped to modules but appeared as a set of activities and not as exercises.	■ On-board content helped a great deal with the students knowing where to go. Appealed more to their chapter-focused experiences in countless other courses. ■ Grading open response was a lot of work for instructors. In addition, many students struggled with writing coherent responses in the foreign language of inference. ■ Some instructors began changing the open response into well-formed multiple choice. ■ Students still struggled with the format of the activities and how to map those to the relevant content.
Period of Light	■ Material organized into chapters. ■ Activities were changed to so-named exercises following each chapter. ■ Material is no longer laid out as modules, but rather in a format that appears more traditional. ■ Exercises are now a combination of open response and multiple choice.	■ Expect this format to be much easier for students and instructors. ■ Instructors can require the students to provide detailed explanations for their multiple choice answers. But they also have the option of just keeping them multiple choice which will make grading in large lecture classrooms realistic again. ■ Significantly more material has been added so that instructors now have more to choose from as they map out a semester plan. ■ All homework questions—multiple choice and open response—are available in rtf format for quick uploading (e.g., with Respondus) to standard course management systems such as Blackboard or Canvas.

Exercises Old and New

If you had the opportunity to use one of the earlier editions of this book, then you will be pleased to know that nearly all of the activities have remained the same. A few have been absorbed into the chapter material, but most are simply positioned in the format of exercises following the appropriate content.

There have been quite a few new exercises added. All of these are in the same student-centered format that has always characterized this book. Every chapter has at least two project activities as well, and two different capstone projects emerge one step at a time as different chapters are covered.

Relevancy Still a Priority

If you are going to ask students to be actively involved in their own learning, it is important you *engage* them. With examples that are relevant to their day-to-day encounters, students are more likely to relate to the material. Once the connection is made, students tend to be drawn into the content and consequently become more involved in their own learning. To help make these critical connections, we have been careful in our choice of exercises. Throughout the book we address current topics that are not only illustrative in content, but are also relatable to the student. Some examples include the following:

- A February 2014 study on gay marriage to facilitate the construction and interpretation of a confidence interval, when the margin of error is given.
- A 2014 Gallup Organization–Purdue University report on the effects of one's college choice to introduce the idea of using confidence intervals to make hypothesis-testing like decisions.
- A study on the efficacy of social networking systems as instructional tools to demonstrate the empirical rule.

Another key to engagement is the presentation of the material. We have deliberately chosen a chapter design that is more open, friendly, and less formal than other textbooks that cover similar material. Don't be fooled; the material is there. We have opted for a format that is more student-appealing.

Software as Part of a College Education

All students, from liberal arts majors to engineers, need to know how to use basic software suites to manipulate numbers, perform calculations, and create graphs. Sobered by how few students learned these basic skills as they progressed through college, we have long endeavored to integrate them naturally into the material in this book. While we encourage our students to use any software, even online applets, we illustrate some graphical and summary computation throughout with Microsoft Excel. The specific software doesn't really matter, but students should probably have some basic exposure before graduating.

About the Author

Dr. William Rayens is Professor and the Dr. Bing Zhang Endowed Department Chair in the Dr. Bing Zhang Department of Statistics at the University of Kentucky. Rayens has an extensive research record focused primarily on the development of multivariate and multi-way statistical methodologies mostly related to problems in chemistry and the neurosciences. He has mentored several Ph.D. students and has been honored at both the College and the University level as an outstanding teacher. Rayens also served as Assistant Provost for General Education during which time he was tasked with implementing new general education reforms at the University of Kentucky, the first changes to that program in almost 30 years. Rayens created STA 210: Introduction to Statistical Reasoning in 2010, and designed the one-of-a-kind Technologically Enhanced Active Learning rooms in the Jacobs Science Building where he and his instructors are genuinely privileged to teach STA 210 and use this book.

1

CHAPTER 1
Number Sense: Basic Numeracy

Introduction

To a statistician, the word "inference" refers to a prominent subfield of statistical science responsible for creating the mathematical tools that make possible many of the statistical arguments that we encounter each day. Those tools can be pretty complicated. However, the one thing they all have in common is that they are designed to facilitate our reasoning *from* data *to* something we can't know with certainty.

This reasoning process—from the known to the unknown—is more commonly what the word inference is used to refer to and not to the set of sophisticated mathematical tools that statisticians have developed to facilitate that process in a particular way.

For example, when a defendant is on trial, the jury is going to have to infer from the evidence presented whether the defendant is guilty or not. A politician trying to decide if she should run for office can only know what a sample of polled voters thinks of her. She's left to infer from the results of that sample what the entire population of voters thinks.

This chapter is not concerned with the formal tools of statistical inference. Rather, our immediate goal is to explore the challenges of what one might call "human inference."

Human Inference

We will use the phrase "human inference" to describe the reflexive inferences we make when we consume simple statistical constructs, like charts, graphs, and numerical summaries. Human inference is how we naturally mobilize that sort of information. We react with our own interpretation of "this is what these data seem to be telling me."

Let's look at some examples:

In August of 2007, the Associated Press ran an article that said "'Harry Potter and the Deathly Hallows' sells 8.3 million in first 24 hours … more than 50,000 a minute." One obvious statistical construct that emerges is the rate "50,000 a minute." Our immediate human inference might be the conclusion that this book was incredibly popular.

This is simple enough, and a legitimate inference, if the rate actually turns out to make sense. We will revisit Mr. Potter later on.

Charts also prompt us to form reflexive inferences. For instance, the graph shown here uses a decorative bar chart to display the decrease in the purchasing power in the Canadian dollar between 1980 and 2000. We are likely to infer that the decrease has been enormous. Exactly how much is an easy question to answer (54%) if we take the time to read the numbers on the chart. However, if we only look at the size of the icons used in the chart, the decrease looks much more dramatic (perhaps on the order of 75%). That would definitely be the wrong human inference to make.

1980	1985	1990	1995	2000
$1.00	$0.70	$0.56	$0.50	$0.46

Purchasing power of the Canadian dollar, 1980 to 2000

Finally, look at the bar chart below. It records the performance of students on a freshman biology exam, broken down by whether the students were breakfast eaters or not. If you look carefully at the graph, you will see that students who ate breakfast tended to have better grades on the exam. Hence, we might be tempted to infer from this graph that eating breakfast is important to good exam performance in college. The question is whether this is a fair inference to make from the information we are given. While the more formal trappings of statistical inference are embodied in so-called confidence intervals and statistical hypothesis testing, the point for now is much simpler: when we see a rate, a graph, or an argument based on statistics, we reflexively form inferences as we absorb and assimilate the information being conveyed. We want to form these inferences correctly.

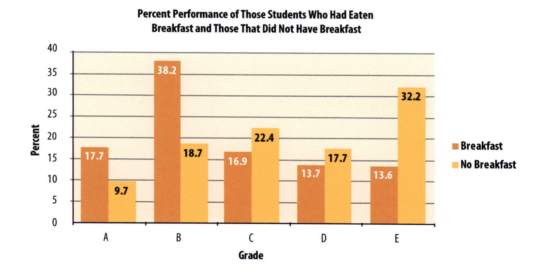

Percent Performance of Those Students Who Had Eaten Breakfast and Those That Did Not Have Breakfast

Decimal Points and Deceptive Charts

The validity of our reflexive human inferences can be compromised in lots of ways. The goal of this chapter is to learn how to recognize and avoid some of the more common pitfalls. Some of what we'll see is pretty basic, while other encounters will be more subtle. Numeracy is not something you can really teach by listing a set of rules required to achieve it! Rather, our goal is to learn by example some ways to practice effective, basic numeracy. In this chapter, we will look at some examples of two very basic numeracy issues:

- Decimal points that were overlooked or in the wrong place
- Deceptive graphs

Decimal Points

Decimal points? Really? We admit this doesn't sound like a college-level topic. But it turns out that decimal points are notorious for causing glitches in human inferences, perhaps even as far back as Plato. It has been conjectured that the whole myth of Atlantis can be traced to his misinterpreting a particular hieroglyphic for east-west distance in which he may have confused the number 100 for 1,000. If he had recorded a particular distance as 100, then his records would have pointed directly to Crete as the location of the civilization lost to a "great cataclysm." That's a nice theory at least!

More troubling is the claim made on BlackWomensHealth.com that over 4 million women in the U.S. are battered to death by a spouse or boyfriend each year. Does that seem plausible to you? Other sources, such as the Feminist Majority Foundation, suggest this gruesome statistic is closer to 4 thousand. Could this discrepancy be due to a decimal point error? We will have more to say about this in a later chapter.

Customers of Verizon once coined a word—Verizonmath—to describe some numeracy problems they were having with customer service representatives of that company. Most of the issues they had seem to have revolved around data rates. For example, one customer was quoted a rate of 0.015 cents per kilobyte as his data rate. That first month he used 3,000 kilobytes and expected a bill for 0.015 × 3,000 cents, or about 45 cents. Instead, his data charge was $45, and he was not happy.

What do you think happened? It turns out that his customer service representative meant to say ".015 *dollars* per kilobyte" or "1.5 cents per kilobyte." Customers were even more frustrated when mid-level management insisted that there was nothing different about those two statements! These were very basic mistakes involving units (arguably decimal points), each of which created a 100-fold difference in the bills customers received.

Cheap data, outrageous partner abuse rates, and even possibly the location of Atlantis. All human inferences that may have been misdirected by misplaced decimal points!

Deceptive Graphs

We saw above that a poorly designed bar graph can be deceptive. Surprisingly, versions of those kinds of graphs appear even in highly reputable places. For example, consider the 2012 graphic from the *New York Times Magazine* shown below. If our eyes are disciplined enough to just focus on the vertical axis, which is where we should be focusing, then our inferences from this graph will be correct. Indeed, about 72% of high school seniors in 1980 had consumed alcohol, and that had dropped to about 40% in 2011.

The problem is that the decorative graphics used to make the plot are also allowed to vary in width, giving us a jug for the seniors of 1980 capable of holding about 128 ounces of liquid and a glass for seniors of 2011 that would hold about 8 ounces. Hence, the pictures themselves suggest a 16-fold difference between 2011 and 1980, when in fact the difference is less than 2-fold.

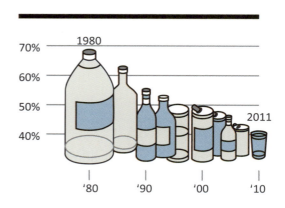

Alcohol use by high-school seniors over time.

This is not illegal and certainly not unattractive. However, it might be a challenge to the immediate inference we form when we see the huge jug for those wild-and-crazy seniors in 1980 and that little glass for those straight-and-narrow seniors in 2011.

Another example of a deceptive graph is shown below. A version of this graph was presented on Fox News in 2012. We see a steady increase in gas prices with a fairly steep incline at the beginning of the graph. Our natural inference is that gas prices have been consistently trending upwards.

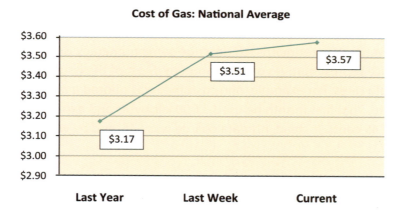

Source: AAA Fuel Gauge Report

It turns out those data were taken from an AAA chart, a version of which you see below, but the data were picked in such a way that the unfortunate outcome was three points—see the blue dots superimposed on the chart—that formed a steadily increasing graph and left out all the downward activity that happened between February 2011 and January 2012.

It is also worth noting that in the Fox chart the horizontal axis gives equal spacing to time intervals that are very different in size (approximately 11 months versus 1 week). This has the effect of making the jump from "Last Year" to "Last Week" appear steeper than it actually was.

These are not deep ideas or difficult issues. But as long as deceptive graphs continue to appear in our media either by accident, ignorance, or design, there will be an on-going need to make sure we stay alert so that our common-sense inferences are correctly formed.

12 Month Average for Self-Serve Regular

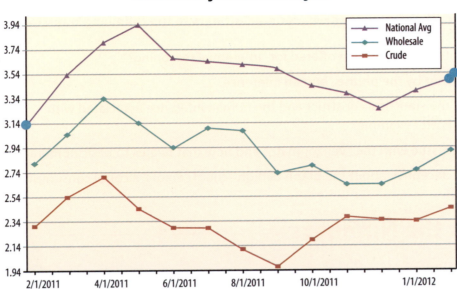

TAKE-HOME POINT

Remember that basic numeracy is a critical step toward the goal of correctly forming human inferences from statistical constructs.

Chapter 1 Exercises

Reading Check

Short Answer: Please provide brief, concise answers to the following questions.

1. What was the first definition offered for "inferential reasoning?"

2. What is meant by the phrase "human inference?"

3. What was the statistical construct and the human inference associated with the Harry Potter example?

4. Explain the issue with the "Verizonmath" and how that relates to basic numeracy.

5. What was deceptive about the graph of high school seniors' alcohol consumption over time?

Beyond the Numbers

1. Cocaine and Kids

Read the *Times* excerpt below.

Title: Cocaine Floods the Playground: Use of the Addictive Drug by Children Doubles in a Year

Authors: Richard Ford and Sean O'Neill

Source: Last accessed 2021: *Times*, March 25, 2006, https://www.thetimes.co.uk/article/cocaine-floods-the-playground-k2659spkxtd

Cocaine use among children has doubled in a year as the fashionable drug of the middle classes extends its reach from the dinner party to the playground. Hundreds of thousands of 11- to 15-year-olds are being offered the Class A drug, which is flooding into the country, according to government figures published yesterday.

An estimated total of 65,000—one in 50 children aged 11–15—said they had taken cocaine, which is known by euphemisms such as "zip" and "tickets" by youngsters who are increasingly experimenting with the drug.

…

It is more available and socially acceptable. Once cocaine use reached acceptability among the 20- and 30-year-olds, it was only a matter of time before it started to become fashionable among youngsters.

…

Figures published yesterday showed that cocaine use among 11- to 15-year-olds doubled from 1% to 2% between 2004 and 2005. The official study questioned more than 9,000 pupils in 305 schools in England at the start of the autumn term last year. There are an estimated 3.2 million pupils aged 11–15. Based on survey findings, an estimated 65,000 pupils took cocaine last year.

A total of 320,000 were offered the drug, including almost 100,000 15-year-olds and 35,000 11-year-olds.

Overall, 11% of 11- to 15-year-olds had taken an illegal drug in the month before they were questioned.

TABLE 1.1 Proportions of Pupils Who Took Individual Drugs in the Last Year: 2001 to 2006[1,2,3]

Type of Drug	Percentages					
	2001	2002	2003	2004	2005	2006
Cannabis	13.4	13.2	13.3	11.3	11.7	10.1
Any Stimulants	5.6	6.2	6.1	5.4	6.2	6.2
Cocaine	1.2	1.3	1.3	1.4	1.9	1.6
Crack	1.1	1.0	1.2	1.1	1.0	0.8
Ecstasy	1.6	1.5	1.4	1.4	1.5	1.6
Amphetamines[4]	1.1	1.2	1.2	1.3	1.2	1.2
Poppers	3.4	4.3	4.0	3.4	3.9	4.2

a. The word "doubles" is used in the subheading to describe the increase in cocaine use among children. Look carefully at Table 1.1. Where does the word "doubles" come from?

i. The article says it doubled from 1% to 2% between 2004 and 2005. But the entry for 2004 is 1.4 and 1.9 for 2005. Good guess that the authors just rounded the 1.4 to 1 and the 1.9 to 2.

ii. This was just an off-hand way of saying it "increased a lot."

iii. The rate for crack was 0.8 in 2006, but 1.60 for cocaine. 1.6 is double 0.8 so that is where the use of the word "doubles" came from.

iv. All stimulants were at a rate of about 5.6 in 2001, but cannabis use in 2004 was at a rate of 11.3. 11.3 is essentially double 5.6, and this is where the use of the word "doubles" came from.

---

1   Source: Beverley Bates, et al., "Smoking, Drinking, and Drug Use among Young People in England in 2006." Edited by Elizabeth Fuller. Copyright © 2011, Re-used with the permission of The Health and Social Care Information Centre. All rights reserved. http://www.hscic.gov.uk/catalogue/PUB11334/smok-drin-drug-youn-peop-eng-2012-repo.pdf. Only a portion of Table 4.6C is recreated here.

2   Estimates from 1998 to 2000 not shown, as they are not comparable with estimates from 2001 onwards because of the changes in the way that drug use was measured.

3   Estimates are shown to one decimal place because of generally low prevalence rates.

4   Surveys from 2004 onward asked about "speed and other amphetamines."

b. Why might your human inference regarding the cocaine problem among children be misled by the headline on the article? How might you revise it?

2. Was Kobe Streaking?

Kobe Bryant's field goal shots for the first game of the 2001 Western Conference finals between the Los Angeles Lakers and the San Antonio Spurs are as follows:

O O O X X X X X X O O X O O O (halftime)
X X O X O O X X O X X O O O X O X O X X O
(O = missed shot, X = hit shot)

The announcer's first-half comment included sentiments such as "What an awful start" after Bryant missed three in a row, "The basket is looking mighty big right now" after he hit five in a row, and "This is an unstoppable run" after he hit the seventh in a row (and coincidentally, right before the run did, in fact, stop).

a. Looking at Kobe's first game field goals, do you think one can argue he got a "hot hand" starting on his fourth field goal attempt and first basket? Why or why not?

b. For our purposes, define a "run" as two or more shots that turned out the same way. That is, two or more hits in a row, or two or more misses in a row, can be considered a run. How many runs were there in Kobe's first game per the results above? Statisticians often use the number of runs (defined a little differently than we have here) to formally assess randomness.

3. You Are So Random. Or Maybe Not.

This is an in-class group activity. Follow your instructor's direction on what to do and how. Get into groups as directed by your instructor. Every person in a group will have the same answers to this assignment, but for record keeping, you may be instructed to have every person fill out her own sheet with those common answers.

Your instructor will facilitate the following sequence of events but must be out of the room while they are taking place.

- Every group of students will be assigned at random to one of two tasks.
- One set of groups (Task 1) will each be given a die and asked to roll it 20 times. If your group is assigned Task 1, then you record the sequence of odds and evens in Table 1.2.
- The other set of groups (Task 2) will each be asked to make up what they think would be a convincing random sequence of odd and even outcomes. If your group is assigned Task 2, then you record those answers in Table 1.2.
- Do not indicate anywhere on this sheet what task your group was assigned.
- Your instructor will return to the room and proceed to guess with amazing accuracy which task was assigned to which group.

TABLE 1.2 Roll Number …(Each of these 20 squares should have either an "O" or an "E" entered.)

1	2	3	4	5	6	7	8	9	10	11	12	13	14	15	16	17	18	19	20

a. After experiencing this demonstration, what have you learned about the concept of randomness?

4. Cutting the Corpus Callosum

Answer the following questions while watching the "Split Brain" video at http://www.youtube.com/watch?v=ZMLzP1VCANo (video last accessed 2021).

a. When Joe was shown the pan in his left visual field, what did he say he saw?

   i. He said he saw a hammer.

   ii. He said he didn't see anything.

   iii. He said he saw a saw.

   iv. He said he saw a pan.

b. When Joe was shown the saw in his left visual field and a hammer in his right visual field, what did he say he saw?

   i. He said he saw a hammer.

   ii. He said he didn't see anything.

   iii. He said he saw a saw.

   iv. He said he saw a pan.

c. When Joe was shown the saw in his left visual field and the hammer in his right, what was he able to draw with his left hand?

   i. He was able to draw a hammer.

   ii. He was not able to draw anything.

   iii. He was able to draw a saw.

   iv. He was able to draw a pan.

5. Have I Got a Story for You!

Title: The Split Brain Revisited

Author: Michael S. Gazzaniga

Source: *Scientific American* 279, no. 1 (1998): 50–55

About the Study: The author, psychology professor Dr. Michael Gazzaniga, describes part of the split brain study he conducted. On page 53 he writes:

> My colleagues and I studied this phenomenon by testing the narrative ability of the left hemisphere. Each hemisphere was shown four small pictures, one of which related to a larger picture also presented to that hemisphere. The patient had to choose the most appropriate small picture. [T]he right hemisphere—that is, the left hand—correctly picked the shovel for the snowstorm; the right hand, controlled by the left hemisphere, correctly picked the chicken to go with the bird's foot. Then we asked the patient why the left hand—or right hemisphere—was pointing to the shovel. Because only the left hemisphere retains the ability to talk, it answered. But because it could not know why the right hemisphere was doing what it was doing, it made up a story about what it could see—namely, the chicken. It said the right hemisphere chose the shovel to clean out a chicken shed.

a. In the study described above, was the snowstorm shown in the left visual field or right visual field? Why?

b. The real import of this activity is contained neatly in the last two sentences of the Gazzaniga excerpt. Explain why the results Gazzaniga cites above are evidence that we are hardwired toward slippery thinking.

c. Explain why the results Gazzaniga cites above are evidence that human inference is potentially a far more complex task than many realize.

Projects

1. Very Lucky Project

Introduction

Please listen to a fascinating radio program that describes some unusual events: "A Very Lucky Wind" is available at Radiolab (https://www.wnycstudios.org/podcasts/radiolab/segments/91686-a-very-lucky-wind). Audio was last accessed in 2021. If you can't find it at that link, search for it. We want to focus on two main themes from the audio:

- The "blade of grass" way of thinking about probabilities
- The "background" piece about how we tend to focus on confirmatory information and leave out other (non-confirmatory) information

The Assignment

After listening to the radio program, describe a situation from your own experiences where something happened that was simply too unusual for you to believe it happened by accident. Take a second to look at that situation based on what you learned from this audio piece. Make sure your comments are confined to the context of one or both of the two broad themes bulleted above. Comment on whether you still view your story as unusual. Your paper should have three paragraphs, and each paragraph should be three or four meaningful sentences *at least.*

Paragraph 1—A concise summary of what the audio is saying about human inferences and slippery evidence. Include a minimum of two examples from the audio.

Paragraph 2—A description of something that happened to you that seemed very unusual.

Paragraph 3—An examination of your experience in light of what you learned from the audio and in the context of one or both of the two themes bulleted above. Make sure you connect your story to one of those two themes. If you are unable to connect to one of the two themes, you may need to reconsider your choice for Paragraph 2.

2. Uncovering Your Own Slippery Evidence

Introduction

The purpose of this assignment is to find an article or story in the news (does not have to be current) that would be a good example of the kind of challenges to human inference that were addressed in this chapter.

The Assignment

Read

You will need to read an article in the news that ultimately allows you to exhibit command of one or more of the topics discussed in this chapter. Your instructor may provide a set of articles for you to choose from, or you may be asked to find the article yourself. In any case, it is important that the article clearly and unambiguously maps to one or more of the topics discussed. For example, you may find an article wherein the wrong inference is made because a number was calculated incorrectly or a graph is inappropriately interpreted. The evidence has to be real and concrete. Don't say "I think this number is wrong because it is just unreasonable." That would not be adequate.

Write

Write two paragraphs summarizing your findings:

Paragraph 1—Provide the source of your article and summarize what your article is about, being careful to reproduce the statistics, tables, and graphs on which you are going to comment.

Paragraph 2—Give a detailed explanation of how this article is an example of one of the topics covered to this point in your class. Make sure you clearly describe what is wrong with the statistical construct you have chosen to discuss. This is an important paragraph and should be at minimum five sentences.

2 CHAPTER 2
Number Sense: Basic Computational Skills and Benchmarks

Introduction

Chapter 1 reminded us that the integrity of the informal inferences we make from statistical constructs is in large part dependent on our numeracy. How can we practice good numeracy? There's no simple answer to such a broad question, but good practical advice would include the following:

- Be ready to look more closely at claims that might seem too good to be true. If it is possible to check simple calculations being presented, then do so.

- Be able to decide if the use of percentages or rates is best, when totals might have been less confusing or deceptive. Or vice-versa.

- Be sure to have a working knowledge of common benchmarks, like how many people there are in your state, how many babies are born each year in the U.S., etc.

- Be sure you have a basic grasp of simple statistical computations such as mean, median, and standard deviation and understand simple statistical displays such as histograms.

This chapter is an opportunity to spend a little time with this advice.

Check Calculations

Don't forget to pay close attention to the statistics people are quoting and the calculations you might be doing. In some cases the numbers may not make any sense.

For example, there was once an article in the respected journal, *Science,* that mentioned a California field that had produced over 750,000 melons per acre. If you just do the math, you'll see that amounts to just over 17 melons per square foot.

That is, one acre is about 43,560 square feet, so 750,000 melons in 43,560 square feet (just divide), produces the roughly 17 melons per square foot.

You don't have to be a farmer to understand how outlandish that is. The point is that it happens, and in this case it got beyond a couple of referees for the paper and an editor. Here, it was just an honest mistake that was rectified later. But if you were not paying close attention when this article came out, you might just infer that the production of this watermelon variety was way more impressive than it actually was.

Sometimes simple math mistakes have devastating consequences. Consider the 2010 case of a well-respected nurse at a Washington State hospital. The nurse had been instructed by a doctor to administer 140 milligrams of calcium chloride to a 9-month-old infant. Here was the nurse's thought process, as recorded in hospital documents:

- There are 10 milligrams of medication in every milliliter
- 140/10 = 14, so the correct dosage is 14 milliliters

The nurse proceeded to administer 14 milliliters of calcium chloride through the infant's IV. What was wrong with the nurse's arithmetic? There are 100 milligrams in a milliliter. So the correct dosage would have been 1.4 milliliters of calcium chloride. That same day, another nurse caught the arithmetic error, but it was not soon enough. The child died four days later; the nurse was fired and died by suicide seven months later. Numeracy is not really an optional skill for all of us to have.

Rate Rates

There are times when it simply makes more sense to compute rates than to look at totals. For example, the table shown here exhibits total flights and on-time flights for five airlines that were flying out of Cincinnati International Airport (CVG) between July 2014 and June 2015. Look at Delta and Frontier. Is it fair to say Frontier, with only 185 delayed flights, is superior to Delta that logged 227?

A quick calculation shows that Delta had 227 delayed flights out of 6478 total flights for a 3.5% outgoing delay rate. Frontier had 185 delayed flights, but that was out of a total of only 1542 flights all together, which makes for a 12.0% outgoing delay rate. So from the perspective of outgoing delay *rate*, Delta is performing much better than Frontier. This kind of calculation is pretty simple, and you'd probably do this without having to be reminded. But it is the kind of mistake that is still made a lot, so it is worth mentioning.

Rates can also be deceptive. There was the famous case in 2013 of the 2½-year-old Mississippi baby who showed no signs of HIV after being born with the disease. She was given high doses of three antiretroviral drugs within 30 hours of her birth and was declared functionally cured in 2013. That's one baby cured, but a 100% cure rate. Both are technically correct summaries, obviously. But they create very different impressions. If the doctors had

TABLE 2.1

CVG Airport July 2014–June 2015		
Airline	Total Flights	Delayed Flights
Delta	6478	227
ExpressJet	4664	578
Frontier	1542	185
American Eagle	5636	744
SkyWest	3387	325

naively reported a 100% cure rate for this early-intervention infusion, it would have invited a very deceptive inference about the success rate. In this case the total made much more sense to report than the rate. Unfortunately, in the summer of 2014, the infection reappeared for this child.

Don't expect it to always be the case that a rate or total will clearly be the most informative. It's typically not that straight forward. Consider the following statements that have been made about maternal health in the U.S.:

> It's more dangerous to give birth in the United States than in 49 other countries. African-American women are at almost four times greater risk than Caucasian women. A safe pregnancy is a human right for every woman regardless of race or income.

> Maternal mortality ratios have increased from 6.6 deaths per 100,000 live births in 1987 to 13.3 deaths per 100,000 live births in 2006.

Death rates did increase *by over 100%* in the U.S., going from 6.6 deaths per 100,000 live births in 1987 to 13.3 deaths per 100,000 live births in 2006. In absolute terms, this was an increase of 6.7 deaths per 100,000 live births. Clearly, neither the total nor the rate is a pleasing number, but neither is particularly deceptive. In this case, having both the percentage increase and the absolute increase seems to be equally useful.

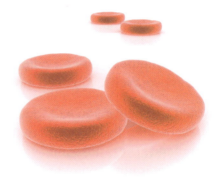

Say the Unsaid

Sometimes numerical arguments are just not that easy to understand without some concerted effort to dig out information you don't have, that you really need. In 2007, former N.Y. City Mayor Giuliani was making a case in opposition to socialized medicine. Citing his own experience with prostate cancer, Giuliani noted that he had an *82% chance of survival in the U.S. but only a 44% chance in England.*

Those are scary and sobering numbers.

The 44% came from data that showed for about every 49 out of 100,000 men in the U.K. who are diagnosed with prostate cancer, 28 will die within five years, for a "five-year survival rate" of about 44%. That is, 28 divided by 49 is roughly 0.44. A similar kind of calculation for the U.S. produced the 82%.

This seems clear enough, but really isn't so clear at all. In the U.S. the prostate sensitive antigen test (PSA) was used widely before 2014, and it probably made prostate cancer detection possible at a much earlier age. So in the U.S., you might have detection coming at age 50 and the patient living to age 70, but in the U.K., detection may not come until age 67 (when external symptoms appear) with patients living to age 70, as well. So it could be that the men are actually living to roughly the same average age. But because the U.K. patients are dying within a five-year window of detection, the five-year survival rate shows up as much stronger for the U.S.

There is no way to know if former Mayor Giuliani was being clever in the construction of his argument or just didn't catch the potential flaw in his reasoning. In any case, it probably is not safe to infer only from those survival rates that under socialized medicine more men would die of prostate cancer.

Know Some Benchmarks

You may never know for sure you've avoided making a deceptive inference from the statistics you are reading or the chart you are looking at. No checklist for making sure this is always avoided could possibly be constructed! So what can you do? Read all statistical arguments critically. If something doesn't seem to make sense, try to figure out why. If enough information is available, check the statistics that are given. When a total (absolute) is given instead of a rate, question whether the rate would be more informative, and vice versa. Even if a statistic makes sense, ask yourself if you need more information for the human inference to be legitimate.

It would also help if you had some benchmarks at your disposal or were willing to look up benchmarks if you needed to. To illustrate what we mean, consider the following ten benchmarks. These are somewhat randomly pulled from any number we could have listed. The larger point is not confined to any specific list of benchmarks, but is simply that having a grasp of some basic benchmarks will be very useful in increasing your numeracy and the accuracy of your human inferences.

1. U.S. population is just over 300 million.
2. Each year about 4 million babies are born in the U.S.
3. About 2.8 million Americans die each year.
4. Roughly 1 in 4 who die do so of heart disease.
5. Roughly 1 in 4 who die, die from cancer, more or less.
6. About 38,000 die in traffic accidents.
7. About 19,000 deaths are homicides.
8. About 13,000 deaths are from AIDS.
9. There are about 40 million black Americans.
10. About 18% of Americans identify themselves as Latino.

How can benchmarks come in handy? Consider the claim that allegedly appeared on BlackWomensHealth.com and can be found archived in *The Critical Analysis: An Overview of African American Progress from Emancipation to Present Day*, by E.D. Johnson. The claim, appearing in Chapter 5 of that book, is as follows:

> More than three million children witness domestic violence, and more than four million women are battered to death by their husbands or boyfriends each year.

Look back at the third benchmark above. Only about 2.4 million people die in the U.S. each year *from all causes*. So the claim that 4 million women die from partner abuse has to be wrong.

There is no way to know for sure why the "4 million" statistic was reported. It has been conjectured this was another example of a decimal point error since some sources have said as many as 4,000 such homicides take place each year. Still others cite statistics that say there are 4 million cases of battery at the hands of a spouse or boyfriend. That could have been the original source, with the addition of "to death" happening inadvertently over time as the statistic was repeated. Regardless, knowing something about the order of magnitude of all deaths in the U.S. would flag a claim like this quickly and reorient an inference that might otherwise go astray.

Certainly, the more we know about benchmarks, the better we will be able to navigate the statistics we hear and read. Consider a report released on *Everytown.org* in December 2018 regarding gun violence. A key statement made is repeated below:

> Data from the Centers for Disease Control and Prevention shows 39,773 people were killed by gun violence in 2017—approximately 1,100 more than were killed by motor vehicle accidents.[1]

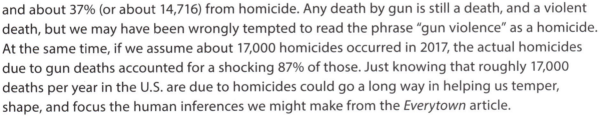

While there is little doubt that the U.S. has serious issues with "gun violence," we know that the 39,773 could not all be homicides. There are only about 17,000 homicides in the U.S. each year total. If you dig a little deeper, you'll find that about 60% of those 39,773 deaths were from suicide and about 37% (or about 14,716) from homicide. Any death by gun is still a death, and a violent death, but we may have been wrongly tempted to read the phrase "gun violence" as a homicide. At the same time, if we assume about 17,000 homicides occurred in 2017, the actual homicides due to gun deaths accounted for a shocking 87% of those. Just knowing that roughly 17,000 deaths per year in the U.S. are due to homicides could go a long way in helping us temper, shape, and focus the human inferences we might make from the *Everytown* article.

We also need to make sure we have a basic grasp of some of the simplest numerical summaries that we are likely to see popping up just about everywhere. Some of what is in the next section will be review, but likely there are some new things there for us all to learn.

---

1   https://everytown.org/press/cdc-data-show-gun-violence-killed-more-people-than-motor-vehicles-in-2017-everytown-moms-demands-action-respond/ (last accessed April, 2021).

Master Background Bugaboos ———————————

Means and Medians

Long before now, you are sure to have learned about means and medians. Let's take a minute to remind ourselves of what they are and how we compute them.

> The **mean** or **average** of a list of numbers is found by adding all the numbers up and dividing by how many you have. The mean is typically denoted by $\overline{x}$ (read "xbar").

Example: Google Salaries

The minimum salary for the ten top-paying jobs at Google (as reported by *Glassdoor* at https://www.glassdoor.com/blog/highest-paying-jobs-at-google/ on November 7th, 2018) are shown in Table 2.2.

TABLE 2.2

Position	Minimum Salary
Senior Vice President	$661,000
Director of Operations	$304,000
Director of Engineering	$300,000
Senior Director, Product Management	$296,000
Director, Global Partnership	$286,000
Senior Director, Talent Management	$282,000
Finance Director	$272,000
Director, Product Management	$268,000
Global Creative Director	$258,000
Director of Marketing	$245,000

To find the average salary, we simply have to add these ten numbers up and divide by ten:

$$\overline{x} = \frac{661,000 + 304,000 + 300,000 + 296,000 + 286,000 + 282,000 + 272,000 + 268,000 + 258,000 + 245,000}{10}$$

$$= \$317,200$$

The mean is the most common way of describing an entire list of data if we are limited to just one number. The median is not far behind in popularity.

> The **median** of a list of numbers is found by first arranging the numbers in increasing or decreasing order, then choosing the one in the middle. If there is an odd number of numbers in the list, then the idea of the "one in the middle" is well defined. If there is an even number of numbers in the list, then the median is the average of the two numbers that define the middle of the list. There is no special notation for the median the way there is for the mean.

Example: Google Salaries Revisited

Refer back to the Google salaries. The data are already in order. There are ten salaries listed, so there is no unique middle. However, the salaries of the Director of Global Partnership and the Senior Director of Talent Management are the two that define the middle unambiguously. So the median is taken to be as follows:

$$\text{median} = \frac{286,000 + 282,000}{2} = \$284,000$$

Back to College Level

Computing the mean and median are very elementary tasks. Only slightly more difficult is the answer to the question "which one should you use to describe a list of data?" In the list of Google salaries, notice that nine of the ten salaries were below the average salary of $317,200. So it is fair to ask, in what sense does the average inform us, granted with only one number, about the larger list? It is bigger than nine of the salaries and smaller than the tenth. So it is somewhat all by itself with respect to the actual data, neither describing well the lower 9 nor the largest value.

The median, on the other hand, is right in the middle of the lower nine salaries and does a much better job informing us about where the bulk of the list of numbers is. Granted, it fails to tell us anything about the elevated pay of a Senior Vice President. It is too much to expect for one number alone to carry a lot of information about a complex list of data, but it is worth noting that the mean and median respond differently to outliers, like the Senior Vice President's pay, in a data set.

TABLE 2.3

Position	Minimum Salary
Senior Vice President	$661,000
Director of Operations	$304,000
Director of Engineering	$300,000
Senior Director, Product Management	$296,000
Director, Global Partnership	**$286,000**
Senior Director, Talent Management	**$282,000**
Finance Director	$272,000
Director, Product Management	$268,000
Global Creative Director	$258,000
Director of Marketing	$245,000

An outlier in a list of data is simply a number in that list that is notably different than the other numbers in the data.

The mean is sensitive to outliers. That is, when the mean is computed on a data set with an outlier, it will tend to be dragged toward that outlier. This risks having the mean be less than adequate as a one-number summary of those data.

The median is less-sensitive to outliers. That is, when the median is computed on a data set with an outlier, it will effectively ignore the outlier and produce an adequate one-number summary of the remaining data.

If the data set has no outliers, then either the mean or the median is fine to use. In these cases, they will often be nearly the same number.

These are worthy things to remind oneself of, even if you already knew this from long ago. Choosing properly between the mean and median has implications for your human inference. With the ten Google salaries, the mean produces an exaggerated view about what typical Google executives make. As we go further into this book, we will find ourselves talking a lot about means. Just be aware that when we get there, we will be talking specifically about data that have symmetric distributions, with no outliers. Hence, the distinction between the mean and the median will have no practical import at that point.

There is often no good reason to compute the mean or median by hand. There are a plethora of good software utilities that will do it for you. Here is how you would do it in Excel. The operative functions are *average* and *median*.

Excel Tip: Computing the Mean and Median

	A	B
1	$661,000	
2	$304,000	
3	$300,000	=average(A1:A10)
4	$296,000	
5	$286,000	
6	$282,000	
7	$272,000	
8	$268,000	
9	$258,000	
10	$245,000	
11		

	A	B
1	$661,000	
2	$304,000	
3	$300,000	=median(A1:A10)
4	$296,000	
5	$286,000	
6	$282,000	
7	$272,000	
8	$268,000	
9	$258,000	
10	$245,000	
11		

Standard Deviation

The mean and median are designed to tell you, roughly, where the "middle" of a set of data is, albeit in different ways as we have just seen. It is also imperative to have some idea of how data are spread out, or how "variable" they are. While there are many ways of quantifying this idea, the most common way is the standard deviation.

The **standard deviation** of a list of n data values is computed as follows:

- Find the mean, $\overline{x}$.
- Subtract $\overline{x}$ from each number in the data set.
- Square the difference you just found.
- Add up all those squared differences and divide by $n - 1$.
- Take the square root of your last calculation.

These steps can be exhibited in a single formula, shown below, where $x_1, \ldots x_n$ denote the n data values. The standard deviation is typically denoted by the letter "s."

$$s = \sqrt{\frac{(x_1 - \overline{x})^2 + (x_2 - \overline{x})^2 + \ldots + (x_n - \overline{x})^2}{n - 1}}$$

The standard deviation is in the same units as the original data. So if the data are in "dollars" then the standard deviation is in "dollars." The variance is also commonly referred to as a measure of variability, and it is simply the square of the standard deviation.

The **variance** of a list of n data values is just the square of the standard deviation s and is typically denoted by s^2.

There really is no good reason to compute the standard deviation by hand. There are a plethora of easy-to-use tools available to do that task. We will do it just once by hand to make sure you have an example of how it is done. Then we'll show you how to do it with common software.

Example: Variation in Google Salaries

Refer back to the Google salaries. We have added columns to the data table to illustrate the computation of the standard deviation.

To complete the computation, we need to add up all the numbers in the third column, divide by $n - 1 = 9$, and take the square root. This results in a standard deviation of $s = \$122,252.74$. The following tip shows you the command needed to find the standard deviation and variance in Excel. The operative functions are *stdev.s* and *var.s*. See below.

TABLE 2.4

Salary	(Salary – $317,200)	(Salary – $317,200)^2
$661,000	$343,800	118198440000.00
$304,000	–$13,200	174240000.00
$300,000	–$17,200	295840000.00
$296,000	–$21,200	449440000.00
$286,000	–$31,200	973440000.00
$282,000	–$35,200	1239040000.00
$272,000	–$45,200	2043040000.00
$268,000	–$49,200	2420640000.00
$258,000	–$59,200	3504640000.00
$245,000	–$72,200	5212840000.00

Excel Tip: Computing the Standard Deviation and Variance

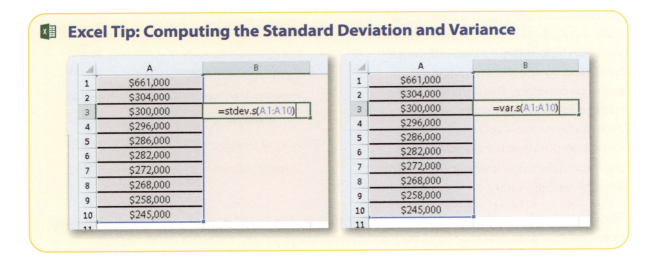

Since the standard deviation is in units of the original data, it is hard to say whether the standard deviation is large or small when computed on a single data set. To make this kind of assessment a little easier, it is not uncommon to divide the standard deviation by the mean. This has the effect not only of cancelling units but producing a number that is in effect the "number of means in the standard deviation" which offers a useful way of interpreting the standard deviation.

> The **coefficient of variation** of a list of data is the standard deviation divided by the mean of those data. It is an effective way of comparing variation across data sets even when their means differ markedly.

Example: Habitat for Humanity Salaries

The table below lists ten salaries for Habitat for Humanity executives (as reported by *Glassdoor* at https://www.glassdoor.com/Salary/Habitat-for-Humanity-Salaries-E5810.htm) by job title.

A comparison of the means, standard deviations, and coefficients of variations between these salaries and those at Google appear in the table below. If we just looked at the standard deviations alone, we see that the Google salaries are more variable than the Habitat for Humanity salaries. The coefficients of variation help quantify that difference by scaling by the means, making it easier to compare. The variation in the Habitat for Humanity salaries is about 26% of the average salary, while the same computation for Google shows the variability there is about 39% of the average Google salary.

TABLE 2.5

Position	Minimum Salary
Executive Director	54,875
Volunteer Coordinator	32,598
Store Manager	35,560
Project Manager	55,751
Director	82,734
Transportation Manager	52,048
Program Manager	48,000
Manager	56,203
Director of Development	46,827
Logistics Manager	57,156

This is yet another way you can increase your numeracy skills. If you are comparing variability between two data sets that have very different average values, then the coefficient of variation is a useful tool to know so that your inferences are correct regarding which data set is more variable.

TABLE 2.6

Statistic	Google	Habitat for Humanity
Mean	$317,200	$52,175.20
Standard Deviation	$122,252.74	$13,745.021
Coefficient of Variation	0.39	0.26

There's no separate function in Excel for the coefficient of variation. You just simply divide the standard deviation by the mean.

Excel Tip: Computing the Coefficient of Variation

	A	B
1	$661,000	
2	$304,000	
3	$300,000	=(STDEV.S(A1:A10))/(AVERAGE(A1:A10))
4	$296,000	
5	$286,000	
6	$282,000	
7	$272,000	
8	$268,000	
9	$258,000	
10	$245,000	
11		

One final thought on the standard deviation. Since it has the sample mean at its core, you won't be surprised to learn that the standard deviation can be very sensitive to outliers. If you look back at the Google data and take out the top salary of $661,000, the standard deviation for the remaining nine drops to $19,937. That's down from $122,253 for all ten salaries!

The standard deviation is sensitive to outliers and that should be taken into account whenever you are making inferences about how spread out data are.

While means, medians, and even standard deviations are easy to compute, particularly with the welcome availability of hand-held software, we still need to make sure we have a grasp on when those computations might lead us astray and risk creating wrong-headed inferences. We are going to end this chapter by looking at a common graphical tool for displaying data.

Histograms

You may have first seen histograms in middle school, where they are often (oddly) referred to as "line plots"—which are really something different entirely. We'll be using those a lot as we move throughout this book, so they are worth reviewing. We also need to be aware of how they can distort our human inferences regarding the data being plotted if we aren't careful.

A histogram is a plot that organizes your data into prescribed groups, called bins, and then records either how many data values are in each bin, or *relatively* how many data values are in each bin.

Next is an example of how you would do that.

Example: Graphing Google

Refer back to the Google salaries. They range from $245,000 to $661,000, so we will need to create groups adequate to capture all of those. Something like Table 2.7 will do.

To construct the histogram, we plot the upper and lower bin limits on the horizontal axis and draw bars above the bins, with heights determined either by the number of data points in each bin (a frequency histogram) or the relative number in each bin (a relative frequency histogram). We have constructed a frequency histogram below. For convenience, the data were divided by $1000 before plotting. To construct this plot, we used a free utility that you may want to be aware of:

> A **free and flexible histogram graphing utility** can be found at http://www.shodor.org/interactivate/activities/Histogram/

It is certainly possible to construct histograms with all kinds of software, including Excel. The steps in Excel are a little more involved, and you don't have the flexibility to quickly change bin sizes the way you do with the above utility. What we are really interested in is what the histogram tells us graphically. We can immediately see that the bulk of the data are down in the $300,000 range, while the Senior Vice President's salary sits as an outlier to the right.

TABLE 2.7

Salary Bin	Number of Salaries in Bin	Relative Number in Bin
$221,000 to $260,000	2	0.2
$261,000 to $300,000	6	0.6
$301,000 to $340,000	1	0.1
$341,000 to $380,000	0	0
$381,000 to $420,000	0	0
$421,000 to $460,000	0	0
$461,000 to $500,000	0	0
$501,000 to $540,000	0	0
$541,000 to $580,000	0	0
$581,000 to $620,000	0	0
$621,000 to $660,000	0	0
$661,000 to $700,000	1	0.1

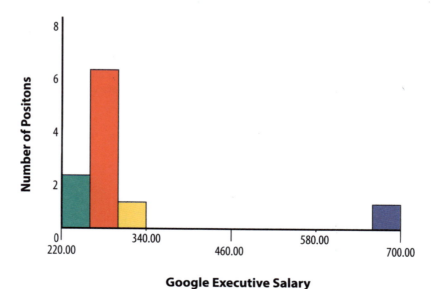

Google Executive Salary

Histograms are simple to construct, and we will be relying on them often later in the book. For now we will end this chapter by recognizing that human inferences we make from any plot, histograms included, can go astray if the plot is not carefully constructed. With histograms the key component to choose is the bin width. Although some books will try to suggest rules for how many bins there should be, and what their widths should be, the fact is this varies by data set. Far better, though more challenging, to just be aware that the impact of the plot may critically depend on those choices. One nice feature of the free graphing utility above is that these choices can be made very quickly and, as a result, you can quickly see how the graphical impact changes as those choices change.

Below we have plotted the same Google salary data, but in the plot on the right, we used a bin width of 400, and in the one on the left, we used a bin width of 100. The one on the left reveals the upper salary as an outlier but disguises any variability in the lower nine. The one on the right fails at doing either.

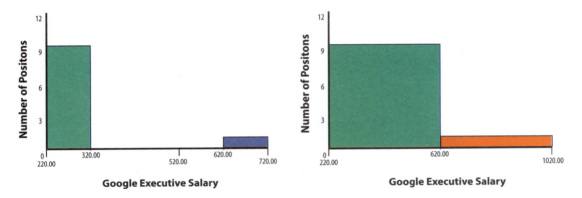

While neither of these plots is incorrect, neither is as informative in its exhibition of the data as is the original one we constructed above. Keep in mind that this is as much an art as a science.

As mentioned above, one can create histograms in Excel. It is just a little more involved than the applet we first used above. On the next page are the steps.

📊 Excel Tip: Creating Histograms in Excel

This is just a tip intended to get you started. Your specific experiences may vary.

- Highlight the data you want to put into a histogram.

- With the data highlighted, go to Insert on the toolbar, click on the arrow on the lower right of the Charts category, and select histogram.

- Click OK and the histogram will be embedded in the active worksheet.

- You can customize titles the usual way titles are edited in Excel.

- You can customize bin sizes by clicking on the horizontal axis and following the directions on the overlay.

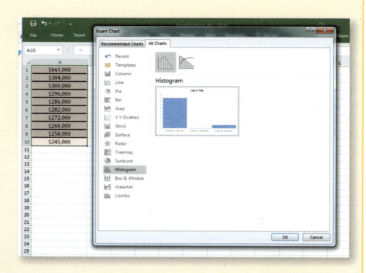

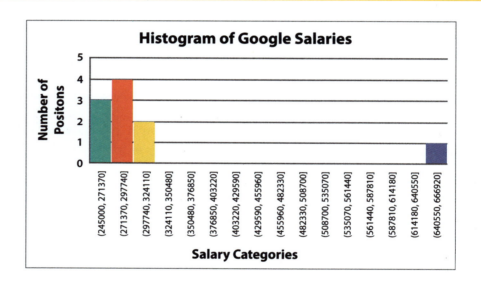

TAKE-HOME POINTS

- If a numerical claim is too good to be true, it may not be. Check simple calculations when possible.

- Percentages and totals might create different impressions. Be prepared to recognize those situations and adjust as necessary.

- Have a good grasp of common benchmarks. These can provide necessary grounding when reading numerical arguments.

- Simple computations and simple graphical summaries are useful. Be prepared to know when they might be accidentally deceptive or misused.

Chapter 2 Exercises

Reading Check

Short Answer: Please provide brief, concise answers to the following questions.

1. Why is it potentially deceptive to say that the U.S. maternal mortality rate increased by 100% between 1987 and 2006?

2. How is Mayor Giuliani creating a confusing picture of health care in the United States and in Europe? Explain.

3. Please find and list five benchmarks that are not included in the ten you encountered in this reading.

4. What is the advantage of using the coefficient of variation to summarize variability, as opposed to just using the standard deviation?

5. Describe a situation where using the median would be more appropriate than using the mean.

Beyond the Numbers

1. Nimble Numbers

a. Read the following excerpt from the Associated Press, following the much-anticipated release of *Deathly Hallows*.

Title: *Harry Potter and the Deathly Hallows* Sells 8.3 Million in First 24 Hours

Source: The Associated Press, July 22, 2007, 6:28 p.m.

It is the richest going-away party in history.

"Harry Potter and the Deathly Hallows," the seventh and final volume of J.K. Rowling's all-conquering fantasy series, sold a mountainous 8.3 million copies in its first 24 hours on sale in the United States, according to Scholastic Inc.

No other book, not even any of the six previous Potters, has been so desired, so quickly. "Deathly Hallows" averaged more than 300,000 copies in sales per hour—more than 50,000 a minute. At a list price of $34.99, it generated more than $250 million of revenue, more than triple the opening weekend take for the latest Potter movie, "Harry Potter and the Order of the Phoenix," which came out July 10.

"The excitement, anticipation, and just plain hysteria that came over the entire country this weekend was a bit like the Beatles' first visit to the U.S.," Scholastic president Lisa Holton said in a statement Sunday.

We breathe about 15 times per minute, and a hummingbird flaps its wings about 3,000 times per minute. So a rate of 50,000 copies per minute would truly take our breath away and be faster than we could discern with our eyes. Is the 50,000 figure right?

i. Yes. 300000/60 = 50000.

ii. No. 500000/60 = 833.33, not 34.99.

iii. Yes. 50,000/300,000 = 0.1667; and 0.1667/20994 = .3499.

iv. No. 300000/60 = 5000, not 50000.

b. A letter to the editor of the *New York Times* complained about a *Times* editorial that quoted the following demoralizing statistic:

An American woman is beaten by her husband or boyfriend every 15 seconds.

The writer of the letter complained: "At that rate, 21 million women would be beaten by their husbands or boyfriends every year. That is simply not the case."

Is the letter writer's numerical reasoning correct?

i. Yes. An incident every 15 seconds is 4 per minute, 240 per hour, 5760 per day, 40320 per week, 2096640 per year, about 21 million.

ii. No. An incident every 15 seconds is 4 per minute, 240 per hour, 5760 per day, 40320 per week, 209640 per year, not 2096640 per year.

iii. Yes. An incident every 15 seconds is 4 per minute, 240 per hour, 5760 per day, 40320 per week, 2096640 per year, about 210.0 million.

iv. No. An incident every 15 seconds is 4 per minute, 240 per hour, 5760 per day, 40320 per week, 2096640 per year which is about 2.1 million, not 21 million.

c. The letter writer in question 1b goes on to cite the National Crime Victimization Survey (NCVS), which estimates about 254,000 cases of violence against women by husbands or boyfriends each year. The NCVS reports cases, not incidents. How does this information support the letter writer's case? Explain.

2. Violence in Chicago

Please read the following excerpt from the *Chicago Tribune*:

Title: Number of CPS Students Shot Rises, as Does Fear of More to Come

Author: Noreen S. Ahmed-Ullah

Source: Last accessed 2021: *Chicago Tribune*, June 26, 2012, https://www.chicagotribune.com/news/ct-xpm-2012-06-26-ct-met-cps-student-violence-0625-20120626-story.html

Twenty-four students were fatally shot during the school year that ended June 15, four fewer than in the 2010–11 year. But the overall shooting toll—319—was the highest in four years and a nearly 22% increase from the previous school year.

a. The "previous school year" being referred to is the 2010–11 year. What was the total number of shootings during the 2010–11 school year?

   i. 261

   ii. 319

   iii. 204

   iv. 563

b. Speaking of violence in Chicago, the table below shows the number of murders in Chicago from 2001 through 2012. Find the mean and the median number of murders over this time period.

TABLE 2.8

Year	2001	2002	2003	2004	2005	2006	2007	2008	2009	2010	2011	2012
Murders	667	656	601	453	451	471	448	513	459	436	435	506

   i. The mean is 508 and median is 459.5.

   ii. The mean is 465 and median is 508.

   iii. The mean is 508 and median is 465.

   iv. The mean and median are each 2006.5.

c. Suppose we construct a histogram of these twelve "Murders" values. We choose the smallest bin to have 420 as a left endpoint and use a bin width of 50. Which of the following has to be the result?

i.
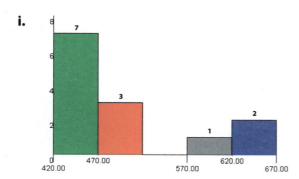

iii.
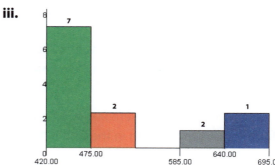

ii.
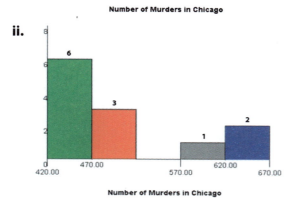

iv.
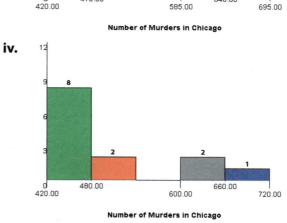

3. Difficult Dilbert Dialogue

The following dialogue is from a Dilbert cartoon published April 17, 1996.

Secretary: Oh my! This is shocking!

Boss: What?

Secretary: *40% of all sick days taken by our staff are Fridays and Mondays!*

Boss: What kind of idiot do they think I am?

Secretary: Not an idiot savant, they can do math.

a. The secretary is trying to poke fun at her pointy-headed boss. Is she successful or is the joke on her?

   i. Successful. Assuming an employee picks a single workday at random to be absent, the chances that the employee will pick a Monday or a Friday are 2/5 which is 40%

   ii. Successful. Assuming an employee picks a single workday at random to be absent, the chances that the employee will pick a Monday or a Friday are $(1/2) \times (1/2)$—since there is a 1 in 2 chance they pick a Monday or they don't—which is 1/4, or about 40%

   iii. Joke's on her. Assuming an employee picks a single workday at random to be absent, the chances that the employee will pick a Monday or a Friday are 2 in 3, not 2 in 5, since once two days are counted, you only have three left.

   iv. Joke's on her. Assuming an employee picks a single workday at random to be absent, the chances that the employee will pick a Monday or a Friday are 2/5 which is 2.5%, so the author of the cartoon was having his own numeracy issues.

Let's put some other words into the secretary's dialogue. What if she had said the following?

Secretary: *When employees choose to take off a pair of sick days in a Tuesday–Monday period of time, 40% of the time they choose Friday and Monday as the two days they pick!*

The following two questions explore in what sense this is or is not a different situation than the one in the actual cartoon.

b. If we denote the five usual workdays as M,T,W,R, and F (using R for Thursday), which of the following lists all pairs of possible workdays.

   i. (M,T), (T,W), (W,R), (R,F), (T,M), (W,T), (R,W), (F,R), (M,F), (F,T)

   ii. (M,T), (M,W), (M,R), (M,F), (T,W), (R,M), (T,F), (R,W), (W,F), (R,F)

   iii. (M,T), (M,W), (M,R), (M,F), (T,W), (T,R), (T,F), (W,R), (W,F), (R,F)

   iv. (M,F), (T,F), (T,M), (W,M), (W,F), (R,M), (R,F), (F,W), (F,R), (M,R)

c. Assume there are ten possible pairs of days an employee could pick from, and that list of the sets of ten pairs suggested in the question above is the right one. If a pair was just picked at random, what is the chance that they'd pick Monday and Friday as the pair, and does this suggest the secretary (in the amended dialogue) has uncovered a bias?

  i. 1 in 10. No evidence of bias since every pair had an equal chance of being chosen.

  ii. 1 in 10. Yes, if 40% are choosing Monday and Friday, that is evidence of bias toward that pair.

  iii. 4 in 10. But that is exactly what the secretary notices was happening, so no evidence of bias.

  iv. 4 in 10. Yes, if 40% are choosing that pair of days, and there is exactly that chance of randomly choosing that pair of days, then this is clear evidence of bias in the days the employees are choosing.

4. Contraception Deception

In October 1995, the U.K. Committee on Safety of Medicines warned that third-generation oral contraceptive pills increased the likelihood of potentially life-threatening blood clots in the legs or lungs twofold—that is, by 100%. This information was passed on in "Dear Doctor" letters to 190,000 general practitioners, pharmacists, and directors of public health as well as in an emergency announcement to the media.

a. Read the ABC News report from 2008 at http://abcnews.go.com/Technology/story?id=6034371&page=1 (link last accessed 2021—make sure you access the entire link). What was the actual increase in the number of blood clots experienced?

  i. Went from 1 in 7,000 to 2 in 7,000

  ii. Went from 1 in 13,000 to 2 in 13,000

  iii. Went from 7,000 to 14,000

  iv. Went from 7,000 in 100,000 to 14,000 in 100,000

b. Read the ABC News report from 2008 at http://abcnews.go.com/Technology/Story?id=6034371&page=1 (link last accessed 2021 – make sure you access the entire link). How many abortions was the scare blamed for?

  i. 7,000

  ii. 13,000

  iii. 190,000

  iv. 1,000

c. Our human inferences are likely to be different depending on whether we read about the percent increase in blood clots or read about the actual change in the number of blood clots. Explain and suggest which one would be the most reasonable in this context.

5. Nursing Knowledge Needed

On June 2, 2007, a nurse in Wales accidentally injected an 85-year-old patient with a lethal amount of insulin during a home visit. While the syringes used to inject insulin are typically marked with insulin units instead of the usual milliliters, this nurse found herself without an insulin syringe on hand. So she retrieved a common syringe from her car and proceeded to do the transformation from milliliters to insulin units in her head.

a. If 1 milliliter equals 100 insulin units, how many milliliters should the patient have been given if she had been prescribed 36 units?

  i. 1 milliliter = 100 units so 0.01 milliliter = 1 unit. It follows that 0.36 milliliters = 3.6 units. So she should have injected the patient with 3.6 milliliters.

  ii. 1 milliliter = 100 units so 0.1 milliliter = 1 unit. It follows that 0.36 milliliters = 36 units. So she should have injected the patient with 0.36 milliliters.

  iii. 1 milliliter = 100 units so 0.01 milliliter = 10 units. It follows that 36 milliliters = 36 units. So she should have injected the patient with 36 milliliters.

  iv. 1 milliliter = 100 units so 0.01 milliliter = 1 unit. It follows that 0.36 milliliters = 36 units. So she should have injected the patient with 0.36 milliliters.

b. The nurse injected the patient four times with a full 0.9 milliliter syringe. What appears to be the nurse's mistake? Defend your answer.

6. Statistical Citizenship

Read the article "Democracy and the Numerate Citizen: Quantitative Literacy in Historical Perspective," by Professor Patricia Cline Cohen: https://www.maa.org/sites/default/files/pdf/QL/pgs7_20.pdf

If the link is broken, search for the article on the web by title and author, or ask for help.

a. What of the following are the three "features of the Constitution that suggest a numerical approach to governance" according to the article?

  A. The size of each state's population determined the composition of the House of Representatives and the number of electors in the electoral college that selected the president.

  B. The post-Civil War era finally brought a full melding of statistical data with the functioning of representative government. A century after the first census of 1790, no one any longer suggested that an expanded census would alarm the people or merely gratify idle curiosity.

  C. The Constitution inaugurated a regular and recurring census based on "actual enumeration," and the results would determine not only apportionment in the House and electoral college, but also apportionment of direct taxes.

    D. The agenda for national statistics collection had a new focus on urban problems, immigration, labor conditions, and standards of living.

    E. The framers handled the thorny problem of noncitizen inhabitants by counting slaves (circumspectly described as "other Persons" in contrast to the category of the "free") at a three-fifths ratio …

      i. B, C, E

      ii. A, C, D

      iii. B, C, D

      iv. A, C, E

b. Initially, the government was reluctant to collect more than the most basic census information of race, sex, and age. During which of the time periods addressed by Cohen did we see a "full melding of statistical data with the functioning of representative government?"

    i. The post-Civil War Era

    ii. The Founding Generation

    iii. The Antebellum Era

    iv. The post-World War II Era

c. Cohen claims that a "full melding of statistical data with the functioning of representative government" eventually happened. He lists several facts supporting that claim. Which of the following is NOT one of the facts he lists?

    i. The census bureau was turned into a permanent federal agency.

    ii. The agenda for national statistics collection had a new focus on urban problems, immigration, labor conditions, and standards of living.

    iii. Censuses began to be run by men and women trained in psychology and education.

    iv. Statistics emerged as a sophisticated field of science.

d. Refer to Appendix A from Cohen's article. Which of the 12 suggestions to promote quantitative literacy would have the greatest impact? Defend your answer.

7. Winging Out Some Computations

Your wingspan is the distance from the tip of the middle finger on your left hand to the tip of the middle finger on your right, with your arms parallel to the ground. The wingspans of 10 hypothetical individuals are recorded in the table.

a. What is the mean wingspan for this group of ten persons?

   i.   65

   ii. 67

   iii. 67.2

   iv. 70

TABLE 2.9

Person	Wingspan (in)
1	70
2	63
3	72
4	67
5	65
6	65
7	68
8	64
9	67
10	71

b. What is the median wingspan for this group of ten persons?

   i.   65

   ii. 67

   iii. 67.2

   iv. 70

c. What is the variance for this group of ten persons?

   i.   9.29

   ii. 4.13

   iii. 17.06

   iv. 3.05

d. What is the coefficient of variation for this group of ten persons?

   i.   0.61

   ii. 1.29

   iii. 0.05

   iv. 0.10

8. A BIGS Change

Suppose the former Kentucky basketball great and subsequent NBA star Anthony Davis drops in for a visit and joins the group of 10 persons in the data set introduced first in the problem above.

a. What is the mean of this new eleven-person data set?

  i. 65

  ii. 67

  iii. 69.1

  iv. 70

b. What is the median of this new data set?

  i. 65

  ii. 67

  iii. 69.1

  iv. 70

TABLE 2.10

Person	Wingspan (in)
1	70
2	63
3	72
4	67
5	65
6	65
7	68
8	64
9	67
10	71
Anthony Davis	88

c. Which changed the most, the mean or the median?

  i. neither changed at all

  ii. the mean changed the most

  iii. they changed the same amount

  iv. the median changed the most

d. What is the coefficient of variation for this group of eleven persons?

  i. 0.61

  ii. 1.29

  iii. 0.05

  iv. 0.10

9. Gates-Proof Inference

Versions of the following completely-made-up story have been told elsewhere. Suppose there are exactly 9 people on a bus in Seattle, and the average of their yearly incomes is $65,000. At the next stop, the highest paid person on the bus (with an income of $120,000 per year) gets off, and Bill Gates gets on. Mr. Gates makes 7.6 billion dollars a year.

a. What is the new average salary of people on the bus?

    i.   The same as before

    ii. $844,496.00

    iii. $844,496,111.10

    iv. $120,000.00

b. What is the new median salary of people on the bus?

    i.   The same as before

    ii. $844,496.00

    iii. $844,496,111.10

    iv. $120,000.00

c. One of the 9 people on the bus was a snoozing journalist who woke up just after Mr. Gates got on. She proceeded to do a survey of salaries for the nine riders and reported the average the next day in the paper. Would this create an inaccurate inference regarding the typical income of Seattle residents? Why?

    i.   Yes. Most residents of Seattle are quite wealthy, and the salaries of the other eight on the bus will pull the mean down.

    ii. Yes. Most residents of Seattle will make a lot less than Mr. Gates, so the reported mean will be an exaggeration.

    iii. No. Most residents of Seattle are quite wealthy, so the mean she is reporting will be right in line with the average in all of Seattle.

    iv. No. Most residents of Seattle will make a lot less than Mr. Gates, so it is quite appropriate that the other eight salaries on the bus pull the reported mean down to a realistic level.

10. The Spice of Life

The wingspans of 18 persons are recorded in the table below, nine being ordinary people with made-up names, and nine being current or former NBA players. We will call this Data Set 1.

a. What is the average wingspan of these 18 persons?

 i. 80.5

 ii. 69

 iii. 88

 iv. 102

TABLE 2.11 Data Set 1

Ordinary Person	Wingspan (in)	NBA Person	Wingspan (in)
Barney	70	Ike Diogu	88
Bee	73	Anthony Davis	88
Floyd	72	Shelden Williams	88
Helen	69	Elton Brand	90
Thelma	69	Shawn Bradley	90
Gomer	68	Bismack Biyombo	90
Andy	68	Saer Sene	93
Opie	64	George Muresan	94
Otis	73	Manute Bol	102

b. What is the median wingspan of these 18 persons?

 i. 80.5

 ii. 69

 iii. 88

 iv. 102

c. Is the mean from question 10a a better summary of the wingspan of Ordinary Persons or of NBA players? Explain.

 i. It's not a very good summary of either since it is right in the middle of the two groups.

 ii. It's a pretty good summary of both since it is right in the middle of the two groups.

 iii. It's not a very good summary of either since it is below both of the two groups.

 iv. It's a pretty good summary of both since it is in the middle of the NBA group, which we would expect.

We have organized the counts of the wingspans from Data Set 1 into the table below and then used those data to produce the histogram you see below. We have left one entry out of the table just so we can make sure you understand how one gets from the table (or the data) to the histogram. The label of "60" on the plot below denotes the first interval from the table—Wingspan ≤ 60. The label of "75" denotes the fourth interval—70 < Wingspan ≤ 75—and so on.

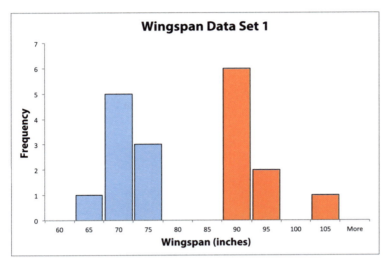

TABLE 2.12

Interval (Bin)	Frequency
Wingspan ≤ 60	0
60 < Wingspan ≤ 65	1
65 < Wingspan ≤ 70	5
70 < Wingspan ≤ 75	3
75 < Wingspan ≤ 80	0
80 < Wingspan ≤ 85	0
85 < Wingspan ≤ 90	–
90 < Wingspan ≤ 95	2
95 < Wingspan ≤ 100	0
100 < Wingspan ≤ 105	1

d. What is the missing value in the table?

   i. 3

   ii. 90

   iii. 0

   iv. 6

e. On the graph to the right, we have the letters A, B, C, and D superimposed on the horizontal axis. Which letter most closely corresponds to the location of the mean of the 18 wingspans in Data Set 1?

   i. A

   ii. B

   iii. C

   iv. D

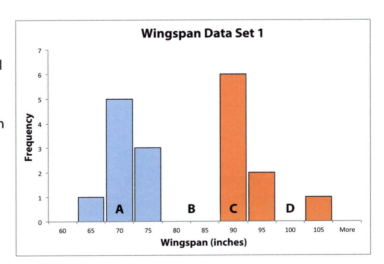

f. Suppose we create a new data set that we will call Data Set 2. In this data set we have eighteen persons—eight with a wingspan of 80.5 inches, five with a wingspan of 75.5, and five with a wingspan of 85.5. What is the mean of Data Set 2?

    i. 75.5

    ii. 80.5

    iii. 85.5

    iv. 90.5

g. Suppose in a data set we call Data Set 2, we have eighteen persons—eight all with a wingspan of 80.5 inches, five all with a wingspan of 75.5, and five all with a wingspan of 85.5. What is the median of Data Set 2?

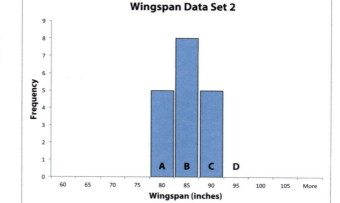

    i. 75.5

    ii. 80.5

    iii. 85.5

    iv. 90.5

h. We have organized the wingspans from Data Set 1 into the histogram you see here. On the histogram, we have the letters A, B, C, and D superimposed on the horizontal axis. Which letter identifies the interval containing the location of the mean of the 18 wingspans in Data Set 2?

    Remember how to read the axis. The bar above the 80, for example, is the count in the interval from 75 to 80.

    i. A

    ii. B

    iii. C

    iv. D

i. How adequate would the mean value alone be for distinguishing the data in Data Set 1 from those in Data Set 2? Explain.

j. Compare the histograms from Data Sets 1 and 2 (both shown above). Specify at least two ways that the histograms are notably different with respect to how their data are spread out.

k. Recall Data Set 2, defined above. This data set has eighteen persons—eight all with a wingspan of 80.5 inches, five all with a wingspan of 75.5, and five all with a wingspan of 85.5. You get to pick two wingspans at random from this data set, eyes closed. Call your choices x_1 and x_2. You will receive a payout of $[(80.5 - x_1)^2 + (80.5 - x_2)^2]$ based on these choices. If it costs you $25.00 to play, what is the maximum profit you can make?

l. Refer to k, above. If you got to pick your data set for the payout game in c., above, would you want to pick Data Set 1 or Data Set 2? Explain. Histograms of those datasets are repeated below.

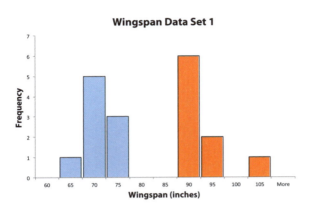

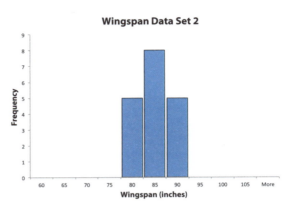

11. Visual Validation

The wingspans of nine ordinary persons and nine current or former NBA players are shown in the table below. This has been called "Data Set 1" in earlier problems in this chapter.

a. Compute the standard deviation of these 18 wingspans. It is recommended that you use a software application to do this.

i. 3.8

ii. 11.8

iii. 14.7

iv. 140.3

b. We defined Data Set 2 as consisting of 18 persons—eight with a wingspan of 80.5 inches, five with a wingspan of 75.5, and five with a wingspan of 85.5. Find the variance of these 18 wingspans. Use of a software application is recommended.

i. 3.8

ii. 11.8

iii. 14.7

iv. 140.3

TABLE 2.11 Data Set 1

Ordinary Person	Wingspan (in)	NBA Person	Wingspan (in)
Barney	70	Ike Diogu	88
Bee	73	Anthony Davis	88
Floyd	72	Shelden Williams	88
Helen	69	Elton Brand	90
Thelma	69	Shawn Bradley	90
Gomer	68	Bismack Biyombo	90
Andy	68	Saer Sene	93
Opie	64	George Muresan	94
Otis	73	Manute Bol	102

c. Histograms of Data Set 1 and Data Set 2 are shown below. Which one has to have the larger standard deviation? Is this how it turned out in parts a. and b., above? Explain.

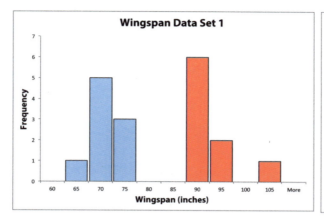

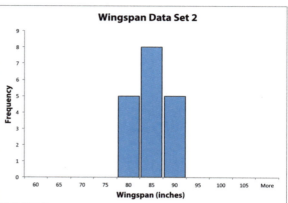

Projects

1. Administration Salaries

Introduction

This chapter reminded us of some basic statistical constructs, and how their computation, use, or display can affect our human inferences.

The Assignment

For this project you are required to do the following:

a. Find the top-ten administrator salaries at your university and at one other comparable university. If your university is a public university, then these should be a part of the public record. Look hard enough and you can find them.

b. Compare the two sets of salaries by:

  i. Computing the means and medians

  ii. Computing the standard deviations and coefficients of variation

  iii. Creating histograms (complete with labels, titles, color)

c. Place all of your findings, including the original data, into a typed, professional paper. The paper should read like a report to your supervisor, not like a bulleted list of answers to a homework problem. Make sure you offer your own detailed comparisons of the salaries and comment if any of the summaries or displays might be misleading.

2. Reality Video

Introduction

This chapter reminded us of some basic statistical constructs, and how their computation, use, or display can affect our human inferences. Now, we want to give you a chance to both show you know what you are doing and to help teach others.

The Assignment

Your instructor may have you do this as individuals or may form you into teams. We'll refer to "teams" below, although you may be a team of one depending on what your instructor wants. Each team is required to do the following:

a. Pick one or more topics from this chapter. Your instructor may want to approve them before you proceed.

b. Construct a three-minute video that teaches the chosen topic to the class. You must carefully and correctly use the language of the chapter in your presentation, and you are required to give an example that is NOT in the chapter.

c. Post the video on YouTube and send the link to your instructor in the manner he stipulates.

3

CHAPTER 3
Statistical Experiments and the Problem of Confounding

A Working Definition

An **experiment** is a test under controlled conditions, with the goal of making cause and effect statements. For example, an experiment may be designed to assess the claim that a popular knee surgery or acupuncture causes a reduction in patient pain. This type of claim is typically formalized into an experimental hypothesis.

An experiment is fundamentally different than a survey because of the extra control that is imposed in the environment from which the data are collected. For example, if you wanted to get a solid idea of whether acupuncture was effective at relieving pain, you could just survey a lot of people, ask them if they had ever had acupuncture, and then ask them how much it helped on a scale of, say, 0 to 5, with 5 being "a lot." But if you collect your data this way, you are not controlling the environment at all. Your subjects may have had acupuncture from all different kinds of practitioners, and they may have had really different levels of baseline pain. This all gets mixed up in the survey results and makes it challenging to see any usefulness to the treatment, if there is indeed any usefulness to be seen. You would have a lot more controls in place if you were trying to collect the same kind of data in an experimental setting. You may recruit subjects who see the same type of acupuncturist, make sure that participants were treated with the same types of needles, had the same ambient music, the same site preparation, etc. Control: it's what makes an experiment an experiment.

The creation of the experimental conditions is critical to the validity of the experimental results. While statistical science has an important role in that phase of the experiment, it has its most visible inferential role when the experimental hypothesis is tested in a mathematical sense.

Troublemaker: Confounding

The focus on control in experimentation produces some of the purest data that can be collected. It is both an art and a science to optimally design an experiment toward that end.

The ways in which you build in these controls are part of what statisticians call the "experimental design." Entire courses are taught on experimental design, but most students are unlikely to be designing their own statistical experiments in college or in their workplaces. More than likely, you, like most of the rest of us, will simply be trying to understand the results of experiments that others have performed.

While experimental data are typically particularly clean data, **confounding** can still compromise inferences from these data. This is of critical interest to us because the integrity of the inferences we make, both formally and informally, from experimental data directly depends on whether we have properly handled confounding. You can read more about the history of confounding in statistical science in the very interesting piece at the following link: http://www.medicine.mcgill.ca/epidemiology/hanley/c607/ch09/Confounding_Hx_Vandenb.pdf

So what is confounding? In the vernacular, "to confound" just means to confuse or to mix up …. You may well hear it used in conversation as a synonym for confusion. That intuitive use is very close to how it is used in statistical science.

If the goal of experimentation is to determine whether a cause and effect relationship exists between two variables (say, presence or absence of acupuncture treatment and level of pain relief), then we want the experiment to be able to speak clearly to that relationship. Confounding is when a third variable, not of any primary interest, distorts that case for causation. This is often because that third variable (say, trust in the procedure to reduce pain, or type of acupuncture needle used) is related in some way to both of the primary variables and may introduce some cause and effect of its own. Therefore, the case for causation becomes confused or muddled.

Example Revisited

Let's revisit the acupuncture example. Let's suppose we want to know if acupuncture treatment causes a reduction in pain. What's wrong with the following design?

- You identify a clinic willing to partner with you for your study. When patients show up for the study, you carefully categorize their baseline pain and only choose study participants who have similar levels of baseline pain in similar physical locations.

- Then you have them fill out a pain survey where they rate their pain coming into the study. Then they are treated at the clinic and asked to rate their pain again, once the treatment is over.

This sounds fine on the surface and was the basic design of countless medical studies once upon a time. The problem is that you won't have any way of knowing if the patients feel better just because they were being treated, regardless of the effectiveness of the treatment. Preposterous you say? Not at all! It turns out we have a rather amazing ability to feel better for reasons that have nothing to do with an active treatment. We will name this important phenomenon below.

Necessary Language

Granted, confounding is a very general term and many types of confounding could be identified. We have chosen to take a very broad, categorical view here and only look at two types:

> Two types of confounding: inadequate or improper comparison in the experimental design, and lack of randomization.

This is an oversimplification to the extent that it would make a lot of serious statisticians cringe, but it is useful for our purposes nonetheless.

We also need a common language for discussing these ideas. Here are the basic definitions we will use. They are likely already familiar to you:

> A "response variable" is the primary variable you are taking measurements on for your experiment.

> The "explanatory variable" is what you are varying in your experiment.

> "Subjects" are who or what you are doing the experiment on.

> A "lurking variable" is another name for that third variable that can cause confounding.

> Finally, the "placebo effect" is a real response from subjects to an inert or inactive treatment. This is what we were alluding to in the brief comments we made above about a hypothetical acupuncture experiment.

The next section in this chapter will discuss several examples of confounding. We will have a chance to practice using this terminology, and soon enough, it will become second nature.

Proper Comparisons and Randomization

Arthroscopic surgery for joint pain or dysfunction has become the most commonly performed orthopedic surgery in the United States. It isn't just athletes, but a diverse group of people, including those suffering from arthritis.

How have doctors traditionally assessed the efficacy of this treatment? See the graphic below.

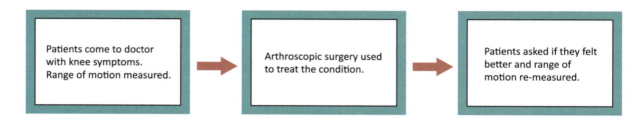

On the one hand, their reasoning seemed to make a lot of sense. Patients presenting with the appropriate symptoms had their level of function measured and were asked to rate their level of pain, both before and after the surgery. By and large, patients felt they had less pain and exhibited an increase in their range of motion. Simple enough. Seems like arthroscopic surgery works.

The Placebo Effect

What is missing with the kind of comparison just described? What we don't know is how much the results might have been confounded by the placebo effect. That is, is it possible that the patients felt better and had a better range of motion in part just because they had felt their problem was being addressed, even if the actual surgery was not effective?

This is the question that was asked by the team of physicians and scientists from the Houston Veterans Affairs Medical Center at the Baylor College of Medicine. In this study, 180 patients with osteoarthritis of the knee were randomly assigned to the traditional arthroscopic surgery and a placebo surgery. This type of placebo surgery is called a "sham surgery." Those patients received skin incisions and had the sights and sounds of the real surgery simulated. After surgery the patients in both groups were followed for 24 months and measurements both on pain and function were taken.

What this team of researchers found was surprising. The outcomes as measured by these pain and function scales were no better for the real surgery group than for the placebo group. More formally, the differences seen between the two groups could not be relegated to anything other than chance, so they were not statistically significant. In fact, in 2002, recommendations to consider avoiding arthroscopic surgery for knees were released, though the impact of those recommendations on practice is unknown.

Vocabulary and Diagrams

Let's practice using our language some more. In the arthroscopic knee surgery example, there are two response variables mentioned: "level of pain" and "level of function." The explanatory variable is "type of procedure," (real or placebo), and the subjects are the participating patients.

Confounding was present in the original design because the physicians did not allow for a placebo comparison. So there was an absence of necessary comparison in their experimental design. In the Baylor study, this source of confounding was addressed by employing a sham surgery.

It is sometimes helpful to be able to diagram the design of an experiment. Before the placebo-controlled study of arthroscopic surgery, patients' pain and function were compared before and after surgery. So there was a comparison taking place, but comparison alone is not always enough. In this case, as we've just seen, the placebo effect was influencing the post-surgery measurements. So the right kind of comparison is important.

If we diagram the experiment that was conducted at Baylor, then we see that the confounding created by the placebo

Diagramming the Faulty Design

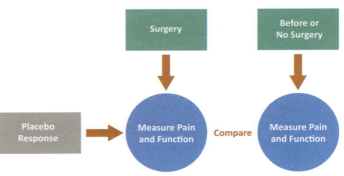

Diagramming the Careful Design

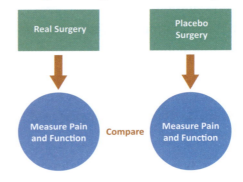

effect was controlled by a direct comparison between the real surgery and an elaborate placebo surgery, also known as a "sham" surgery. Since patients were randomized to these two treatments, it stands to reason that any differences in post treatment pain and function that exist between the two groups can be attributed to the treatments being different.

Randomization

Let's turn now from comparison to randomization. In the mid-to-late 1980s, the disease AIDS was just making its way prominently into the public consciousness. It came as no surprise that pharmaceutical companies were in a frantic rush to be the first to offer a drug that would treat patients with early symptoms of the disease and profit accordingly. Milan Panic, then the flamboyant chairman of ICN Pharmaceuticals in California, was the first to claim success in a January 1987 news conference. The FDA disagreed.

Here is a summary of the results that ICN reported from their medical experiments. About the same number of patients participated in three different treatments, two with different doses of the drug ribavirin, and the third, a placebo treatment. On the surface, nothing seems amiss. In fact, a close look at the table of results shows that the drug, especially in 800 mg form, seems to be highly effective compared to a placebo. What was the FDA's objection?

To quote a June 5th, 1990, article by Michael Lev in the *New York Times*, "The agency questioned the methods used in the test. Dr. Frank E. Young, then Commissioner of the F.D.A., publicly challenged the tests at an AIDS conference in Washington, D.C., because the group receiving a placebo in the study might have contained more patients considered seriously ill than were in the group that received ribavirin, skewing the results in ribavirin's favor."

TABLE 3.1

Treatment	Number of Patients	How Many Developed AIDS
800 mg	52	0
600 mg	55	6
Placebo	56	10

In short, there was evidence, or at least suspicion, that ICN had not randomized their subjects to the three treatments. Hence, the confounding owing to health of the patients at the time they entered the study compromised the integrity of the inference about "effective treatment" that Mr. Panic had been hoping for.

It's worth reviewing the formal language one more time. In the ribavirin study, the response variable was the "development of AIDS (yes or no)," the explanatory variable was the "type of intervention (level of Ribavirin, placebo)," and the subjects were "the 163 patients participating in the study."

The confounding, as mentioned already, was caused by the "lack of randomization into the treatments." This could have been easily avoided with a purposeful randomization of patients to the various treatment groups.

Randomization helps to keep the treatment groups as much alike as possible. That is, if a large group of patients is randomized into two treatment groups then, odds are, the variation inherent in that larger group will be somewhat represented in both of the two treatment groups.

Statistical Significance

As we have noted, experimental data are among the purest data that can be collected because of the control that can be designed into a well-thought out experiment. But even if confounding is not an issue affecting inferences from experimental data, there is still a hurdle to overcome. Let's look at an example.

Flibanserin was originally created as a drug designed to treat depression. While it proved unsuccessful for that purpose, researchers noticed it seemed to have a positive effect on female sex drive. Subsequently, studies with flibanserin were conducted involving pre-menopausal women with generalized acquired Hypoactive Sexual Desire Disorder.

The studies were mostly well-designed experiments. One key study involved 1,378 women who were randomized to one of two treatments—either they received flibanserin or a placebo. All the women were required to keep a record of whether they had sex, and if they did, whether it was satisfying in their view. The participants were screened for depression and other medical problems, eliminating a series of possible confounding variables.

Here's what happened.

- At the end of the study period, the flibanserin women reported an average of 4.5 sexually satisfying events, up from 2.8 at baseline before treatment. That's a 61% increase.

- Women in the placebo group reported an average of about 3.7 sexually satisfying events, up from 2.7 at baseline, a 37% increase.

So 61% increase for the flibanserin group and 37% increase for the placebo group. That's sounds pretty convincing.

What's the Issue?

While the FDA wanted to know about this actual difference, they also wanted to know something much more technical and probabilistic. They wanted to know *how likely* that difference between 61% and 37% was to have happened by chance alone. The FDA needed to know if the results of the experiment were statistically significant. A useful, non-technical definition of statistical significance is shown below:

> Experimental results are **statistically significant** if the observed difference among the treatments is large enough to have been unlikely to have happened by chance.

This is an unapologetically probabilistic statement. It is perhaps the one phrase that consumers of statistical inference encounter most often. It's a kind of "badging" of results, saying something very technical about their importance.

In this chapter we don't address any of the mathematics behind this important concept, opting instead for just this informal introduction.

What about the flibanserin study? Indeed, the difference between 61% and 37% was judged to be statistically significant. That is, the experiment was able to conclude that this difference between the flibanserin-treated patient group and the placebo group was large enough to have been unlikely to have occurred by chance. That is, there was no reason to doubt that the difference was due to flibanserin. This was a very important hurdle to get over for the drug company.

This story ended in an unusual fashion, however. The practical benefit of flibanserin was very small, and there were possible side effects. So on June 18, 2010, a federal advisory panel to the FDA unanimously voted against recommending approval in spite of the statistical significance. The struggle for approval continued until early August 2015 when the FDA finally approved the drug. The small company that had been able to bring flibanserin through these last few, important regulatory hurdles was quickly bought by a large pharmaceutical company for nearly a billion dollars.

TAKE-HOME POINTS

- Credible inferences from experimental data have to be free from confounding, which has been categorized as having two primary sources—lack of proper comparison and improper randomization of subjects to treatments.
- The placebo effect and lack of randomization can create very real obstacles to making credible inferences from experimental data.
- Credible inferences from experimental data have to ultimately be held to a mathematically formal standard of statistical significance in order to assess whether the results were likely to have occurred by chance.

Chapter 3 Exercises

Reading Check

Short Answer: Please provide brief, concise answers to the following questions.

1. What is the statistical definition of "confounding?"

2. What are the two primary sources of confounding discussed in the assigned reading?

3. What is a "response variable?" Give an example not covered in the assigned reading.

4. What is an "explanatory variable?" Give an example not covered in the assigned reading.

5. Describe a simple experiment that measures one variable's effect on another. Name at least one potential confounding variable.

6. How did an understanding of the placebo effect change arthroscopic knee surgery?

7. When done correctly, explain how comparison addresses confounding.

8. Suppose a social experiment is being done to compare two ways that a police officer can handle a domestic violence call. Effectiveness is measured by number of call-backs to that address within a week. How might the treatment randomization be handled here?

9. What critical mistake did the manufacturers of ribavirin make in their early drug studies?

10. What percentage of all the patients studied in the ribavirin trial developed AIDS.

11. What were the treatments in the flibanserin study?

12. What percent increase from baseline in the number of sexually satisfying events was reported by women in the placebo group?

13. What is the definition of "statistical significance?"

14. Were the flibanserin results statistically significant?

15. Why didn't the FDA approve flibanserin?

Beyond the Numbers

Your instructor may choose to have you answer the multiple choice exercises as open response, or you may be required to show your work. Follow your instructor's instructions.

1. Slippery Evidence and Confounding

Please read the excerpts shown from one of the early studies on the effectiveness of online instruction. The notation "s.d." in the article is used for "standard deviation."

Title: Learning in an Online Format versus an In-Class Format: An Experimental Study

Authors: Allan H. Schulman and Randi L. Sims

Source: *T.H.E. Journal* 26, no. 11 (1999): 54–56

Methodology Students enrolled in five different undergraduate online courses during the Fall semester 1997 participated in a voluntary test-retest study designed to measure their learning of the course material. These students were compared with students enrolled in traditional in-class courses taught by the same instructors.

Subjects In total, 40 undergraduate students were enrolled in the online courses and 59 undergraduate students were enrolled in the in-class courses during the testing period.

Pre-tests Instructors designed pre-tests to measure the level of knowledge students had of the course content prior to the start of the course. The average pre-test score for online students was 40.70 (s.d. = 24.03). The average pre-test score for in-class students was 27.64 (s.d. = 21.62).

Post-tests Instructors designed post-tests on a 100-point scale to test students' knowledge at the end of the course. The average post-test score for online students was 77.80 (s.d. = 18.64). The average post-test score for in-class students was 77.58 (s.d. = 16.93).

Results [O]ur results indicate that there were no significant differences for post-test scores.

a. Looking at the information given in the **Results** section, why might you infer that in-class instruction is no better than online instruction?

   i.   The standard deviations for both sets of pre-test scores were nearly equal.

   ii. The pre-test scores weren't very different from the corresponding standard deviations.

   iii. There was no significant difference in the post-test scores.

   iv. The pre-test scores were not deemed significantly different.

b. What is one obvious source of confounding that threatens the article's conclusion that online learning is no less effective than face-to-face learning?

   i. Students got to self-select; that is, they got to decide for themselves which course (online or face-to-face) they enrolled in.

   ii. Students were not able to decide for themselves which course (online or face-to-face) they enrolled in.

   iii. There were significantly more males in the online class than in the face-to-face class.

   iv. The pre-test scores of the two groups were identical, making it virtually impossible to say one type of class was better than the other.

c. Diagram this experiment similar to the examples given in the reading.

2. Make Mine a Large

In the 2009 *New York Times* piece "Excess Pounds, but Not Too Many, May Lead to Longer Life," author Roni Caryn Rabin reported:

> Being overweight won't kill you—it may even help you live longer. That's the latest from a study that analyzed data on 11,326 Canadian adults, ages 25 and older, who were followed over a 12-year period. The report … found that overall, people who were overweight but not obese—defined as a body mass index of 25 to 29.9— were actually less likely to die than people of normal weight, defined as a B.M.I. of 18.5 to 24.9.
>
> By contrast, people who were underweight, with a B.M.I. under 18.5, were more likely to die than those of average weight. Their risk of dying was 73% higher than that of normal weight people.

a. Although this article doesn't describe an experiment, it does imply that being a little overweight may lead to a longer life. Identify at least one confounding variable that may compromise the validity of this inference and support your choice.

3. Brains and Beats

Do children who study music perform better in school? One of the studies supporting this claim was conducted by University of California professor Gordon Shaw and reported in a 1999 edition of the *Deseret News*. In this study, students in the 95th Street School, one of Los Angeles' 100 poorest-performing institutions, received both piano lessons and automated mathematics

training. Their ability to understand and analyze ratios and fractions was then compared to a 1997 study involving students from under-achieving schools in Orange County who were given automated and traditional mathematics instruction (but no musical training).

The *News* reported that "[t]he Los Angeles students scored 2% higher than their Orange County counterparts in their ability to understand and analyze ratios and fractions—concepts usually not introduced until sixth grade."

a. How do you know that the subjects in this study could not have been randomized to the treatments that were compared? This was one of the more widely-reported studies that supported music training as positively impacting general academic performance.

   i. The article excerpt says that the students at the 95th Street School self-selected which treatment they wanted to be in.

   ii. The article excerpt implies that random assignment was against fair-treatment policies in the Orange County system and, therefore, could not be utilized.

   iii. The treatments were paired, so every student ended up getting both treatments and, hence, no randomization was needed.

   iv. The treatments were at two separate schools at two separate points in time.

b. Aside from randomization issues, what is another possible source of confounding (other than lack of randomization) that might challenge a conclusion that exposure to music caused the L.A. students to do better.

   i. The group in Orange County had both traditional and automated mathematics instruction (not just automated).

   ii. The group in L.A. had both traditional and automated mathematics instruction (not just automated).

   iii. The group in Orange County had both traditional and automated mathematics instruction (not just traditional).

   iv. The group in L.A. had both traditional and automated mathematics instruction (not just traditional).

4. Fuzzy Quasi Is a Bear

Quasi experiments are studies that are unable to use randomization to evaluate effectiveness of interventions. This can make it difficult to tease out possible confounders and assess the integrity of any cause and effect claims.

Consider this example adapted from a paper in *Clinical Infectious Diseases* ("The Use and Interpretation of Quasi-Experimental Studies in Infectious Diseases," Vol. 38, Issue 1, pp. 1586–1591). A hospital wants to know if providing alcohol-based hand cleaners for staff will reduce the rate at which bacterial infections occur in the patient population. The hospital needs to design a study that will address this question. Access the article, read it, and answer the following.

a. Suppose a researcher suggests to place these cleaners in randomly chosen patient rooms and not in others. At the end of the study, the infection rates for each group could be compared. Why does the referenced paper say this simple design might be an impossible design to implement in practice?

   i. To quote the paper, "most hospitals have strict rules about the use of randomization where medical treatments are concerned."

   ii. To quote the paper, "it is difficult, politically, to implement use of an alcohol-based disinfectant only in certain parts of a hospital or only on certain sides of a ward."

   iii. To quote the paper, "it would be difficult medically to have cleaners in rooms where patients have or may have serious allergies to ingredients that most of us would find innocuous."

   iv. To quote the paper, "all nursing professionals are trained in the proper use of quasi-experimental data and, as such, would find a randomized trial to be awkward and untenable."

b. The hospital will more likely adopt a quasi-design that collects data for an extended period of time before hand cleaner dispensers are installed, and then compares those findings with data collected after the dispensers are installed. What are two possible sources of confounding with this design that the referenced article mentions?

   i. Severity of patient illness and quality of hand cleaner.

   ii. Quality of medical care and severity of internal, ambient pollution.

   iii. Quality of medical care and how years of experience of hospital administration.

   iv. Severity of patient illness and quality of medical care.

c. In the absence of random assignment, arguments addressing confounding often have to be made some other way. What are two arguments that one might be able to make to support results obtained from a quasi-design like the one in Question 2?

5. Experimentation Takes Flight

This exercise is appropriate for an in-class activity or a mini-project. Follow your instructor's lead on how she wants it to be completed. A set of subjects needs to be divided into two groups denoted as Group A and Group B. Members of Group A will build the paper airplane Design A. Members of Group B will build the paper airplane Design B.

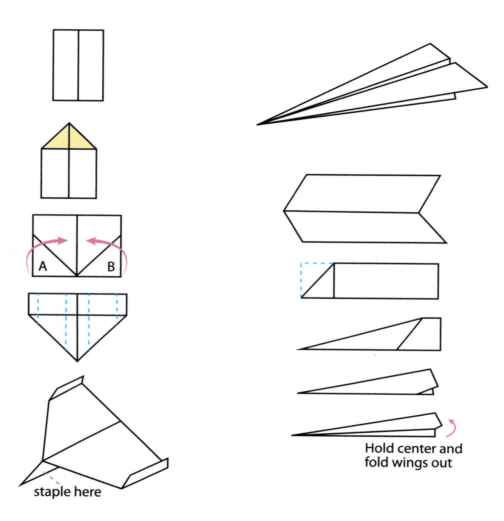

staple here

Hold center and fold wings out

Planes will be flown per instructor's directions. The linear distance traveled will be measured. This can be done very precisely or can be done in a clever and quick way (for example, the number of tiles or stairs passed). Your instructor will explain what she wants. Record the measurements in a table like the one shown below and compute the average distance flown for each design.

TABLE 3.2 Airplane Distance Traveled

Design A	Design B
Average:	Average:

a. Identify the explanatory and response variables in this experiment.

b. Identify at least one potential source of confounding and suggest how it might be rectified.

c. By just looking at the two sets of data points and their averages, does there seem to be a difference due to design? Elaborate.

6. Catching on to Experimentation

This exercise is appropriate for an in-class activity or a mini-project. Follow your instructor's lead on how he wants it to be completed. A set of subjects needs to be divided into two groups denoted Group R and Group L. Group R will perform this experiment with their right hands and Group L will use their left hands. Here's what you need to do.

■ Choose a subject to begin. Make sure that he or she knows which hand to use.

■ The subject extends his or her hand and holds out the thumb and forefinger in a pinching stance (about 1" gap).

- A group member will hold a ruler above the subject's hand so that when the ruler is let go, the subject will be able to catch it by pinching his or her fingers together. The zero centimeter mark on the ruler should be level in between the subject's fingers.

- The subject will be instructed to catch the ruler with his or her two fingers as soon as possible after it is dropped. No advance notice about release will be given.

- Record the position of the subject's fingers on the ruler when he or she catches it. Convert that number to a reaction time in seconds using Table 3.3, and record it in a table similar to the one shown as Table 3.4.

- Repeat until all subjects have performed.

TABLE 3.3

Distance (cm)	Time (sec)	Distance (cm)	Time (sec)
1	0.045	16	0.181
2	0.064	17	0.186
3	0.078	18	0.192
4	**0.090**	19	0.197
5	0.101	20	0.202
6	0.111	21	0.207
7	0.119	22	0.212
8	0.128	23	0.217
9	0.135	24	0.221
10	0.143	25	0.226
11	0.150	26	0.230
12	0.156	27	0.235
13	0.163	28	0.239
14	0.169	29	0.243
15	0.175	30	0.247

a. The data in Table 3.4 come from the formula:

$$\text{Distance} = (0.5)\, at^2$$

where "t" is time, in seconds, and "a" is acceleration due to gravity (9.8 meters/sec). Verify that the time t corresponding to a distance of 4 cm is 0.090 sec. Show all your work.

b. Diagram this experiment similar to the examples given in the reading.

c. What is the response variable in this experiment?

d. What is the explanatory variable?

e. What is a potential confounding variable? How might this have affected the experiment?

f. Find the means of both groups using the data you collect (ostensibly in Table 3.4, or similar). Based on those two values, is there evidence of a difference between the reaction times of Group L and Group R? Defend your answer.

TABLE 3.4

Group R	Time (sec)	Group L	Time (sec)
1		1	
2		2	
3		3	
4		4	
5		5	
6		6	

g. What role would the variance of the measurements in each group play in refining the precision of the assessment you just gave that only considered means? Explain.

h. Instead of having some people use their left hands and some use their right hands, the experiment could have been designed so that all of the subjects used both their right and left hands, in random order. State and defend two reasons why this might have been a better design to have used.

7. Cancer Carafe

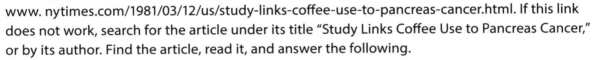

On March 12, 1981, the *New York Times* reported on a Harvard study that linked coffee consumption and pancreatic cancer. Article author Harold Schmeck Jr. noted that "[t]he report estimated that more than half of the pancreatic cancer cases that occurred in the United States might be attributable to coffee drinking ...." To complete the assignment, begin by reading Schmeck's article: http://www.nytimes.com/1981/03/12/us/study-links-coffee-use-to-pancreas-cancer.html. If this link does not work, search for the article under its title "Study Links Coffee Use to Pancreas Cancer," or by its author. Find the article, read it, and answer the following.

a. What two groups were being compared in this experiment?

i. 369 pancreatic cancer patients at 11 hospitals in the Boston metropolitan area were compared to 369 patients similar in age and sex who were hospitalized for reasons unrelated to the pancreas.

ii. 644 pancreatic cancer patients at 11 hospitals in the Boston metropolitan area were compared to 369 patients similar in age and sex who were hospitalized for reasons unrelated to the pancreas.

iii. 369 pancreatic cancer patients at 11 hospitals in the Boston metropolitan area were compared to 644 patients similar in age and sex who were hospitalized for reasons unrelated to the pancreas.

iv. 644 pancreatic cancer patients at 11 hospitals in the Boston metropolitan area were compared to 644 patients similar in age and sex who were hospitalized for reasons unrelated to the pancreas.

b. At least two other sources of circumstantial evidence were cited in the article as further support of a link between pancreatic cancer and coffee. Which one of the following is one of the two mentioned?

i.   The apparent increase in frequency of cancer of the pancreas in the United States in recent decades and the low rates observed in certain British groups who do not drink coffee.

ii. The apparent increase in frequency of cancer of the pancreas in the United States in recent decades, and the low rates observed in such groups as Mormons and Seventh Day Adventists, who do not drink coffee.

iii. The apparent increase in frequency of cancer of the pancreas in such groups as Mormons and Seventh Day Adventists, who do not drink coffee, with no such increases seen in the population at large.

iv. The fact that the general population and such groups as Mormons and Seventh Day Adventists, who do not drink coffee, all seemed to have roughly the same pancreatic cancer rates.

c. Schmeck followed up on this article with another one titled "Critics Say Coffee Study Was Flawed." List three potential sources of confounding mentioned in this article and comment on why these could potentially destroy any claim to cause and effect? You can find the article online at https://www.nytimes.com/1981/06/30/science/critics-say-coffee-study-was-flawed.html?searchResultPosition=1. If this link does not work, search for the article under its title or by its author.

8. Of Mice and People

Read the article "Misleading Mouse Studies Waste Medical Resources" by Erika Hayden, which appeared in Nature on March 26, 2014. You may find it at https://www.nature.com/news/misleading-mouse-studies-waste-medical-resources-1.14938. If this link does not work, search for the article under its title or by its author.

a. The article addresses amyotrophic lateral sclerosis (ALS), suggesting that mice studies might be misleading. The article gives two reasons why. What are they?

i.   The odd fact that the mice and humans with ALS tended to die from the same cause and the lack of reproducibility in the mice studies.

ii. The fact that mice in the study tended to die from a different cause than humans with ALS and the lack of reproducibility in the mice studies.

iii. The fact that mice in the study tended to die from a different cause than humans with ALS and the lack of reproducibility in the human studies.

iv. The fact that mice in the study tended to die from bowel obstruction, unlike the humans who died from ALS and the lack of reproducibility in the mice studies.

b. A 1958 amendment to the Food, Drugs, and Cosmetic Act of 1938 called the "Delaney Clause" has been instrumental in the banning of food additives since its enactment. Find the exact text of the one-sentence Delaney Clause. Which one of these is it?

i. "… [T]he Secretary of the Food and Drug Administration shall not approve for use in food any chemical additive found to induce cancer in man, or, after tests, found to induce cancer in animals."

ii. "… [T]he President of the United States shall not approve for use in food any chemical additive found to induce cancer in man, or, after tests, found to induce cancer in mice."

iii. "… [T]he Secretary of the Food and Drug Administration shall not approve for use in food any chemical additive found to induce cancer in man, or, after tests, found to induce cancer in nonhuman primates."

iv. "… [T]he President of the United States shall not approve for use in food any chemical additive found to induce cancer in man, or, after tests, found to induce cancer in animals."

c. In February 2014, sandwich giant Subway announced it would stop using azodicarbonamide in its breads. Research this decision and comment on the indirect role that the Delaney Amendment had in Subway's decision. Find the CNN article by Elizabeth Landau about this issue. It has been available at this link: https://www.cnn.com/2014/02/06/health/subway-bread-chemical/index.html. Why did Subway stop using azodicarbonamide in its breads?

i. Because it is a substance that is used in yoga mats and tennis shoe soles.

ii. Because it was found to be statistically negligible in increasing desirable bread elasticity.

iii. A derivative was found to cause cancers of the lung and blood vessels in mice.

iv. It was part of a common agreement on the part of all commercial bakers to make their bread dough more natural.

9. Random Reflections

It is easy to confuse different sources of randomization and the reasons as to why each is important to experimentation. The following table lists the likely effect when random assignment is used (or not) and random sampling is used (or not). Six entries have been left out of the table and scrambled in the list below. Match the effect descriptions to their corresponding letters in Table 3.5.

1. Confounding not addressed; results generalize to population _____
2. Experimental conclusions do not generalize to population _____
3. Confounding not addressed; so weak claim for causation with results _____
4. Confounding addressed; results generalize to population _____
5. Experimental conclusions generalize to population _____
6. Confounding addressed; results don't generalize beyond sample _____

TABLE 3.5

	Random Assignment to Treatments	No Random Assignment to Treatments	
Subjects Randomly Sampled from Population	A	B	C
Subjects Not Randomly Sampled from Population	D	Confounding not addressed; results don't generalize beyond sample	E
	Confounding addressed; so solid claim for causation with results	F	

10. Random Opposition

Not everyone thinks that random assignment is ethical or even sensible. A full treatment of that discussion is beyond the scope of our mission here, but it is instructive to consider a problem that was the topic of a debate in the 1920s between two prominent statisticians, Sir Ronald Fisher and William Gossett.

a. Suppose you have a field that is divided into 24 rectangular plots as shown here. Two crop varieties (A and B) are to be assigned at random to those 24 plots and their yields compared after a season. An online randomizer (Research Randomizer©) was used to make the assignments. The result is shown to the right.

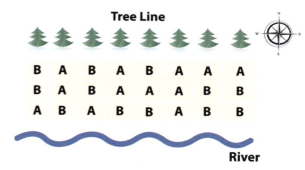

Suppose you know before you plant that the ground decreases in fertility as you move from the river to the tree line. This random assignment has resulted in nine of the twelve B varieties appearing in the lower two rows of plots, and nine of the twelve A varieties are in the upper two rows. Hence, B has a distinct advantage with respect to fertility.

Design a purposely *non-random* distribution of A and B that would effectively balance out any North-South *and* East-West variation in soil quality, sun exposure, etc. Enter your letters in the empty plot below. You should end up with 12 As and 12 Bs. Make sure you offer support for your design.

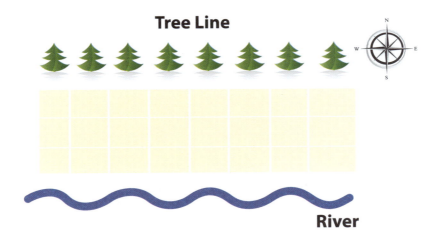

Tree Line

River

11. A Badge of Big

Statistical science endeavors to answer the question "are treatment results different enough that they are unlikely to have occurred by chance?" If so, the results are said to be statistically significant. In another exercise you may have completed an activity where twelve reaction times were computed—six for a group using their left hands and six for a group using their right hands. Your instructor may want you to use those values rather than the twelve results recorded in Table 3.6. Follow her instructions.

TABLE 3.6

Group R	Time (sec)	Group L	Time (sec)
1	0.090	1	0.111
2	0.119	2	0.181
3	0.143	3	0.090
4	0.169	4	0.186
5	0.064	5	0.045
6	0.150	6	0.143

a. What is the mean of the six Group R measurements?

   i. 0.0016

   ii. 0.1260

   iii. 0.0030

   iv. 0.1225

b. What is the mean of the six Group L measurements?

   i. 0.0016

   ii. 0.1260

   iii. 0.0030

   iv. 0.1225

c. What is the variance of the six Group R measurements? You will probably want to use a software package. The variance is the square of the standard deviation. See Chapter 2.

   i. 0.0016

   ii. 0.1260

   iii. 0.0030

   iv. 0.1225

d. What is the variance of the six Group L measurements? You will probably want to use a software package. The variance is the square of the standard deviation. See Chapter 2.

    i. 0.0016

    ii. 0.1260

    iii. 0.0030

    iv. 0.1225

e. Compute the following value:

$$z = \frac{(\text{mean of Group R}) - (\text{mean of Group L})}{\sqrt{\frac{\text{variance of Group R}}{6} + \frac{\text{variance of Group L}}{6}}}$$

The difference between the left-hand reaction times and right-hand reaction times is not statistically significant if $-2.23 < z < 2.23$. Are the results from the data table in this exercise statistically significant?

    i. No, they are not; z is computed to be -0.12689, which is inside the interval given.

    ii. No, they are not; z is computed to be 0.12689, which is inside the interval given.

    iii. Yes, they are; z is computed to be 1.2689, which is outside the interval given.

    iv. Yes, they are; z is computed to be -1.2689, which is outside the interval given.

f. What does statistical significance mean in the context of this activity?

12. Designer Thoughts

Background

Whether an experiment is found to produce statistically significant results is not just a function of the effectiveness of the treatments being evaluated. The design of the experiment and how the data are analyzed can make a huge difference as well. The data on the right represent the reaction times of subjects playing an online game designed to test hand-eye coordination.

There are two ways to look at these data.

Scenario A: As 24 individuals randomly assigned to two groups (left and right hand usage)

Scenario B: As 12 individuals who each had to use both their left and right hands in a paired experiment.

Let's see why it matters. The decompositions below determine whether the results are statistically significant or not. Some technical details have been compromised so that the presentation remains accessible.

TABLE 3.7

Left	Right
1	1.05
0.74	0.76
0.66	0.71
0.78	0.79
0.68	0.69
0.65	0.72
0.75	0.75
0.69	0.72
0.94	0.99
0.79	0.8
0.81	0.82
0.62	0.67

If which hand is being used is going to show up as statistically significant in this experiment, then the ratio of the "Variance Attributed to the Hand" to the "Variance Left Unexplained" is going to have to be 0.45 or larger.

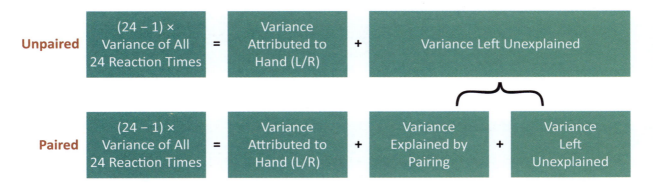

a. Verify that the variance of all 24 observations in Table 3.7 is 0.013234. You may want to use software to do this.

b. Look at the data from the perspective of Scenario A. If the Variance Attributed to Hand is 0.005400, will which hand is being used show up as statistically significant?

   i. Yes. Ratio of the "Variance Attributed to the Hand" to the "Variance Left Unexplained" is 0.005400.

   ii. Yes. Ratio of the "Variance Attributed to the Hand" to the "Variance Left Unexplained" is 0.0181.

   iii. No. Ratio of the "Variance Attributed to the Hand" to the "Variance Left Unexplained" is only 0.005400.

   iv. No. Ratio of the "Variance Attributed to the Hand" to the "Variance Left Unexplained" is only 0.0181.

c. Let's look at the data from the perspective of Scenario B. If the Variance Attributed to Hand is 0.005400 and the variance explained by the pairing is 0.29608, what is the ratio of "Variance Attributed to Hand" to the "Variance Left Unexplained?" Is this ratio big enough to say there is a statistically significant difference in reaction times?

   i. Yes. Ratio of the "Variance Attributed to the Hand" to the "Variance Left Unexplained" is 1.86.

   ii. Yes. Ratio of the "Variance Attributed to the Hand" to the "Variance Left Unexplained" is 0.296.

   iii. No. Ratio of the "Variance Attributed to the Hand" to the "Variance Left Unexplained" is only 0.296.

   iv. No. Ratio of the "Variance Attributed to the Hand" to the "Variance Left Unexplained" is only 1.86.

d. Take a few moments to reflect on what you've shown in this exercise. What are the implications for human inference if paired data were not analyzed as paired (or an experiment wasn't designed as paired when it could have been)?

13. What to Believe?

Background

When it comes to experimentation, principled implementation and ethical reporting are obviously critical to the integrity of all associated human inference. Unfortunately, there are a surprisingly large number of cases where dishonesty or exploitation—not just carelessness or confounding—tainted or nullified the significance of the findings. You will investigate three examples of research misconduct in this activity.

a. Piltdown Meltdown, 1912

In 1912, Charles Dawson discovered two skulls found in the Piltdown quarry in Sussex, England. These skulls were said to be from a primitive hominid and were hailed as the "missing link" between man and ape. Research the Piltdown Man and write a well-formed summary of what ultimately was revealed about the skulls. Give specific details about the nature of the deception that was uncovered.

b. Marker Mice, 1974

William T. Summerlin used to work at Memorial Sloan Kettering Cancer Center in New York City, conducting research in transplantation immunology. His work could have had major implications for reducing the rejection rates of transplanted tissue. Research Dr. Summerlin and write a well-formed summary of what ultimately was revealed about his work. Give specific details about the nature of the deception that was uncovered.

c. Doing the Dishes, 2010

Dr. Vipul Bhrigu was a researcher at the University of Michigan when he allegedly began to feel professionally threatened by the work of a graduate student in the same lab, Heather Ames. It was alleged that, desperate for his work not to be overshadowed, he concocted a plan to keep his student from moving ahead. Research Dr. Bhrigu and write a well-formed summary of what ultimately was revealed about his deception. Give specific details about the nature of the sabotage that was uncovered.

Projects

1. Designing Your Own Experiment: Part I

Introduction

Your instructor may choose to continue this project throughout other parts of the book once you acquire the foundation required to analyze your results.

The scope of what is required depends on whether you are being asked to do this individually or as part of a group. Your instructor will clarify for your class.

The Assignment

Experiment or Observational Study? They aren't the same. In an observational study subjects are observed and data are collected, but no manipulation of the subjects' environment occurs. The treatment that each subject receives is not determined by the investigator. In a randomized experiment, the manipulation of the subjects' environment is the purpose of the treatments, and those treatments are applied at random by the investigator. Your instructor may be willing to accept either one for this project, but you will want to make sure at the outset. We just use the word "experiment" below without distinction.

Designing Your Experiment: You need an experiment that meets the following requirements:

Individual Project

Have a minimum of 50 subjects randomized to two groups, formed from a single change in a **single** variable.

- ▸ E.g., suppose the topic is "Does listening to music affect one's ability to memorize information?"

Apply the treatments being compared to the two groups.

- ▸ E.g., Group 1 might try to memorize a short list of words while listening to music on headphones. Group 2 has the same task but is allowed to do the memorization in a quiet atmosphere. Both groups have the same amount of time to look at the list. The variable being changes is the presence (or not) of music.

Collect the data you are interested in.

- ▸ E.g., you could collect the number of words each person remembered from their lists. Or you could collect the proportion of words each person remembered from their lists.

Group Project

Same as the Individual Project, but with a minimum of 150 subjects.

The Submission

Your final project report should be professionally prepared, typed, and proofread. There should be three sections.

a. **Background Research:** You need to do extensive, college-level background research into the question you are investigating in your experiment. Find out what is already known about the topic from similar experiments. For example, if you were to look at the effect of distractions on memorization, you should look at what other *experiments* have found about this or a similar topic. Don't just look for non- experimental opinions from experts. Be sure to not rely on anecdotal opinions as research (e.g., "The guys in my fraternity all say they can concentrate better when they listen to music …").

b. **Subject Recruitment:** It is very unlikely that you can select a genuine random sample of subjects to participate in the experiment. This also happens in real experimentation. One often ends up with whomever they end up with as subjects. Still protocols are needed. Give some thought to whether you want to put limits on age (minimum and maximum), gender (what proportion of each), etc. for participants. In this section, describe in detail how you recruited your subjects and document any issues that arose.

c. **Data Collection:** Describe in detail how your data were collected and why you chose to collect them that way. Document all data that are have been collected in a computer generated table with subject IDs, and their responses.

2. Reality Video

Introduction

This chapter discussed how confounding can create problems with our human inferences. Now, we want to give you a chance to both show you know what you are doing and to help teach others.

The Assignment

Your instructor may have you do this as individuals or may form you into teams. We'll refer to "teams" below, although you may be a team of one depending on what your instructor wants. The following is required by each team:

a. Pick one or more topics from this chapter. Your instructor may want to approve them before you proceed.

b. Construct a three-minute video that teaches the chosen topic to the class. You must carefully and correctly use the language of the chapter in your presentation, and you are required to give an example that is NOT in the chapter.

c. Post the video on YouTube and send the link to your instructor in the manner he stipulates.

4 CHAPTER 4
Sampling: Purpose and Challenges

Introduction

The 18th century British essayist Samuel Johnson was purported to have said "You don't have to eat the whole ox to know the meat is tough." Similarly, if you know how to sample the right way, then you don't have to interview everyone in the whole country to have a very good idea of what they think about a particular issue.

That is the goal of sampling: to make inferences about a population from what we know about a much smaller sample. While it is not something one can do perfectly, it is something that one can do probabilistically. That is, it is something one can do and offer a mathematical defense of one's confidence in the inference.

Let us be a little more specific. We often are confronted with situations where we want to know some number in a larger population that we just can't know perfectly. For example, what proportion of all Americans agree with the 2013 Supreme Court ruling that the Defense of Marriage Act was unconstitutional? What's the average amount of money that Americans over 50 have saved for retirement?

To know these kinds of numbers perfectly could entail interviewing millions of people, which is a completely impossible task. What we can do is take a sample of some of those millions and compute those numbers of interest just from the sample. We would then like to infer that the corresponding number in the population is close to the number we have been able to compute from the sample.

It turns out we cannot always be assured that the number we want to know is close to the one we have been able to compute. However, we are able to make meaningful statements about how likely a given range around that sample number is to contain the one we do not know in the population. In one sense it is a mathematical way of proclaiming how much confidence we have in our estimation procedure.

Not surprisingly, identifying who (or what) to include in the sample is of critical importance. We will have a lot to say about that. Once the sample is determined, contact has to be made and data collected. There are many ways to do this. Door-to-door surveys, telephone surveys, or internet surveys are all commonly used. But let us not get ahead of ourselves. Before we learn formal rules, we want to make sure we understand the role of common sense in sample selection. Let us look first at a famous example that helped shape that common sense while the field of statistical science was still quite young.

How NOT to Sample

In 1936, New Deal incumbent Franklin D. Roosevelt was running for re-election against Kansas Governor Alf Landon. While it was generally thought the race would be tight, the *Literary Digest* predicted that Landon would win the popular vote by a margin of 57% to 43%. The *Digest* had called every election correctly for twenty years, and its predictions were highly respected.

In the end FDR won 60.8% of the popular vote and carried the electoral vote in all states except two. It was the largest electoral landslide in the history of the two-party system at that point. What did the *Digest* do wrong?

As they had always done in the past, the *Digest* used telephone lists and lists of automobile owners to select their sample. This meant, however, that they ended up targeting mostly middle and upper-class citizens, who did, in the end, vote for Landon. However, the lower classes, thanks in large part to the New Deal, voted overwhelmingly for Roosevelt. This was the first time in modern history that an election split this decisively along socio-economic lines.

The *Digest* tried to apologize and even laugh at itself, but it never recovered from this public embarrassment and went bankrupt shortly thereafter. This was not just an issue about how the sample was taken, but about the lists available to sample from. Remember, the science of statistics was really very new in 1936, some would say dating only to the 1920s, and there were no polling organizations the way we think of them today. The first take-away point is that lists from which samples are chosen have to be unbiased.

Secondly, some samples are just as a result of voluntary response and don't involve sampling from any kind of list. Data collected this way can be very misleading. Here is one famous example. Ann Landers once asked, "If you had it to do over again, would you have children?" She ended up with over 10,000 responses and about 70% of those recorded a resounding "NO!" Wisely, Ms. Landers acknowledged not long after that survey that people with the strongest feelings about a topic are those most likely to respond to a voluntary poll question.

Just how biased is a voluntary response poll? Typically, you just don't know. In this case we are able to get some idea of how biased the Ann Landers poll might have been because *Newsday* commissioned a poll of 1,373 parents, being careful to employ statistically correct sampling methods. *Newsday* found that 91% of parents said they would have children again. That's a pretty remarkable difference from the 30% that Ann Landers found.

Finally, bias is sometimes intentional. So-called "push polls" are designed to push respondents in a particular direction. Push polls find their way into all major political parties, it seems, at one time or another. A push poll will often ask a very biased question, one that, even if completely untrue, might plant a seed of doubt in the minds of the recipients. Also, if the poll was successful in pushing responses in a particular direction, it is a convenient way to generate a biased headline the next day in the local papers.

An often quoted push poll came from the 2000 Republican primary when the Bush camp asked voters in South Carolina, "Would you be more likely or less likely to vote for John McCain for president if you knew he had fathered an illegitimate black child?" Never mind that McCain had no so-called "illegitimate" children. The damage was done and a headline was generated.

Frugging is a similar kind of idea. Frugging is fundraising under the guise of legitimate research. The goal really is to raise money; asking questions that are very biased toward the positions of the target audience seems to increase contributions. Our common sense rightly tells us that inferences from push poll data or frugging data will likely be nonsense.

Inappropriate samples are sometimes easy to avoid and recognize, but not always. Aside from the common sense assessment that biased samples are not that useful, a more subtle and more important fact is that with biased samples, we cannot say anything mathematically meaningful about how good the inferences from the samples are. The science of statistics is largely concerned with how one can make these important mathematical statements when the samples have been taken the right way.

Simple Language

Now that we have had an introduction to the idea of statistical sampling, let us turn to some simple but critical language that will facilitate a more meaningful, more complex discussion about sample-based inferences. We have already used some of this language above because it is language in our vernacular. We need to make sure we fully absorb it now, however, if we are to develop a real appreciation of how statistical inference intersects with our daily lives.

> The **population** is that larger collection of subjects or items that you are interested in understanding, which is way too large or too complex for you to examine each subject or item individually.

> A **sample** is a collection of subjects or items that you select from the population to examine.

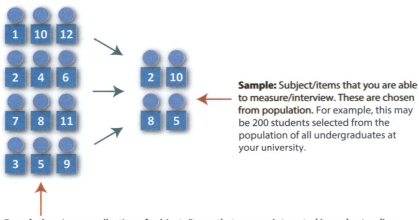

Sample: Subject/items that you are able to measure/interview. These are chosen from population. For example, this may be 200 students selected from the population of all undergraduates at your university.

Population: Larger collection of subjects/items that you are interested in understanding something about. For example, this may be all undergraduates enrolled at your university.

We have already noted above that there are some obviously wrong ways to select a sample from the population. At the same time, there are many reasonable ways. Be aware, however, that correct statistical sampling means correct in a mathematical or probabilistic sense. It is quite a bit more complicated than just some common-sense idea of "fair" or "reasonable."

As we have said, the population is that larger collection of subjects or items that you are interested in understanding. Let us be a bit more specific about what we might be interested in:

> A population **parameter** is a number that describes some population characteristic we are interested in.

It may be that it is a *proportion*, such as the true proportion of all undergraduate students at your university who would answer "Yes," to the question "Do you support same-sex marriage?" if all were asked. Or it could be an *average*, such as the average student debt for all students at your university. We almost never will know population parameters. That's why we sample.

In a well-chosen sample from the population:

> The sample **statistic** is a number of interest to the researcher that describes the sample.

A statistic you may be interested in might be the sample proportion of all students in your sample who responded "Yes," to the question: "Do you support same-sex marriage?" Or it could be the sample mean student debt for all students in your sample. Refer to the graphic below. **Remember S-S: Statistic/Sample and P-P: Parameter/Population.**

The purpose of statistical sampling is to be able to estimate the population parameter using the sample statistic. Determining this estimated value, along with an expression of some mathematical assessment of confidence in the estimate, is the most basic act of formal statistical inference.

Beware: It is naïve to just assume that the parameter will more or less be the same as the statistic, or else the sample must have been somehow messed up when it was taken. If it were only that simple.

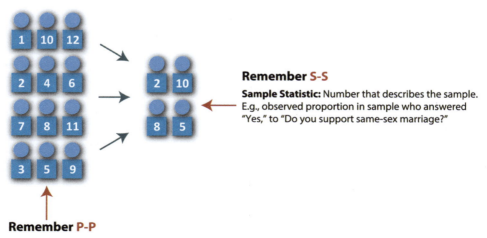

Remember S-S
Sample Statistic: Number that describes the sample. E.g., observed proportion in sample who answered "Yes," to "Do you support same-sex marriage?"

Remember P-P
Population Parameter: Number that describes the population. E.g., true proportion of all students at your university who would answer "Yes," to "Do you support same-sex marriage?"

Fortunately, statistical inference is based on some really beautiful mathematics that ultimately allow us to both give coherent expression to the estimate and to say how reliable that estimate is. Our first step toward understanding how all of this works is to make sure we understand how essential it is that the sample be taken the right probabilistic way. Much more has to be said about this. First, however, let us look at an example from the news and make sure we are comfortable with the language that has been introduced.

Example: Gun Laws

Here is an excerpt from a CBS News/*New York Times* poll concerning gun laws.

Title: Poll: Majority of Americans Back Stricter Gun Laws

authors: Sarah Dutton, Jennifer De Pinto, Anthony Salvanto, Fred Backus, Leigh Ann Caldwell

Source: CBS News, January 17, 2013. https://www.cbsnews.com/news/poll-majority-of-americans-back-stricter-gun-laws/

As the president outlined sweeping new proposals aimed to reduce gun violence, a new CBS News/*New York Times* poll found that Americans back the central components of the President's proposals, including background checks, a national gun sale database, limits on high capacity magazines and a ban on semi-automatic weapons. Asked if they generally back stricter gun laws, more than half of respondents—54%—support stricter gun laws … That is a jump from April—before the Newtown and Aurora shootings—when only 39% backed stricter gun laws but about the same as ten years ago.

…

This poll was conducted by telephone from January 11–15, 2013, among 1,110 adults nationwide. Phone numbers were dialed from samples of both standard land-line and cell phones. The error due to sampling for results based on the entire sample could be plus or minus three percentage points.

In this article:

- The **sample** is comprised of the 1,110 adults contacted in this telephone survey taken in early January 2013. What was of interest in this sample? The poll wanted to know whether the person being interviewed backed the central components of President Obama's gun proposals.

- The survey yielded a **statistic** of 54% who said they did back those components.

- To identify the **population,** you have to ask yourself what larger group did the researchers want to address? In this case, that group seems to be "All Americans."

■ Finally, the population **parameter** is the true but unknown proportion (or percentage) of all Americans who would have said they back the central components of the President's gun proposals, had it been possible to ask all Americans. This is just a number between 0 and 1 (or percentage between 0% and 100%). We just don't know what the value of that number is.

Simple Random Sampling

If we don't take our sample the right way, then it is virtually impossible to say anything mathematically meaningful about how good a statistic is as an estimate of the parameter of interest. "Right way" means more than "fairly" or "carefully." It is a probabilistic sense of "right way." The concept of a "simple random sample" is perhaps the easiest way to think about what this means, though it not such an easy sample to take in practice.

> A **simple random sample** (shorthand "SRS") is a sample of size n that is chosen from the population in such a way that every set of n individuals had the same chance of being chosen.

This is not a very intuitive definition! Notice that it doesn't say anything about making sure you have the same number of men as women in your sample, or some equal distribution across different age groups. It is unapologetically probabilistic in nature. The following example will help clarify the concept.

Example: Simple Random Sampling

Suppose there are five students who qualify academically to be valedictorian speakers at their upcoming graduation. Unfortunately, it is only possible for two of them to speak at the formal ceremony. The Dean of Students decides the best way to choose those two is to select a simple random sample of two from the five student names.

| Waylon | Thalia | Marcia | Angela | Abdulla |

With only five students in the population, it is easy to just list all possible samples of two students:

1. Waylon and Thalia
2. Waylon and Marcia
3. Waylon and Angela
4. Waylon and Abdulla
5. Thalia and Marcia
6. Thalia and Angela
7. Thalia and Abdulla
8. Marcia and Angela
9. Marcia and Abdulla
10. Angela and Abdulla

If the Dean is going to select a simple random sample of size two, then she has to sample in such a way that all samples of size 2—and there are ten of them here—are equally likely to be chosen. So each sample of size two will have to have the same one-in-ten chance of being chosen.

You have to admit this is a very formal way to think about sampling. You would also be right to imagine that the counting we did here would get difficult very quickly. For example, suppose there are 100 students in your class, and you wanted to select an SRS of five of those. If we continue to think as we have just above, then we'd have 75,287,520 such samples to list. If you live in a little town of 30000 residents with drivers' licenses and want to choose an SRS of size 100 from those, then there are 4.6915×10^{289} possible subsets of size 100. Imagine the Gallup Poll choosing an SRS of 1200 "likely voters" from 300 million. The number of possible subsets of size 1200 can't even be computed, though there is a well-known formula for it.

Fortunately, we do not have to ever actually specify all possible samples in order to select an SRS. We just need some accessible list of the population elements. This list is often called a "frame" since it may not correspond exactly to the population of interest. It may be a list of all voter registrations, all drivers' license numbers, or all possible landline and cell phone numbers in a certain location. Still, we should prepare ourselves to be impressed! The beautiful mathematics mentioned above allow statistical science to say something very concrete about *all those possible samples* of a given size, *without having to list them all*. We will start to unpack that connection in the next chapter. But first, before we look at how one might select an SRS in practice, let us make sure we understand that by no means are all seemingly rational samples simple random samples.

Suppose we take a sample of two students from a high school class where you are student teaching. You decide to do this by starting at a randomly chosen spot and then alternating, left-to-right, choosing the first male and female student encountered. That might seem fair in sense, and it does have a random element to it. But it is not an SRS. Why not? Certainly, a possible sample might be two males sitting side-by-side in the same row, but the chances of selecting that sample of size 2 using this alternating sex strategy is zero. However, any male and female students sitting side-by-side have a non-zero chance of being chosen as the sample. So those two different samples of size 2 have *different chances of being chosen*, so the sampling plan could not have been an SRS. See the graphic.

This sample of size 2 has a zero chance of being chosen.

This sample of size 2 has a non-zero chance of being chosen.

Practically Speaking

The best way to think about selecting a simple random sample is to imagine "mixing up" your population and reaching in with your eyes closed and pulling out a subject one at a time until you have your sample of size *n*. This ends up satisfying the mathematical definition of an SRS. However, rarely will you be able to physically mix up your population and select a sample in this way. Still yet, there are many ways of actually identifying a sample of size *n* if you are able to label the objects in your population and have access to those labels. Typical labels might be names, house numbers, social security numbers, drivers' license registrations, etc.

One useful tool for selecting a simple random sample is "Research Randomizer" (https://www.randomizer.org/). To use Research Randomizer, you have to have your population of *N* objects numbered 1 to *N*. Then it is easy to identify a sample of size *n*. You will get a chance to practice with Research Randomizer in the exercises. Here's how it works. If you had access to a list of 20,000 voter registration IDs and wanted to select a simple random sample of 100 of those, you'd do the following:

- Number the voter registrations from 1 to 20,000 (order in which that is done is not important).
- Instruct Research Randomizer to give you 100 unique numbers from 1 to 20,000.
- Map the 100 numbers returned to the corresponding voter registrations, and that is your sample.

The sample of size 100 is chosen in such a way that all samples of size 100 are equally likely to be chosen. That is, it is a simple random sample.

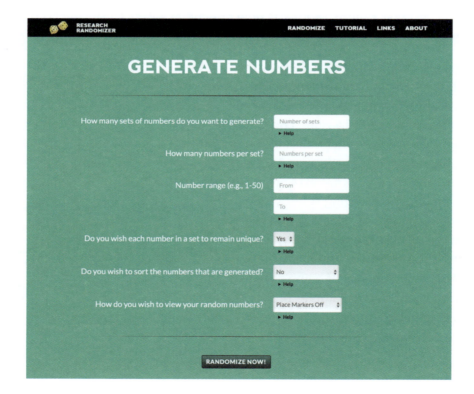

Key Idea Revisited

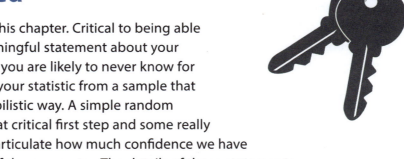

Let us revisit the main idea of this chapter. Critical to being able to make a *mathematically* meaningful statement about your population parameter—which you are likely to never know for certain—is to have computed your statistic from a sample that was chosen in a correct probabilistic way. A simple random sample is one such sample. That critical first step and some really neat mathematics allow us to articulate how much confidence we have in the statistic as an estimate of the parameter. The details of those statements occupy the next chapter.

Keep in mind that a correct or fair sample in this probabilistic sense is not a cross-sectional sample. That is, you are not guaranteed that a simple random sample will be some kind of "cross section" of the population on all features you deem important (race, income, education level, etc.). This may be the most misunderstood issue with probabilistic sampling there is. Let's look at a famous example.

Dewey Defeats Truman

Often called the world's most famous newspaper error, the banner proclaiming "Dewey Defeats Truman" was plastered across the front page of the *Chicago Tribune* on November 3, 1948, the day after the incumbent President defeated the New York Governor, Thomas Dewey, by a popular-vote margin of 50% to 45%. There are several reasons the *Tribune* got it wrong, including going to press early. But several polling organizations also got it wrong, including Gallup (predicted 44% Truman to 50% Dewey), Crossley, and Roper.

What happened to the polls is generally thought of as a stark reminder that cross-sectional samples can't do for you what probabilistic samples can, especially since you'll never know if they were actually an accurate cross-section. You have to give Gallup credit for really trying to create a reasonable cross-section. For example, in St. Louis a Gallup poll interviewer was required to interview 13 subjects:

- exactly six live in suburbs, seven in the central city;
- exactly seven were to be men, six women;
- of the seven men, three under 40 and one black, six white;
- rental prices paid had to be one $44.01 or more, three $18.01 to $44.00, and two under $18.00.

This is very nicely thought out in one sense, but as a sampling plan, it clearly missed the mark ultimately. That is, it failed to capture a true cross-section of the population in the sample, or otherwise as went the sample, so had to go the population. This clearly did not happen.

This way of thinking about sampling, which was typical in the early days of polling, is still widely confused with probability sampling. On the one hand, if you are able to pull off a true cross-sectional sample, then you don't need statistical science to tell you about how good your inferences are going to be. Indeed, if the cross-sectional sample is perfect, then there is no uncertainty in that sample with respect to how the population would respond.

The issue, of course, as illustrated by the famous example above, is that you are never going to know if your best effort at a cross-sectional sample has been good enough. If it is not, then your inferences about the population could be very wrong. And you can't make any meaningful mathematical statements about how wrong you might be! If your sample is a probability sample, like an SRS, your inferences can most definitely still be wrong. But you have the added benefit of being able to quantify, in a sense, how likely that is to happen. That's the trade-off! Mainstream statistical science has long taken the random sampling approach, preferring to be able to offer concrete mathematical statements about the uncertainty of sample statistics.

TAKE-HOME POINTS

- Avoiding biased samples is just good common sense, whereas making meaningful statements from proper samples is the business of statistical science.
- A probabilistic sample like an SRS is a critical step in the process of estimating a population parameter with a sample statistic and being able to make a mathematical statement about how well that has gone.

Chapter 4 Exercises

Practice and Basic Comprehension ───────

Short Answer: Please provide brief, concise answers to the following questions.

1. What is the goal of sampling?

2. Define and distinguish "sample" and "population." Give an example of each.

3. Define and distinguish "statistic" and "parameter." Give an example of each.

4. Explain what a simple random sample is. Use probabilistic language.

5. Suppose you had a population with only six items in it. List all samples of size two that could be chosen.

6. Suppose you want to sample opinion in your fraternity about a proposed alcohol ban on campus. You alphabetize your membership roster and then you choose every 10th name on that list until you have your sample. Is this an SRS? Explain.

7. What is the tool Research Randomizer useful for?

8. What is the key to being able to mathematically express confidence in your parameter estimate?

9. What was the likely mistake made by the Gallup organization during polling for the 1948 Presidential election?

10. What is the difference between a cross-sectional sample and a simple random sample? Give examples of each.

Beyond the Numbers ─────────────────

1. Texting Error

Please read the following excerpt from the *New York Times*:

Title: Poll Finds Support for Ban on Texting at the Wheel

Author: Marjorie Connelly

Source: *New York Times*, September 27, 2009, http://www.nytimes.com/2009/09/28/technology/28truckerside.html

Read the following extract from the above article and answer the related questions to see if you understand the data.

> The public overwhelmingly supports the prohibition of text messaging while driving, the latest *New York Times*/CBS News poll finds. Ninety percent of adults say sending a text message while driving should be illegal, and only 8% disagree.
>
> …

The *Times*/CBS News telephone poll was conducted September 19–23 with 1,042 adults nationwide and has a margin of sampling error of plus or minus three percentage points.

a. How was the sample taken, and what was the result of the survey?

   i. The poll was a telephone poll and 90% of the population said sending a text message while driving should be illegal.

   ii. The poll was texted to 90% of the population and all said sending a text message while driving should be illegal.

   iii. The poll was a telephone poll and 90% of the sample said sending a text message while driving should be illegal.

   iv. The poll was texted to 90% of the population and 8% said sending a text message while driving should be illegal.

b. Suppose someone said to you, "Sure, of the 1,042 surveyed by the poll, 90% agreed, but I bet if you interviewed all American adults, you could potentially find only 50% agreeing!" Is this a possible scenario and, if so, why?

   i. Yes, it is possible. The 90% was just from a sample of American adults, not results from all American adults.

   ii. No, it is not possible. You could never see 90% in a sample when the result for the entire population was 50%.

   iii. Yes, it is possible. The 50% was just from a sample of American adults, not results from all American adults.

   iv. No, it is not possible. You could never see 50% in a sample when the result for the entire population was 90%.

2. A Weak Majority

Please read the following excerpt from the *New York Times*:

Title: Poll Finds Slim Majority Back More Afghanistan Troops

Author: Adam Nagourney and Dalia Sussman

Source: *New York Times*, December 9, 2009. http://www.nytimes.com/2009/12/10/world/asia/10poll.html

Read the following extract from the above article and answer the related questions.

A bare majority of Americans support President Obama's plan to send 30,000 more troops to Afghanistan, but many are skeptical that the United States can count on Afghanistan as a partner in the fight or that the escalation would reduce the chances of a domestic terrorist attack, according to the latest *New York Times*/CBS News poll.

…

The support for Mr. Obama's Afghanistan policy is decidedly ambivalent, and the nation's appetite for any intervention is limited. Over all, Americans support sending the troops in by 51% to 43%, while 55% said setting a date to begin troop withdrawals was a bad idea.

…

The poll was conducted by telephone from Friday through Tuesday night, with 1,031 respondents, and has a margin of sampling error of plus or minus three percentage points.

a. About how many people in the sample supported sending more troops to Afghanistan?

   i. 51% of 30,000 which is about 15,300 people

   ii. 30% of 1,031 which is about 309 people

   iii. 51% of 1,031 which is about 53 people

   iv. 51% of 1,031 which is about 526 people

b. Suppose the headline had read as follows: "Poll Finds Slim Majority of All Americans Back More Afghanistan Troops." Offer some objections to that and support your position.

3. Push and Pull Opinions

Please read the following excerpt to learn more about so-called "push polls."

Title: RNC Distributes "Obama Agenda" Push Poll to Raise Money

Author: Evan McMorris-Santoro

Source: Last accessed 2021: *Talking Points Memo*, November 6, 2009,

https://talkingpointsmemo.com/dc/rnc-distributes-obama-agenda-push-poll-to-raise-money

In an article posted to talkingpointsmemo.com, writer Evan McMorris-Santoro discusses an article published previously in the *Washington Post* about frugging at the Republican National Committee. Frugging is fundraising under the guise of marketing research and is widely considered by marketers as an unethical marketing practice. Read an extract from McMorris-Santoro's article below.

The *Washington Post* published a story about a version of the survey after one of its reporters received one in the mail in Virginia. From the paper's report:

Surveys designed to persuade rather than survey are a common though dirty tactic in the political arena, the text equivalent of telephone push-polls. The sending of polls for fundraising purposes is also widely considered unethical, a practice known as "frugging"—fundraising under the guise of research. In August, the RNC suggested in a similarly formatted "Future of American Health Care Survey" that "GOP voters might be discriminated against for medical treatment in a Democrat-imposed health care rationing system." Following on outcry from Democrats, a Republican Party spokesperson called that survey "inartfully worded."

The paper called out some of the questions in the survey as "overstating" the truth about Democratic views:

Obama has made no moves as president in support of reinstituting the draft. Rep. Charles Rangel (D-N.Y.) has several times introduced a measure to bring it back, but his bill has little support among Democrats.

RNC spokesperson Gail Gitcho declined to comment on the questions found in the survey, nor would she comment on how many of the surveys have been distributed.

"The document surveys Republican opinion," she said when asked about the survey. "And raises a little money."

The RNC is not the only organization to use these types of surveys. Political candidates from both parties, as well as organizations like the League of Women Voters have also used them. Following is an extract from a poll sent out by the League of Women Voters. Read through the questions and the shaded box at the end of the poll.

NATIONAL OPINION SURVEY

7. In your opinion, what is the best way to limit the impact of special interest money on the political process? Please choose one answer only.
   - ☐ Public financing of congressional elections.
   - ☐ Lower limits on the amount individuals can donate to candidates.
   - ☐ Better disclosure requirements for all political contributions.
   - ☐ Other _____

8. Do you support or oppose the ruling by the U.S. Supreme Court that says corporations and unions can spend as much money as they want to elect or defeat political candidates?
   - ☐ Strongly support
   - ☐ Support
   - ☐ Oppose
   - ☐ Strongly oppose

9. In view of the Supreme Court's ruling, would you support or oppose an effort by Congress to reinstate limits on corporate and union spending on election campaigns?
   - ☐ Strongly support
   - ☐ Support
   - ☐ Oppose
   - ☐ Strongly oppose

10. In your opinion, do you believe partisan redistricting determines the result of too many Congressional elections before any votes are cast?
   - ☐ Strongly agree
   - ☐ Agree
   - ☐ Disagree
   - ☐ Strongly disagree

11. How likely are you to vote in the 2012 Presidential election?
   - ☐ Very likely
   - ☐ Somewhat likely
   - ☐ Not likely

12. Do you support or oppose laws that require voters to show photographic identification before casting their ballots?
   - ☐ Support – These laws help combat voter fraud.
   - ☐ Oppose – These laws reduce turnout and disenfranchise youth as well as ethnic and racial minority voters.

The survey continues on the next page ...

13. Do you support or oppose laws that allow people to register to vote and cast a ballot at the polls on election day?
   - ☐ Support – These laws help increase participation in our elections.
   - ☐ Oppose – These laws increase the likelihood of voter fraud.

14. Which of the following issues is the most important to you? Please rank from 1-12 with "1" being the most important issue to you.
   - ___ Education
   - ___ Wars in Iraq/Afghanistan
   - ___ Immigration
   - ___ Health care
   - ___ Ethics in government
   - ___ Tax fairness
   - ___ Voting rights
   - ___ Jobs/Economy
   - ___ Energy and Environmental policy
   - ___ Campaign finance reform
   - ___ Homeland security
   - ___ Reproductive rights

Thank you for your participation.
Please sign below to submit your responses for tabulation.

☐ Yes, include my responses in the survey tabulation _____ SIGNATURE

☐ Yes, to help strengthen our democracy and put political power back in the hands of the American people, I will join the League of Women Voters. Enclosed is my contribution in the amount of:
   - ☐ $20   ☐ $25   ☐ $35   ☐ $55   ☐ Other $___

Please charge my contribution to my:
☐ Visa   ☐ MasterCard   ☐ American Express   ☐ Discover

CARD NUMBER _____

EXPIRATION DATE ___/___

SIGNATURE _____

EMAIL ADDRESS _____

Female Voter
1234 Main Street
Cincinnati, OH 45230-1234

Survey No. FVXXX1234

Please make your check payable to the League of Women Voters of the United States (LWVUS) and return it with this form.

Have a look at the League of Women Voters' survey above.

a. Is this an example of frugging? Explain.

b. Are there any obvious biases in the questions that you think would affect the accuracy of the poll results?

4. Monkey Surveys

SurveyMonkey®, a free online survey tool, offers us examples of five common survey question mistakes (see https://www.surveymonkey.com/mp/5-common-survey-mistakes-ruin-your-data/). We have repeated some of their examples below. In each case, describe what is wrong with the question being asked and give some suggestions on how to fix the problem. You are encouraged to visit the site referenced above for help.

a. Should concerned parents use infant car seats?

b. Where do you enjoy drinking beer?

c. How satisfied or dissatisfied are you with the pay and work benefits of your current job?

d. Do you always eat breakfast? (Yes/No)

e. Do you own a tablet PC?

5. Random Evolution

The definition of a simple random sample (SRS) can be confusing: An SRS of size n is a sample of size n, chosen in such a way that all samples of size n have the same chance of being chosen. It doesn't help that the word random is used in many different ways, but when it comes to selecting a simple random sample, we have to be very careful to know its technical meaning. We will explore these issues in this next set of exercises.

On November 30th, 2012, National Public Radio ran a short segment entitled "That's So Random: The Evolution of an Odd Word." You may find this segment at http://www.npr.org/2012/11/30/166240531/thats-so-random-the-evolution-of-an-odd-word.

a. List two uses of the word "random" from the audio that are different from the technical definition given above.

6. Careful Counting

The audio segment above ends with Charlie McDonnell (of the British "Fun Science" videos) noting that "every now and then, at random, you end up with something awesome." We might take that to mean that every now and then, a simple random sample is representative of a population with respect to a certain list of demographics. Let's look at a simple example to see how likely that might be. Suppose you have a population with two men—one a Republican and one a Democrat; and two women—one a Republican and one a Democrat.

a. There are six possible distinct samples of size two from this four-person population. If M = male, F = female, R = Republican, and D = Democrat, then the possible samples are:

i. (MR, MR), (MR, FD), (MR, FD), (MR, FR), (MD, FD), (FD,FD)

ii. (MR, MD), (MR, FR), (MR, FD), (MD, FR), (MD, FD), (FR,FD)

iii. (MD, MD), (MD, FR), (MR, FR), (MD, FR), (MD, FD), (FD,FD)

iv. (MR, MD), (MR, FR), (MR, FR), (MR, FR), (MD, FD), (FD,FD)

b. For a simple random sample of size two, all samples of size two have the same chance of being chosen. What would the likelihood be of choosing any one of these samples?

i. 1 in 2

ii. 2 in 6

iii. 4 in 10

iv. 1 in 6

c. Suppose for a sample of size two to be "representative" of the population, it has to have exactly one man and one woman, and one Democrat and one Republican. What is the chance of selecting a simple random sample of size two from this population that is representative (in this sense of the word)?

i. 1 in 2

ii. 2 in 6

iii. 4 in 10

iv. 1 in 6

7. Social Media Sampling

Suppose you have 113 Friends on Facebook and you want to choose a simple random sample of 20 of them.

a. What is your population?

i. Your 113 Friends on Facebook

ii. The 20 Friends selected from Facebook

iii. All registered users of Facebook

iv. The proportion of all Facebook users who are your "Friend"

b. Describe in detail how you would select your simple random sample.

8. Bravos for Bucks

The VIP brand Kindle Fire cover received 4,945 reviews on Amazon by early 2012, averaging a nearly perfect 4.9 stars out of five. That's quite impressive. It is tempting to think that online reviews, especially those posted at major sites like Amazon, are representative of consumer experiences. We know, however, that voluntary responses are often biased. But are product reviews even less accurate than previously thought? In his 2012 *Time* article "9 Reasons Why You Shouldn't Trust Online Reviews," Brad Tuttle writes, "You shouldn't believe everything you read. And if you're reading online reviews of products, hotels, restaurants, or local businesses or services Is this cited correctly? Then you should believe even less." You can find Tuttle's article online at http://business.time.com/2012/02/03/9-reasons-why-you-shouldnt-trust-online-reviews/.

a. There are several reasons listed in the article as to why you should be very cautious about online reviews. Which one of the following is not one of the first three reasons listed in the article?

    i. The marketplace for fake reviews operates fairly openly.

    ii. The Russians are known to plant positive reviews for socially detrimental consumer goods.

    iii. Companies give freebies in exchange for reviews.

    iv. Even if you think you can spot fake reviews, you probably can't.

b. What was VIP doing to boost the ratings of its Kindle Fire cover?

    i. Supplying a discount code for use in a Kindle Fire purchase if they posted a review.

    ii. Reimbursing customers for the cover if they posted a review.

    iii. Falsely upgrading customers' credit ratings if they posted a review.

    iv. Sending customers gift cards for iTunes if they posted a review.

c. How well did people perform in spotting fake reviews?

    i. They were right about 90% of the time.

    ii. They were right about 10% of the time.

    iii. They were right about 17% of the time.

    iv. They were right about 50% of the time.

9. Bezos Bosom Buddy

Fixing the problem of biased online product reviews is not going to be easy to do. Let's pretend for a moment that the founder and CEO of Amazon, Jeff Bezos, hired you to devise a solution that captures what you know about statistical sampling.

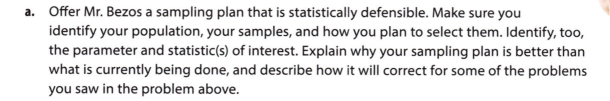

a. Offer Mr. Bezos a sampling plan that is statistically defensible. Make sure you identify your population, your samples, and how you plan to select them. Identify, too, the parameter and statistic(s) of interest. Explain why your sampling plan is better than what is currently being done, and describe how it will correct for some of the problems you saw in the problem above.

10. Gulliver Travels

900 people live in Gulliver, a small town in Michigan's Upper Peninsula. You want to know what proportion of Gulliver's population supports legalizing marijuana. Suppose you already know the following demographic information about Gulliver's 900 citizens:

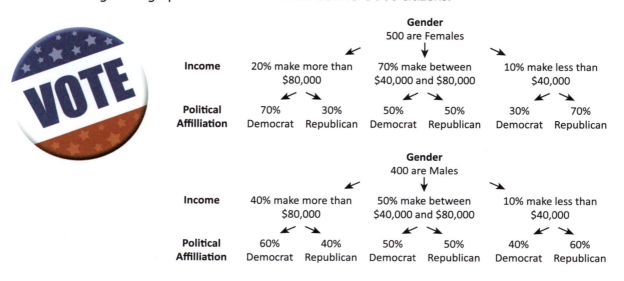

a. You have enough money to interview 90 residents. Working much the way Gallup did in the 1930s, you want your sample of 90 to mirror the distribution of subjects in the population exactly (at least along the lines of gender, income, and political affiliation). How many people would your sample place in the groups shown in the table below? If a calculation results in a partial person (e.g., 6.4 persons), leave the number as it is—don't round.

TABLE 4.1

Category	Number of Persons
Males	i.
Females	ii.
Males making between $40,000 and $80,000 yearly	iii.
Females making less than $40,000 per year who are Democrats	iv.
Male Republicans making over $80,000 per year	v.

b. Suppose the cross-sectional sample taken above represents a perfect microcosm of the larger population with respect to the legalization of marijuana. Is there any uncertainty involved in using this sample to represent the proportion of people in Gulliver who favor the legalization of marijuana?

    i. No. If truly random, then as goes the sample so goes the population.

    ii. Yes. Marijuana, while having no long-term ill health effects, is known to increase short-term anxiety and uncertainty.

    iii. No. If truly cross-sectional, then as goes the sample so goes the population.

    iv. Yes. Even if truly a perfect cross-section, sampling variability will create uncertainty.

c. Suppose you decided, instead, to take a simple random sample of Gulliver's population. Explain how you could take an SRS of size 90 from this population.

d. Will a carefully chosen simple random sample be representative of the population?

    i. Absolutely. That's what it means to be "random."

    ii. Absolutely. Provided it is, indeed, "carefully chosen."

    iii. Not necessarily. An SRS is not designed to be representative in a cross-sectional sense.

    iv. Not necessarily. If even one has tried to be really careful taking the SRS, it is still relatively easy to make mistakes in how it was selected.

11. Fondue Folly

Background

Two researchers at the Kellogg School of Management at Northwestern recently identified a new form of bias in surveys. Professors Gal and Rucker called this bias "response substitution" and referred to it as the "process whereby some respondents frame their answers to questions in a survey in such a way as to express their views on issues outside the survey's scope—issues on which they have strong opinions." What follows is one of three experiments they ran in an effort to see if the phenomenon was real.

> One-hundred and thirty-seven undergraduates were assigned to one of four treatments. Roughly a quarter were presented with Scenario 1 (below) and asked first to rate Anne's wastefulness, and then rate her intelligence (Treatment 1). Another quarter were also presented with Scenario 1, but were asked to rate Anne's intelligence before rating her wastefulness (Treatment 2).

> **Scenario 1:** Anne has a dinner party for some friends. She buys a fondue set and ingredients for a fondue dinner. The dinner party costs her $250. Her friends enjoy the meal and have a good time. She never uses the fondue set again.

> A quarter of the undergraduates were presented with Scenario 2 (below) and asked to rank Jane's intelligence first, then her wastefulness (Treatment 3). The remaining students, also presented with Scenario 2, were asked to rank Jane's wastefulness first, then her intelligence (Treatment 4).

Scenario 2: Jane has a dinner party for some friends. She reserves dinner at a fondue restaurant. The dinner party costs her $250. Her friends enjoy the meal and have a good time. All rankings were recorded on a five-point scale, with a score of 5 representing the highest perception of intelligent wastefulness. A plot of the average ratings in each treatment appears to the right.

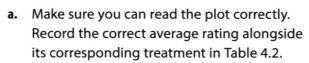
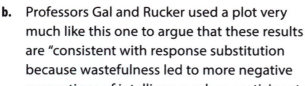

a. Make sure you can read the plot correctly. Record the correct average rating alongside its corresponding treatment in Table 4.2.

b. Professors Gal and Rucker used a plot very much like this one to argue that these results are "consistent with response substitution because wastefulness led to more negative perceptions of intelligence when participants did not have an opportunity to provide their attitude toward wastefulness." How does the plot support that conclusion?

TABLE 4.2

Treatment	Average Rating
1	i.
2	ii.
3	iii.
4	iv.

i. Anne was perceived as wasteful, and if her intelligence was rated first, then she received a higher intelligence score than when it was rated second.

ii. Anne was perceived as wasteful, and if her intelligence was rated second, then she received a higher intelligence score than when it was rated first.

iii. Anne was perceived as wasteful, and if her intelligence was rated second, then she received a lower intelligence score than when it was rated first.

iv. Jane was perceived as wasteful, and if her intelligence was rated second, then she received a higher intelligence score than when it was rated first.

c. How might you try to address response substitution when designing a survey? Briefly describe a study design that could be used to test the effectiveness of your suggestion.

12. No-Stumble Sampling

Background

A simple random sample is the easiest kind of statistically viable sample to select and measure. But how do you actually select an SRS? One useful tool that you were introduced to in this chapter is Research Randomizer, available at http://www.randomizer.org/. The following questions are designed to allow you to get familiar with this tool.

Data from the NHTSA's 1998 San Diego field sobriety test validation study can be found in this workbook's appendix. This data set is also available at www.statconcepts.com/datasets. There are 296 participants in this study, so there are 296 case numbers displayed. Note, though, that these case numbers do not run sequentially from 1 to 296. Your job is to use Research Randomizer to select a sample of 20 cases from this data set.

a. Explain how you plan to identify the cases for Research Randomizer

b. What entries did you use for the following Research Randomizer fields?

TABLE 4.3

How many sets of numbers do you want to generate?	How many numbers per set?	Number Range (e.g., 1–50)

c. For the 20 cases selected, fill out the following chart:

TABLE 4.4

Case Number	Actual BAC	Case Number	Actual BAC

d. What is the average BAC ("Blood Alcohol Content") of the 20 selected cases?

e. What proportion of cases in your sample had BACs at or above the legal limit of 0.04?

13. Social Media Sampling—Revisited

Suppose you have 113 friends on Facebook and you want to choose a simple random sample of 20 of them to ask a survey question you have constructed. Carefully explain how you could use Research Randomizer to select your sample.

14. Purposive Sampling

A simple random sample is the type of probability sample we encountered in the reading. The upside of using probability samples is that they allow mathematical assessments of the accuracy of sample-based parameter estimates. Purposive sampling is another type of sampling. It is not probabilistic, meaning that it does not adhere to probability theory. As a result, numerical assessments of sampling integrity are typically not possible. Still, one often sees these types of samples used—particularly in social science studies— so it is important to know what they are. You will look at four types of purposive sampling in this activity:

- Convenience Sampling
- Snowball Sampling
- Heterogeneity Sampling
- Expert Sampling

Use your research skills to look up each of these types of samples and answer the questions below.

a. What is meant by a *convenience sample?*

    i. A sample where the initially chosen subjects are asked to collect data from other subjects they choose, and then those other subjects have to recruit others.

    ii. A sample designed to capture opinions that are especially informed on the topic of interest.

    iii. A sample that is chosen because the subjects are easily accessed.

    iv. A sample designed to capture diversity across subgroups.

b. What is meant by a *snowball sample?*

    i. A sample where the initially chosen subjects are asked to collect data from other subjects they choose, and then those other subjects have to recruit others.

    ii. A sample designed to capture opinions that are especially informed on the topic of interest.

    iii. A sample that is chosen because the subjects are easily accessed.

    iv. A sample designed to capture diversity across subgroups.

c. What is meant by a *heterogeneity sample?*

    i. A sample where the initially chosen subjects are asked to collect data from other subjects they choose, and then those other subjects have to recruit others.

    ii. A sample designed to capture opinions that are especially informed on the topic of interest.

    iii. A sample that is chosen because the subjects are easily accessed.

    iv. A sample designed to capture diversity across subgroups.

d. Define what is meant by an *expert sample?*

  i. A sample where the initially chosen subjects are asked to collect data from other subjects they choose, and then those other subjects have to recruit others.

  ii. A sample designed to capture opinions that are especially informed on the topic of interest.

  iii. A sample that is chosen because the subjects are easily accessed.

  iv. A sample designed to capture diversity across subgroups.

e. Give some well-stated reasons why you think it would be difficult to infer from these types of samples to a larger population.

15. Getting Gallup

The following is an excerpt from Gallup's website describing their Daily Tracking Survey. The page was last available at http://www.gallup.com/174155/gallup-daily-tracking-methodology.aspx

> Gallup interviews U.S. adults aged 18 and older living in all 50 states and the District of Columbia using a dual-frame design, which includes both landline and cellphone numbers. Gallup samples landline and cell phone numbers using random-digit-dial methods. Gallup purchases samples for this study from Survey Sampling International (SSI). Gallup chooses landline respondents at random within each household based on which member had the next birthday. Each sample of national adults includes a minimum quota of 50% cellphone respondents and 50% landline respondents, with additional minimum quotas by time zone within region. Gallup conducts interviews in Spanish for respondents who are primarily Spanish-speaking.

a. What is the actual population being addressed by a Gallup telephone survey?

  i. Adults 18 years and older, living in the United States.

  ii. Adults 21 years and older.

  iii. All Adults 18 years and older, with access to at least one landline.

  iv. All Americans.

b. In what sense can a random-digit-dial sample be thought of as a simple random sample?

  i. The numbers are selected at random from the list of all people who have drivers licenses.

  ii. The numbers are selected at random, by a computer, for all working exchanges.

  iii. The numbers are carefully spread out over all exchanges, using a computer.

  iv. The survey expert is careful to make up telephone numbers at random and calls until one works.

16. Weighting Room

In 2012, the *National Journal* reported: "Critics allege that pollsters are interviewing too many Democrats—and too few Republicans or independents—and artificially inflating the Democratic candidates' performance." Suppose a simple random sample of voters yielded the following poll results:

TABLE 4.5 Results of a Random Sample of 100 Likely Voters

	Planned to Vote "Obama"	Planned to Vote "Romney"
80 Democrats	70%	30%
20 Republicans	20%	80%

a. What proportion of likely voters overall (Democrats and Republicans combined) planned to vote for Barack Obama?

   i.   56%

   ii. 45%

   iii. 60%

   iv. 35%

b. Do you think the proportion you provided above is an underestimate, overestimate, or neither?

   i.   It is not possible to know.

   ii. Overestimate

   iii. Underestimate

   iv. Neither an overestimate or an underestimate

c. Suppose that in the larger population, half of all likely voters are Democrats and half are Republicans. Now imagine that our simple random sample was comprised of 50 Democrats and 50 Republicans (instead of 80 Democrats and 20 Republicans), with the same percentages in each category (e.g. 70% of Democrats planning to vote for Obama) as seen in the actual results. Re-compute the proportion of likely voters who planned to vote for Barack Obama. What percent change is this from the computation you did above?

   i.   15% increase

   ii. 15% decrease

   iii. 25% decrease

   iv. 25% increase

Projects ————————————————————

1. Conducting a Formal Survey: Part I

Introduction

Your instructor may choose to continue this project throughout other parts of the book once you acquire the foundation required to analyze your results.

The scope of what is required depends on whether you are being asked to do this individually or as part of a group. Your instructor will clarify for your class.

The Assignment

Designing Your Survey: You need to construct a survey that meets the following requirements:

Individual Project

Choose a population you are interested in knowing more about (e.g., all your Facebook friends, everyone in your living community.) This needs to be a group of individuals you'll have direct access to since this is where you'll be taking your sample. The population should have at least 150 people in it.

Construct a single Yes/No question for your survey (required).

▸ E.g., "Do you think college food service programs should partner with local farmers?"

Select an authentic, simple random sample (e.g., using Research Randomizer) from your population. The minimum sample size is 50 individuals. The sample MUST be a simple random sample unless otherwise approved by your instructor.

▸ E.g. You acquire a list of all 500 students living in your dorm, listed in alphabetical order (1 to 500). You use Research Randomizer to select 50 from that list.

Contact each individual chosen in the sample and ask the question you have chosen to ask. Have a call-back plan in case some individuals chosen do not reply the first time. Record each sampled individual (by ID, not by name) in a table with a record of whether they answered "Yes" or "No" to your question.

Group Project

Choose a population you are interested in knowing more about (e.g., all your Facebook friends, everyone in your living community.) This needs to be a group of individuals you'll have direct access to since this is where you'll be taking your sample. The population should have at least 500 people in it.

Construct a single Yes/No question for your survey (required).

 ▸ E.g., "Do you think college food service programs should partner with local farmers?"

Select an authentic, simple random sample (e.g., using Research Randomizer) from your population. The minimum sample size is 150 individuals. The sample MUST be a simple random sample unless otherwise approved by your instructor.

 ▸ E.g. You acquire a list of all 500 students living in your dorm, listed in alphabetical order (1 to 500). You use Research Randomizer to select 150 from that list.

Contact each individual chosen in the sample and ask the question you have chosen to ask. Have a call-back plan in case some individuals chosen do not reply the first time. Record each sampled individual (by ID, not by name) in a table with a record of whether they answered "Yes" or "No" to your question.

Project Requirements

Your final project report should be professionally prepared, typed, and proofread. There should be three sections.

1. **Background Research:** You need to do extensive, college-level background research into the question you are investigating with your survey. Find out what is already known about public response to those or similar questions. For example, if you ask about college food service partnering with local farmers, you should look at what other surveys similar to this one have found about public opinion on this or a similar topic. Don't just look for a count of the number of colleges that are currently partnering with local farmers. Look for the opinions that other surveys have surfaced about this question (or a related question.) Be sure not to just use anecdotal opinion as research (e.g., "Everyone I talk to in my fraternity thinks …").

2. **Sampling Plan:** Describe in detail how your sample was selected, why it is a simple random sample, and what your call-back plan was in the event of non-responders.

3. **Data Summary:** Document all responses that have been collected in a computer-generated table with subject IDs and their Yes/No responses. Calculate the proportion who said "Yes" and compare and contrast that with what similar studies found that are referenced in your Background Research

2. Reality Video

Introduction

This chapter helped us understand the basic language of sampling. Now, we want to give you a chance to both show you know what you are doing and to help teach others.

The Assignment

Your instructor may have you do this as individuals or may form you into teams. We'll refer to "teams" below, although you may be a team of one depending on what your instructor wants. Each team is required to:

a. Pick one or more topics from this chapter. Your instructor may want to approve them before you proceed.

b. Construct a three-minute video that teaches the chosen topic to the class. You must carefully and correctly use the language of the chapter in your presentation, and you are required to give an example that is NOT in the chapter.

c. Post the video on YouTube and send the link to your instructor in the manner he stipulates.

5 CHAPTER 5
Confidence Intervals: What They Are and How We Use Them

Introduction

The last chapter promised that committing to take a probabilistic sample, such as a simple random sample, would pay dividends. The payoff would be our ability to construct a meaningful mathematical statement about the likely location of the unknown parameter based on the observed statistic. This chapter begins to deliver on that promise.

To understand what kind of statement we are talking about, let's take a look at a CBS News/*New York Times* poll concerning gun laws. Here is what the article says. The underlining is ours.

Title: Poll: Majority of Americans Back Stricter Gun Laws

Authors: Sarah Dutton, Jennifer De Pinto, Anthony Salvanto, Fred Backus and Leigh Ann Caldwell

Source: CBS News January 17, 2013. http://www.cbsnews.com/8301-250_162-57564597/ poll-majority-of-americans-back-stricter-gun-laws/

> As the president outlined sweeping new proposals aimed to reduce gun violence, a new CBS News/*New York Times* poll found that Americans back the central components of the president's proposals, including background checks, a national gun sale database, limits on high capacity magazines and a ban on semi-automatic weapons. Asked if they generally back stricter gun laws, more than half of respondents—54%—support stricter gun laws … That is a jump from April—before the Newtown and Aurora shootings—when only 39% backed stricter gun laws but about the same as ten years ago.
>
> …
>
> This poll was conducted by telephone from January 11–15, 2013, among 1,110 adults nationwide. Phone numbers were dialed from samples of both standard land-line and cell phones. <u>The error due to sampling for results based on the entire sample could be plus or minus three percentage points</u>.

Pay particular attention to the underlined sentence. This idea of "error due to sampling" is what you are going to see stated in one form or another almost every time poll results are reported. It is also commonly known as the "margin of error" for the survey. The margin of error is used in conjunction with the reported statistic to form a "confidence interval." In the example above, the confidence interval is .54 ± .03, or, expressed as percentages 54% ± 3%.

> A **confidence interval** has the following form: *statistic ± margin of error*

A confidence interval is a probabilistic statement about where we think the parameter might be. How one should think about this statement is a little more tedious than it might first appear. To get that just right, we need to first make sure we understand what random sampling error is and why it is something that demands attention.

Random Sampling Error

Let's start by acknowledging a fundamental short-coming of sampling. Think back to the example above. Another, equally well-chosen sample of 1,110 adult Americans asked the same question would almost surely not yield the same sample percentage (54% who favor stricter gun laws). Nor would a third, or a fourth.

In fact, we have no way of knowing for sure if the true proportion of all adult Americans who favor stricter gun laws—the parameter of interest—is even close to 0.54. That variability is what we call sampling variability or sampling error. It's not really an error in the sense of "mistake," but that language is common.

> **Sampling variability** is the variability that will be seen in a statistic from sample to sample. It is also commonly called random sampling error.

We might ask, if a sample statistic is going to vary from sample to sample, then which one is the correct one? As long as every one of the samples was taken correctly—as probabilistic samples— then the answer is that all the statistics are correct. A better question would be, how can we do a decent job of estimating the parameter of interest in the face of this kind of variability? Before we answer that question, we need to make sure we appreciate that sampling variability is real and may, or may not, be small.

Example: Simulated Poll Data

In a 2018 Gallup poll, a probability sample of 1,041 adults, aged 18 and older and living in all 50 states and the District of Columbia, were asked if the government is doing enough to protect the environment. Only 38% said "yes." (last accessed 5/25/2021 - https://news.gallup.com/poll/232007/americans-want-government-more-environment.aspx). An artificial population of 250 voters was created with 94 of those assigned the opinion of "yes" for this question and the rest "no." So the actual proportion of people in this population who would have said "yes" to the question, had they all been asked, was designed to be 94/250 or 0.376. That is, we took the statistic from a real survey and made it the parameter in a simulated population.

One-hundred and fifteen simple random samples of size 50 were drawn from this population, and for each sample, the sample percentage of "yes" answers in the sample was recorded. The data are shown in Table 5.1. If you doubted sampling variability was real, you probably shouldn't any longer! Notice that the sample percentages range from a low of 20% to a maximum of 60%.

Remember, if you are a pollster, you only have one of those sample results because you are only going to be willing to pay for one sample (in this case, of size 50). Therefore, it is very important to have a way of quantifying where that statistic likely is, relative to the parameter. Given how much the statistic can bounce around from sample-to-sample, owing to sampling variability, this is clearly a worthy task! That's what the margin of error does for us.

TABLE 5.1 Statistics Based on Samples of Size 50

36	36	42	30	28		34	36	34	48	32
38	46	48	38	36		32	40	38	34	40
42	38	28	36	40		36	44	42	36	38
32	32	38	32	36		36	36	36	42	40
34	46	32	36	36		56	42	28	52	36
46	48	36	48	44		40	38	26	36	34
36	40	38	50	40		30	38	36	36	32
22	46	44	36	38		36	48	36	34	28
26	30	34	34	36		32	38	38	36	48
20	38	36	36	60		40	32	42	40	42
48	28	46	34	34		30	32	44	44	36
40	36	32	50	32						

Anatomy of the Confidence Interval

Margin of Error

In the CBS News article above, we read "the error due to sampling for results based on the entire sample could be plus or minus three percentage points." As we have already noted, the authors of the article are telling us that the margin of error associated with the survey is ± 3%, that is, ± 3 percentage points. Nearly everyone has encountered this idea well before they reach college. But to understand how we are to actually use the margin of error and the associated confidence interval, we need to first understand that the margin of error has two parts.

> The **standard error** of a statistic is the standard deviation of all possible statistics based on random samples of size n.

> The **confidence coefficient** associated with a statistic reflects how confident you want to be that the subsequent confidence interval has captured the parameter.

> The **margin of error** has the form *(confidence coefficient)* × *(standard error)*.

This is where the story can start to get a little complicated. The actual form of the standard error will depend on two things: the parameter you are interested in estimating, and the kind of probability sample you have taken. But as long as you have taken a probability sample, then the good news is that you will be able to determine a margin of error, and,

TABLE 5.2 Confidence Coefficients

C% If You Want a Level of Confidence of:	z* Then Use This Confidence Coefficient
50%	0.67
60%	0.84
70%	1.04
80%	1.28
90%	1.64
95%	2
99%	2.58
99.9%	3.29

hence, a confidence interval. The confidence coefficient is typically denoted by z^* and typical values of z^* can be easily accessed from Table 5.2. We will better understand in the next chapter where z^* actually comes from.

Results for Sample Proportion

In the CBS news article above, the natural statistic is a sample proportion, which is typically denoted by $\hat{p}$ and the poll was ultimately interested in a population proportion. That is, the poll was concerned about the true proportion of all Americans who would have said they would back stricter gun laws, had they all been asked. You can think of this as a proportion or a percentage, either one. You just have to be consistent with that thinking. If the sample were a simple random sample, then the margin of error associated with the sample is given by the following:

> The **standard error for a sample proportion based on a simple random sample of size n** is estimated as $\frac{1}{2}\left[\frac{1}{\sqrt{n}}\right]$.

> The **C% margin of error for a sample proportion based on a simple random sample of size n** is estimated as $\text{MOE} = \frac{z^*}{2}\left[\frac{1}{\sqrt{n}}\right]$, where z^* is the corresponding confidence coefficient.

> The **C% confidence interval for a population proportion based on a simple random sample of size n** is $\hat{p} \pm \frac{z^*}{2}\left[\frac{1}{\sqrt{n}}\right]$, and $\hat{p}$ is the sample proportion (statistic) from the sample.

Results for Sample Mean

Suppose the CBS poll had asked respondents to indicate how much of a property tax increase, in dollars, they would be willing to pay in order to have tighter gun control regulations in their local communities. In this case, the natural statistic is a sample mean, which is typically denoted by $\overline{x}$, and the corresponding parametric of interest is the average of all responses in the population, had it been possible to ask everyone that question. In this case, similar to above, if the sample were a simple random sample, then the margin of error associated with the sample is given by the following:

> The **standard error for a sample mean based on a simple random sample of size n** is estimated as $\left[\frac{\sigma}{\sqrt{n}}\right]$, where σ is the **standard deviation of all the measurements in the entire population. This is typically never known,** and s, the standard deviation of the n responses given by those surveyed, is used instead.

> The **C% margin of error for a sample mean based on a simple random sample of size n** is estimated as $\text{MOE} = z^*\left[\frac{\sigma}{\sqrt{n}}\right]$, where z^* is the corresponding confidence coefficient.

> The **C% confidence interval for a population mean based on a simple random sample of size n** is $\overline{x} \pm z^*\left[\frac{\sigma}{\sqrt{n}}\right]$, and $\overline{x}$ is the sample mean (statistic) from the sample.

For simplicity of message, this book tends to focus on the sample proportion. However, the larger point is that as long as you started with a probability sample, of which a simple random sample is an example, then there is likely a path to a formula for a margin of error. If you did not choose a probability sample, but rather chose a kind of convenience sample, or tried to choose a cross-sectional sample, then there is no such path available.

Example: CBS Guns Poll Revisited

In the CBS poll above, the sample size was reported to be 1,110. We can be sure the sampling was some type of probability sample since the poll was conducted by a professional source, but it was unlikely a simple random sample since those are so hard to take in practice. We also are not told explicitly what level of confidence to attach to the reported margin of error. However, it is always safe to assume that the confidence level is 95% if it is unreported, and the polling organization is reputable. Let's collect what we know:

- $n = 1110$
- $C\% = 95\%$, so $z^* = 2$

Therefore, the 95% margin of error should be about $\frac{2}{2}\left[\frac{1}{\sqrt{1110}}\right] = 0.03$. This is what was reported in the article.

What Kind of Confidence? ————————————

To this point we have left unsaid what kind of confidence a confidence interval gives us. Now that we understand sampling variability, it may come as no surprise that the explanation has to come in those terms. Let's go back to the simulated example that we encountered above. In Table 5.3 we have computed a 95% confidence interval around every one of the 115 sample percentages. Recall the parameter in this example was 37.60%. If you count the number of samples in Table 5.3 that produced a confidence interval that contains 37.60%, you will find that 110 do. Notice that 110/115 is 95.7%. This is no accident, and it illustrates nicely the sense in which a confidence interval gives us confidence about where the parameter is.

> The **interpretation of a C% confidence interval** is as follows: if many samples of size n were taken from the population of interest, and for each, a C% confidence interval constructed, then about C% of all those intervals would contain the unknown parameter.

Thus, confidence as conveyed by a confidence interval is confidence in a repeated sampling sense. Of course, you don't actually do the repeated sampling, but when you form a confidence interval from a single sample of size n, that is the sense in which the confidence interval is telling you something useful, but probabilistic, about your parameter. You want to be careful not to say there is a C% chance that the parameter is in the interval. The parameter is a fixed, just unknown number associated with the population. Far better to say, there is a C% chance that the interval you have computed contains the parameter.

TABLE 5.3 **Confidence Intervals for Repeated Samples of Size 50**

Sample % "Yes"	Confidence Interval	Sample % "Yes"	Confidence Interval	Sample % "Yes"	Confidence Interval	Sample % "Yes"	Confidence Interval	Sample % "Yes"	Confidence Interval
36	(21.86, 50.14)	36	(21.86, 50.14)	42	(27.86, 56.14)	30	(15.86, 44.14)	28	(13.86, 42.14)
38	(23.86, 52.14)	46	(31.86, 60.14)	48	(33.86, 62.14)	38	(23.86, 52.14)	36	(21.86, 50.14)
42	(27.86, 56.14)	38	(23.86, 52.14)	28	(13.86, 42.14)	36	(21.86, 50.14)	40	(25.86, 54.14)
32	(17.86, 46.14)	32	(17.86, 46.14)	38	(23.86, 52.14)	32	(17.86, 46.14)	36	(21.86, 50.14)
34	(19.86, 48.14)	46	(31.86, 60.14)	32	(17.86, 46.14)	36	(21.86, 50.14)	36	(21.86, 50.14)
46	(31.86, 60.14)	48	(33.86, 62.14)	36	(21.86, 50.14)	48	(33.86, 62.14)	44	(29.86, 58.14)
36	(21.86, 50.14)	40	(25.86, 54.14)	38	(23.86, 52.14)	50	(35.86, 64.14)	40	(25.86, 54.14)
22	(7.86, 36.14)	46	(31.86, 60.14)	44	(29.86, 58.14)	36	(21.86, 50.14)	38	(23.86, 52.14)
26	(11.86, 40.14)	30	(15.86, 44.14)	34	(19.86, 48.14)	34	(19.86, 48.14)	36	(21.86, 50.14)
20	(5.86, 34.14)	38	(23.86, 52.14)	36	(21.86, 50.14)	36	(21.86, 50.14)	60	(45.86, 74.14)
48	(33.86, 62.14)	28	(13.86, 42.14)	46	(31.86, 60.14)	34	(19.86, 48.14)	34	(19.86, 48.14)
40	(25.86, 54.14)	36	(21.86, 50.14)	32	(17.86, 46.14)	50	(35.86, 64.14)	32	(17.86, 46.14)
34	(19.86, 48.14)	36	(21.86, 50.14)	34	(19.86, 48.14)	48	(33.86, 62.14)	32	(17.86, 46.14)
32	(17.86, 46.14)	40	(25.86, 54.14)	38	(23.86, 52.14)	34	(19.86, 48.14)	40	(25.86, 54.14)
36	(21.86, 50.14)	44	(29.86, 58.14)	42	(27.86, 56.14)	36	(21.86, 50.14)	38	(23.86, 52.14)
36	(21.86, 50.14)	36	(21.86, 50.14)	36	(21.86, 50.14)	42	(27.86, 56.14)	40	(25.86, 54.14)
56	(41.86, 70.14)	42	(27.86, 56.14)	28	(13.86, 42.14)	52	(37.86, 66.14)	36	(21.86, 50.14)
40	(25.86, 54.14)	38	(23.86, 52.14)	26	(11.86, 40.14)	36	(21.86, 50.14)	34	(19.86, 48.14)
30	(15.86, 44.14)	38	(23.86, 52.14)	36	(21.86, 50.14)	36	(21.86, 50.14)	32	(17.86, 46.14)
36	(21.86, 50.14)	48	(33.86, 62.14)	36	(21.86, 50.14)	34	(19.86, 48.14)	28	(13.86, 42.14)
32	(17.86, 46.14)	38	(23.86, 52.14)	38	(23.86, 52.14)	36	(21.86, 50.14)	48	(33.86, 62.14)
40	(25.86, 54.14)	32	(17.86, 46.14)	42	(27.86, 56.14)	40	(25.86, 54.14)	42	(27.86, 56.14)
30	(15.86, 44.14)	32	(17.86, 46.14)	44	(29.86, 58.14)	44	(29.86, 58.14)	36	(21.86, 50.14)

Example: Interpreting Gun Poll

In the CBS poll above, a 95% confidence interval was easily computed from the information in the article to be 54% ± 3%, or (51%, 57%). Therefore, if we were to imagine lots and lots of samples of size 1,110 being taken from this population, and each time a 95% confidence interval formed around the statistic that resulted, then about 95% of those intervals would cover the parameter. In that sense, we could say there is a 95% chance that (51%, 57%) is one such interval.

Suppose we had wanted to construct a 60% interval instead. Referring to Table 5.2, we find that z^* is 0.84. Therefore, the 60% margin of error is $\frac{0.84}{2}\left[\frac{1}{\sqrt{1110}}\right] = 0.01$, and the corresponding 60% confidence interval is (53%, 55%). Notice that with less confidence, the margin of error is smaller, and, hence, the confidence interval narrower than for the 95% case. This is completely reasonable. That is, while we may be pleased to have a tighter interval—less room to look around for the parameter—we are far less confident we have even captured the parameter in this interval. There are no free lunches, even in statistical science!

TAKE-HOME POINTS

- Simple formulas are available for the margin of error and associated confidence intervals, provided the data were collected as a simple random sample, or similarly statistically correct fashion.

- The confidence conveyed by a confidence interval is most precisely described in a repeated sampling sense.

Chapter 5 Exercises

Reading Check

Short Answer: Please provide brief, concise answers to the following questions.

1. What is sampling variability and why is it important to understand?

2. What is the formula for an 80% confidence interval for a population proportion?

3. What is the correct interpretation of a 95% confidence interval?

4. How does a standard error differ from the margin of error?

5. What is a 99.9% margin of error for a sample mean based on a sample of size 200? Your answer will be in terms of the population standard deviation.

Beyond the Numbers

1. MOE Information Needed

Title: A Shifting Landscape: A Decade of Change in American Attitudes about Same-sex Marriage and LGBT Issues (Feb. 26th, 2014)

Authors: Robert Jones, Daniel Cox, and Juhem Navarro-Rivera

Source: http://publicreligion.org/site/wp-content/uploads/2014/02/2014.LGBT_REPORT.pdf

The following is an excerpt from the *Executive Summary* of this report:

> Support for same-sex marriage jumped 21 percentage points from 2003, when Massachusetts became the first state to legalize same-sex marriage, to 2013. Currently, a majority (53%) of Americans favor allowing gay and lesbian couples to legally marry, compared to 41% who oppose. In 2003, less than one-third (32%) of Americans supported allowing gay and lesbian people to legally marry, compared to nearly 6-in-10 (59%) who opposed.

Near the end of the report, the authors add:

> Results of the survey were based on bilingual (Spanish and English) telephone interviews conducted between November 12, 2013, and December 18, 2013, by professional interviewers under the direction of Princeton Survey Research Associates. Interviews were conducted by telephone among a random sample of 4,509 adults 18 years of age or older in the entire United States (1,801 respondents were interviewed on a cell phone, including 977 without a landline phone).

The margin of error is +/– 1.7 percentage points for the general sample at the 95% confidence level. In addition to sampling error, surveys may also be subject to error or bias due to question wording, context, and order effects.

a. Using the MOE given in the article, what is the 95% confidence interval for the true proportion of Americans in favor of allowing gay and lesbian couples to marry.

   i.   30.3% to 33.7%

   ii. 51.3% to 54.7%

   iii. 93.3% to 96.7%

   iv. 39.3% to 42.7%

b. In the Executive Summary, the authors state that "a majority (53%) of Americans favor allowing gay and lesbian couples to legally marry…" Were all Americans asked? How could the statement be changed so that it more precisely reflects the inference based on the sample data?

   i.   Yes. Could say a majority of Americans sampled favor ….

   ii. No. Could say the sample suggests a majority of Americans favor ….

   iii. Yes. Could say a majority of Americans may or may not favor ….

   iv. No. Could say a minority of Americans sampled favor ….

c. Suppose you were to take another random sample of Americans at the same point in time and ask the same question. Would you find—without question—that more than 50% of the respondents were in favor of allowing gay and lesbian couples to marry? Why or why not?

   i.   No. That's sampling variability.

   ii. No. There's no way to guarantee the samples were correctly taken both times.

   iii. Yes. As long as each sample is a probability sample, they will yield very similar results.

   iv. Yes. It is guaranteed that the next sample statistic has to be within 1.7 percentage points of the first sample. That's what the margin of error does for us.

d. Develop and defend some reasons as to why the following statement appears in this article: "1,801 respondents were interviewed on a cell phone, including 977 without a landline phone."

2. Is It Warm in Here?

Title: Americans Do Care About Climate Change

Author: Annie Leonard

Source: *New York Times* May 8, 2014. https://www.nytimes.com/roomfordebate/2014/05/08/climate-debate-isnt-so-heated-in-the-us/americans-do-care-about-climate-change

The following is an excerpt from the *New York Times* article:

> Americans do care about climate change. Polls showing lower levels of concern than in some countries don't tell the whole story. I travel widely around the U.S., attending meetings at schools, churches and community gatherings. Everywhere I go, I see people who are not only concerned about climate change, but are actively working on solutions.
>
> Nearly two-thirds (67%) of Americans accept the scientific evidence of global warming; fewer than one in six remain in denial.

The full report referenced by Leonard's article tells us that the original survey was conducted by the Pew Research Center in October 2013. There was a (95%) margin of sampling error of about 2.9% associated with the entire sample.

a. What are the Pew Center poll's sample and statistic?

   i. Sample is all Americans. Statistic would be the true proportion of all Americans who accept scientific evidence of global warming, if it were possible to ask all Americans.

   ii. We don't know the size of the sample, but the implication is that it is a sample of Americans. The statistic reported is that 67% accept scientific evidence of global warming.

   iii. Sample is all Americans. Statistic would be those subjects who actually answered the question.

   iv. We don't know the size of the population, but the implication is that it is a sample of people working at the Pew Research Center. The statistic reported is that 67% accept scientific evidence of global warming.

b. What are the population and the parameter?

   i. Population is all Americans. Parameter would be the true proportion of all Americans who accept scientific evidence of global warming, if it were possible to ask all Americans.

   ii. We don't know the size of the population, but the implication is that it is a population mostly of Americans. The parameter reported is that 67% accept scientific evidence of global warming.

   iii. We don't know the size of the population, but the implication is that it is comprised of the people working at the Pew Research Center. The parameter reported is that 67% accept scientific evidence of global warming.

   iv. Population is all Americans. The parameter is unknown but would be computed from the answers provided by those who actually answered the question.

c. Using the MOE given in the article, construct a confidence interval for the true proportion of all Americans who accept the scientific evidence of global warming.

    i. 67% +/– 0.29

    ii. 67% +/– 2.9%

    iii. 0.67 +/– 2.9%

    iv. 67% +/– 0.29%

d. When the data were broken down into subgroups, such as Republicans and Democrats, the associated MOEs increased. Explain why that makes sense.

3. Americans and Their Guns

Title: Poll: Majority of Americans Back Stricter Gun Laws

Authors: Sarah Dutton, Jennifer De Pinto, Anthony Salvanto, Fred Backus, and Leigh Ann Caldwell

Source: CBS News January 17, 2013. http://www.cbsnews.com/8301-250_162-57564597/poll-majority-of-americans-back-stricter-gun-laws/

> As the president outlined sweeping new proposals aimed to reduce gun violence, a new CBS News/*New York Times* poll found that Americans back the central components of the president's proposals, including background checks, a national gun sale database, limits on high capacity magazines and a ban on semi-automatic weapons. Asked if they generally back stricter gun laws, more than half of respondents—54%—support stricter gun laws … That is a jump from April—before the Newtown and Aurora shootings—when only 39% backed stricter gun laws but about the same as ten years ago.
>
> …
>
> This poll was conducted by telephone from January 11–15, 2013, among 1,110 adults nationwide. Phone numbers were dialed from samples of both standard land-line and cell phones. The error due to sampling for results based on the entire sample could be plus or minus three percentage points.

a. What are the sample and the statistic for the CBS News poll?

    i. The sample is the 1,110 adults contacted by telephone, and the statistic is the proportion of all adults who would have said they back stricter gun laws, had it been possible to ask all of them.

    ii. The sample is the 1,110 adults contacted by telephone, and the statistic is the 54% who say they support stricter gun laws.

    iii. The sample is all U.S. adults, and the statistic is the proportion of all adults who would have said they back stricter gun laws, had it been possible to ask all of them.

    iv. The sample is all U.S. adults, and the statistic is the 54% who said they back stricter gun laws.

b. What are the population and the parameter?

    i. The population is the 1,110 adults contacted by telephone, and the parameter is the proportion of all adults who would have said they back stricter gun laws, had it been possible to ask all of them.

    ii. The population is the 1,110 adults contacted by telephone, and the parameter is the 54% who say they support stricter gun laws.

    iii. The population is all U.S. adults, and the parameter is the proportion of all adults who would have said they back stricter gun laws, had it been possible to ask all of them.

    iv. The population is all U.S. adults, and the parameter is the 54% who said they back stricter gun laws.

c. This survey reports a margin of sampling error of about 3%. Confirm that margin with the formula you learned in this class for a 95% margin of error. Show your work.

d. Compute an 80% confidence interval for the true proportion of Americans who "generally back stricter gun laws." What percent decrease in width is this interval from a 95% interval?

    i. The width of the interval went from 0.06 to 0.04. That is a 2.0% decrease in the interval width.

    ii. The width of the interval went from 0.02 to 0.01. That is a 1.0% decrease in the interval width.

    iii. The width of the interval went from 0.02 to 0.01. That is a 50% decrease in the interval width.

    iv. The width of the interval went from 0.06 to 0.04. That is a 33.3% decrease in the interval width.

e. The headline on this article refers to the "majority of Americans." Carefully explain if the reference is to the population or the sample. Is the headline statistically defensible? Be sure to offer a clear defense of your answer by citing appropriate statistical evidence from the article.

4. Sticker Shock

Title: Great Jobs, Great Lives: The 2014 Gallup-Purdue Index Report

Authors: Gallup Organization and Purdue University

Source: https://www.luminafoundation.org/files/resources/galluppurdueindex-report-2014.pdf

Does it really matter where you go to college? Not so much, according to this Gallup-Purdue study. Five areas of well-being were measured among recent college graduates: purpose well-being, social well-being, financial well-being, community well-being, and physical well-being. The study concluded "that the type of schools these college graduates attended—public or private, small or large, very selective or less selective—hardly matters at all. … Just as

many graduates of public as not-for-profit private institutions are thriving—
which Gallup defines as strong, consistent, and progressing—in all areas of their
well-being." Percentages of thriving graduates among each institution type are
shown in the bar chart below.

These results were obtained
from internet surveys conducted
between February 4 and March
7, 2014. The margin of error is
estimated to be 1%.

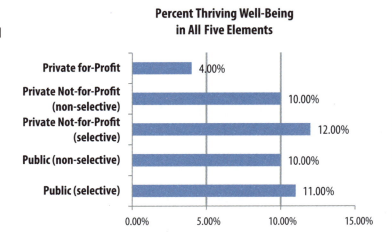

Percent Thriving Well-Being in All Five Elements

a. What is the 95% confidence
interval for the true thriving
rates among graduates of public
selective institutions?

    i.   (0.11, 0.13)

    ii. (0.09, 0.15)

    iii. (0.15, 0.19)

    iv. (0.10, 0.12)

b. What is the 95% confidence interval for the true thriving rates among graduates of private
not-for-profit selective institutions?

    i.   (0.11, 0.13)

    ii. (0.09, 0.15)

    iii. (0.15, 0.19)

    iv. (0.10, 0.12)

c. Do the intervals computed in a and b overlap? Do you think this tells you anything about
whether thriving rates really differ for private not-for-profit selective institutions and public
selective institutions? Explain.

d. What is the 95% confidence interval for the true thriving rates among graduates of public
non-selective institutions?

    i.   (0.03, 0.05)

    ii. (0.09, 0.11)

    iii. (0.01, 0.07)

    iv. (0.07, 0.13)

e. What is the 95% confidence interval for the true thriving rates among graduates of private for-profit selective institutions?

  i. (0.03, 0.05)

  ii. (0.09, 0.11)

  iii. (0.01, 0.07)

  iv. (0.07, 0.13)

f. Do the intervals computed in d and e overlap? Do you think this tells you anything about whether thriving rates really differ for public non-selective and private for-profit selective institutions? Explain.

5. A Common Challenge

A team of psychology researchers was interested in potential misinterpretations of the term "confidence" in a confidence interval. They collected data from 442 undergraduate students, 34 graduate students, and 118 of their research-active colleagues. All subjects were presented with a "fictitious scenario of a professor who conducts an experiment and reports a 95% CI for the [population proportion] that ranges from 0.1 to 0.4. Neither the topic of study nor the underlying statistical model used to compute the CI was specified in the survey." The subjects were then asked to specify whether they agreed or disagreed with each of the following statements as interpretations of that confidence interval (CI). Table 5.4 displays what the investigators found.

TABLE 5.4 **Percentage of Subjects Agreeing with the Statement**[1]

Statement	First Year Students ($n = 442$)	Master Students ($n = 34$)	Researchers ($n = 118$)
1. The probability that the true proportion is greater than 0 is at least 95%.	51%	32%	38%
2. The probability that the true proportion equals 0 is smaller than 5%.	55%	44%	47%
3. There is a 95% probability that the true proportion lies between 0.1 and 0.4.	58%	50%	59%
4. We can be 95% confident that the true proportion lies between 0.1 and 0.4.	49%	50%	55%
5. If we were to repeat the experiment over and over, then 95% of the time the true proportion falls between 0.1 and 0.4.	66%	79%	58%

All of the statements are wrong! Students and professional researchers alike found the interpretation of a confidence interval to be challenging. Yet, it is not acceptable to step away from this challenge. Confidence intervals are used everywhere as a kind of statistical seal of approval for survey and experiment results.

---

1   Hoekstra, R., Morey, R., Rouder, J., and Wagenmakers, E-J. "Robust misinterpretation of confidence intervals," *Psychon Bull Rev*. Published online January 14, 2014. http://www.ejwagenmakers.com/inpress/HoekstraEtAlPBR.pdf. Table has been reformatted, one statement omitted, and "mean" changed to "proportion" throughout.

a. What are two things that are wrong with Statement 1? Look back in the chapter if necessary to make sure that you have the right interpretation.

  i. The statement treats the confidence interval as random, and it treats the parameter as fixed from sample to sample.

  ii. The statement treats the parameter as random, and it refers nonsensically to the value of p relative to 0.

  iii. The statement treats the parameter as random, and it confuses the use of 95% with that of 5%.

  iv. The statement treats the parameter as random, and it treats the confidence interval as fixed from sample to sample.

b. What is wrong with Statement 5? Be very clear. Look back at your content on confidence intervals to make sure that you have the right interpretation.

  i. The statement treats the confidence interval as random, and it treats the parameter as fixed from sample to sample.

  ii. The statement treats the parameter as random, and it refers nonsensically to the value of p relative to 0.

  iii. The statement treats the parameter as random, and it confuses the use of 95% with that of 5%.

  iv. The statement treats the parameter as random, and it treats the confidence interval as fixed from sample to sample.

c. What is the correct interpretation of the 95% confidence interval (0.1, 0.4) used in the study?

  i. If many samples of size 442 were imagined taken from the fictitious population, and for each, a 95% confidence interval constructed, then about 95% of all those intervals would contain the unknown parameter.

  ii. There is a 95% chance that the unknown parameter in this fictitious population is between 0.1 and 0.4.

  iii. If many samples of size n were imagined taken from the fictitious population, and for each, a 95% confidence interval constructed, then about 95% of all those intervals would contain the true proportion of subject who would misinterpret the meaning of a confidence interval.

  iv. If many samples of size n were imagined taken from the fictitious population, and for each, a 95% confidence interval constructed, then about 95% of all those intervals would contain the unknown parameter.

d. What percentage of all participants in the study thought Statement 3 was a correct interpretation of a confidence interval?

   i. About 34.3%

   ii. About 58%

   iii. About 42%

   iv. About 343 participants.

6. Inaugural Intervals

A random sample of 10 U.S. presidents was taken and the age at inauguration recorded. See Table 5.5.

a. What is the sample mean of these 10 presidents' ages **in days,** at the time of their inauguration? You are encouraged to use a software package (such as Microsoft Excel or Apple Numbers) as directed by your instructor.

   i. 19500.8

   ii. 18854.2

   iii. 1246.8

   iv. 646.6

TABLE 5.5

President	Age at Inauguration
James Madison	57 years, 353 days
Martin Van Buren	54 years, 89 days
Millard Fillmore	50 years, 183 days
Warren G. Harding	55 years, 122 days
William McKinley	54 years, 34 days
William Howard Taft	51 years, 170 days
George Washington	57 years, 67 days
Benjamin Harrison	55 years, 196 days
Franklin D. Roosevelt	51 years, 33 days
Ulysses S. Grant	46 years, 311 days

b. Find the sample standard deviation of these 10 presidents' ages in days. You are encouraged to use a software package (such as Microsoft Excel or Apple Numbers) as directed by your instructor.

   i. 19500.8

   ii. 18854.2

   iii. 1246.8

   iv. 646.6

c. What is the confidence coefficient for a 90% confidence interval?

   i. 1.04

   ii. 1.64

   iii. 0.90

   iv. 2.58

d. What is the 90% confidence interval for the true average age (in days) of all U.S. presidents, up to and including President Biden, at the time of inauguration?

   i. 18854 days to 20147 days

   ii. 15036 days to 23215 days

   iii. 11018 days to 15963 days

   iv. 19184 days to 21886 days

e. Carefully interpret the meaning of the interval you just computed in the previous question.

f. Do some research on your own and determine the true average age of *all* U.S. presidents, up to and including President Biden. You are encouraged to use a software package (such as Microsoft Excel or Apple Numbers) as directed by your instructor. What is this average?

   i. About 19035 days

   ii. About 23396 days

   iii. About 20430 days

   iv. About 22067 days

g. See if the interval you computed in part d. contains this true average. Whether it does or doesn't, carefully explain in what sense there was a 95% chance of that outcome happening.

Projects

1. Conducting a Formal Survey: Part II

Introduction

This project is a natural progression from Part I, introduced in Chapter 4. Your instructors may choose to continue this project throughout other parts of the book once you acquire the foundation required to analyze your results.

The scope of what is required depends on whether you are being asked to do this individually or as part of a group. Your instructor will clarify for your class.

From Part I

In Part I, you did the following:

Individual Project

- Selected a simple random sample of 50 individuals
- Constructed a single yes/no question that was asked of all 50 of those individuals.

▸ Recorded each sampled individual (by ID, not by name) in a table with a record of whether they answered "yes" or "no" to your question.

Group Project

▸ Selected a simple random sample of 150 individuals

▸ Constructed a single yes/no question that was asked of all 150 of those individuals

▸ Recorded each sampled individual (by ID, not by name) in a table with a record of whether they answered "yes" or "no" to your question

Confidence Intervals

The goal of this part of the survey project is to construct and explain the value added to your survey project by an assortment of confidence intervals. Here are the specific parts you'll need to gather before you assemble your explanation:

a. Compute the sample proportion of individuals in your study who said "yes" to your yes/no question.

b. Construct a 60% confidence interval around that sample proportion for the population proportion of individuals who would have said "yes," had they all been asked.

c. Construct an 80% confidence interval around that sample proportion for the population proportion of individuals who would have said "yes," had they all been asked.

d. Construct a 95% confidence interval around that sample proportion for the population proportion of individuals who would have said "yes," had they all been asked.

e. Construct a display of those three intervals similar to what you see below. The plot should be mostly self-explanatory, but it features the following:

   i. The interval 0 to 1 in steps of 0.1

   ii. A vertical line segment drawn from the value of the sample proportion

   iii. A plot of each of your confidence intervals from b–d with the endpoints labeled with the numbers you calculate

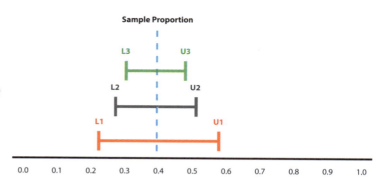

Project Requirements

This installment of your project report should be professionally prepared, typed, and proofread. Your instructor may require that it be added to Part I. Follow his instructions. Your task is to explain to your readers what confidence intervals in your project contribute to the understanding of your survey results. To that end, you are required to do the following:

■ Explain in non-technical language why you are computing confidence intervals.

- Exhibit the plot described above—fully annotated—and explain:

  ▸ What each interval means. Be careful to provide the correct interpretation, awkward though it might be.

  ▸ Why they are nested the way they will necessarily be

  ▸ What the tradeoffs between confidence and interval size are

All of your explanations and comments MUST be in the context of your specific survey. You are not trying to convince your reader of anything. Rather, you are explaining how confidence intervals add to the impact of your survey findings.

2. Designing Your Own Experiment: Part II

Introduction

This project is a natural progression from Part I, introduced in Chapter 3. Your instructors may choose to continue this project throughout other parts of the book once you acquire the foundation required to analyze your results.

The scope of what is required depends on whether you are being asked to do this individually or as part of a group. Your instructor will clarify for your class.

From Part I

In Part I you did the following:

Individual Project

  ▸ Randomized subjects into two groups, formed from a single change in a single variable. You had a minimum of 50 subjects.

  ▸ Applied the treatments in each group and collected the data

  ▸ Collected the response data for each subject

Group Project

  ▸ Randomized subjects into two groups, formed from a single change in a single variable. You had a minimum of 150 subjects.

  ▸ Applied the treatments in each group and collected the data

  ▸ Collected the response data for each subject

Confidence Intervals

The goal of this part of the experiment project is to construct and explain the value added by confidence intervals. Here are the specific parts you'll need to gather before you assemble your explanation:

a. Construct a 95% confidence interval around the statistic (will be a mean or a proportion) for each of the two treatment groups.

b. Construct a display of those two intervals similar to what you see below. The plot should be mostly self-explanatory, but it features the following:

   i. The horizontal axis will be scaled differently depending on if you have means or proportions as your statistic.

   ii. A plot of the two confidence intervals with the endpoints labeled with the numbers you calculate.

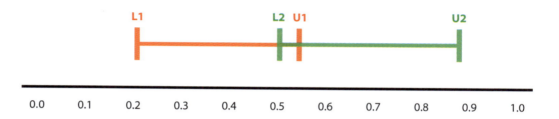

Project Requirements

This installment of your project report should be professionally prepared, typed, and proofread. Your instructor may require that it be added to Part I. Follow her instructions. Your task is to explain to your readers what confidence intervals in your project contribute to the understanding of your survey results. To that end, you are required to do the following:

- Exhibit the plot described above—fully annotated—and explain:

  ▸ What each interval means.

  ▸ Be careful to provide the correct interpretation, awkward though it might be.

- Based on your results, make a case for whether you think your two treatments are different or not.

All of your explanations and comments MUST be in the context of your specific experiment. You are not trying to convince your reader of anything. Rather, you are explaining how confidence intervals add to the impact of your findings.

3. Reality Video

Introduction

This module explained to us how confidence intervals are used and interpreted. Now, we want to give you a chance to both show you know what you are doing and to help teach others.

The Assignment

Your instructor may have you do this as individuals or may form you into teams. We'll refer to "teams" below, although you may be a team of one depending on what your instructor wants. Each team is required to do the following:

a. Pick one or more topics from this chapter. Your instructor may want to approve them before you proceed.

b. Construct a three-minute video that teaches the chosen topic to the class. You must carefully and correctly use the language of the chapter in your presentation, and you are required to give an example that is NOT in the chapter.

c. Post the video on YouTube and send the link to your instructor in the manner he stipulates.

6

CHAPTER 6
Confidence Intervals: Where They Come From

Introduction

Now that we know the basics of how to form and interpret a confidence interval, we are positioned to dig a little deeper into where they actually come from. At the end of this chapter, we will better understand how the probability associated with a confidence level is inherited from the probability associated with how the sample is taken. We will therefore better understand why starting with a probability sample is critical, both to the existence of the formulas that describe confidence intervals and their interpretation. We will start by learning about sampling distributions and the Empirical Rule, and then we will combine the two as we create the full picture of what is going on.

Sampling Distributions

As vexing as sampling variability can be, it is actually quite predictable for probability samples. We can provide a non-mathematical glimpse of this predictability pretty easily as long as you recall what a histogram is. If you don't, you may want to stop here for a minute and remind yourself.

Suppose we were to take many different samples of some fixed size, say *n*, and for each sample we record the statistic of interest, say the proportion in the sample who said "Yes," when asked, "Are you in favor of same-sex marriages?" A plot of all these different possible sample statistics is called a sampling distribution.

> A **sampling distribution** is the plot of values of a statistic (say, a mean or proportion) over lots of possible samples of size *n* taken from the same population. For a finite number of samples, that plot is typically a histogram.

Example: Simulated Poll Data

In Chapter 5 we looked at a simulated population based on a 2018 Gallup poll that asked the question "Is the government doing enough to protect the environment?" To illustrate how vexing sampling variability can be, we took 115 simple random samples from that simulated population of 250 voters, with a designed parameter of 37.6. That is, 37.6% of all voters in the population would have said "yes" to the question, had they all been asked. In Table 5.1 we saw that the sample percentages were all over the place, ranging from 20% to 60%.

If we plot the sampling distribution of those 115 sample percentages, we will see that this seemingly chaotic list of sample percentages actually has some structure lurking within. A histogram of these 115 sample percentages (based on samples of size 50) is shown in Figure 6.1. You should read the axes this way: there were 21 of the 115 percentages between 26 and 32, including 32 but not including 26. Similarly, there were 51 (about 44% of the sample proportions) between 32 and 38, including 38, but not including 32, etc.

What do you notice about the sampling distribution right away? It's bell-shaped. It also has its peak in the interval (32,38], and we know the parameter was 37.6. What we are seeing in this example is no accident. The following is always true:

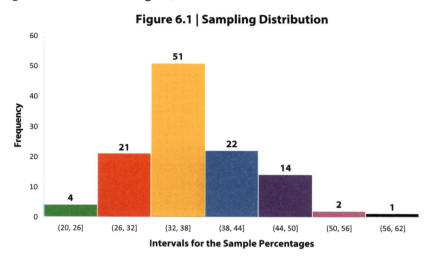

Figure 6.1 | Sampling Distribution

> **The sampling distribution of a sample statistic based on probability samples is guaranteed to be:**
> - Bell-shaped
> - Peak above the parameter from the population
> - Spread out in a manner known in advance and quantifiable

You are guaranteed that the sampling distribution of your statistic will have these properties, provided your samples were probability samples (e.g. simple random samples). If your samples were convenience samples, then this structure is not guaranteed. As we are about to see, knowing these three things about sampling distributions is the key needed to unlock the mystery of where confidence intervals come from.

A word of caution! The message here is *not* that pollsters (or you) should actually take many different samples of size *n*. Rather, the message is that because the mathematics of sampling distributions tells us quite a lot about their properties, we never actually have to do that enormous—typically impossible—task in practice. Amazingly, the claim above about what a sampling distribution will look like applies to any population, and most any statistic, provided the sample was a probability sample (and the sample size was not tiny). The next step for us is to understand that bell-shaped distributions admit useful probability statements about the data they describe.

The Empirical Rule

The U.S. Department of Education sponsors the website "The College Scorecard" that contains all kinds of useful data about graduation rates, cost and value of a college education, median income earned by graduates ten years out of school, etc. We accessed those data (last updated in October of 2018) and pulled the average SAT score for 1384 U.S. colleges and universities. These are plotted in Figure 6.2. The mean of all these averages is 1055.62, and the standard deviation is 18.86.

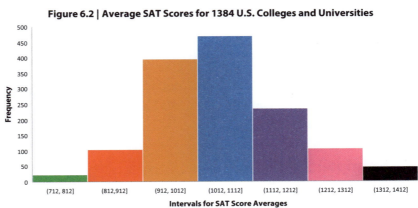

Figure 6.2 | Average SAT Scores for 1384 U.S. Colleges and Universities

Notice that the distribution of the averages is bell-shaped. Indeed, sampling distributions are by no means the only distributions that have this shape. Of interest to us now is the neat probability structure that is always resident in this kind of shape. But before we state the general rule, let's look further at this example. We took all 1384 of these SAT averages and counted how many were within one standard deviation of the mean, how many within two, etc. Make sure you understand what we counted. We summarized our results in Table 6.1.

TABLE 6.1 Counts and Percentages from the College Scorecard Data

Interval	Count	Percentage out of 1384
Between $1055.616 - (1 \times 18.863)$ and $1055.616 + (1 \times 18.863)$	988	71.4%
Between $1055.616 - (2 \times 18.863)$ and $1055.616 + (2 \times 18.863)$	1300	93.9%
Between $1055.616 - (3 \times 18.863)$ and $1055.616 + (3 \times 18.863)$	1384	100%

You should interpret the percentages in the third column this way. If you were to choose one of these colleges or universities at random (from the full group of 1384) and look at the average SAT score for their admitted students, there would be about a 94% chance that the score selected would be within two standard deviations of the full group mean. That is, there would be about a 94% chance that the selected score would be between 1017.89 and 1093.34 (within two standard deviations on either side of 1055.62).

The news in this section is that we know what these percentages have to be, more or less, for any bell-shaped distribution without having to do the actual counting like we just did. Here is the rule we are referring to:

> **The Empirical Rule**: Suppose a bell-shaped distribution has mean denoted by μ and standard deviation denoted by σ. Then:
> - About 68% of all observations represented by that distribution will fall within one standard deviation of the mean.
> - About 95% will fall within two standard deviations of the mean.
> - About 99.7% will fall within three standard deviations of the mean.

Compare the Empirical Rule approximations for the College Scoreboard data with the actual counts. The Empirical Rule predicted about 68% if all observations would fall within one standard deviation either to the left or right of the mean. In the College Scorecard data, we found that actually 71.4% did. Similarly, the Rule predicted that 95% will fall within two standard deviations of the mean, and we saw about 94%. The College Scorecard data were not perfectly bell-shaped, but even so, the actual results were very close to those suggested by the Empirical Rule.

The Empirical Rule is often presented in a graphical form, as you see here. Let's look at a couple of examples.

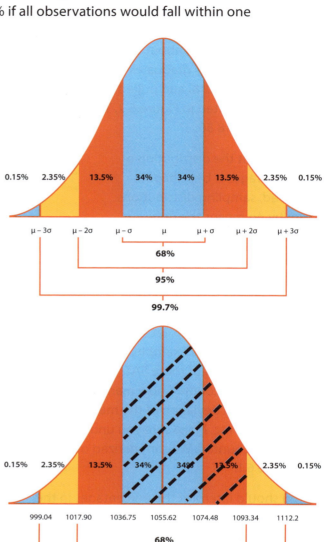

Example: SAT Scores

How would we use the Empirical Rule to estimate the chance of choosing a school at random from the 1435 above that has an average SAT score for admitted students that is between 1036.75 and 1093.34? A quick check of the mean and standard deviation reveals that 1036.75 is one standard deviation below the mean and 1093.34 is two above. That region is shaded on the figure to the right. The answer, using the Empirical Rule approximation, is therefore 34% + 34% + 13.5% = 81.5%

Example: Sports Marketing

Suppose the average score on a sports marketing exam is 70, with a standard deviation of 6. Assuming the distribution of all the test scores is bell-shaped, about what percentage of students scored between 64 and 76? Once again, all we have to do is to superimpose the mean and standard deviation on the Empirical Rule template, and then locate those values that are relevant. In this case, 64 is one standard deviation below, and 76 is one above. If we just read the percentages off the template, it is clear that about 68% of all the students in that class had to score between 64 and 76.

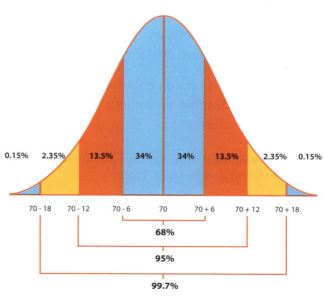

Sampling Distribution of the Sample Proportion ──

With the Empirical Rule in our bag of tricks, we can now apply it to a sampling distribution of a sample proportion, based on repeated probability samples of size n. We already know that we are guaranteed that this distribution is bell-shaped. And we know it peaks above the parameter p. Further, the standard deviation of the sampling distribution of the sample proportion can be accurately estimated in advance, as promised. Let's collect all these facts in one place:

> The sampling distribution of the sample proportion is:
> - **Bell-shaped;**
> - **Has mean, μ = p**, the true proportion of interest for the entire population;
> - **Has a standard deviation of** $\sigma_{\hat{p}} = \frac{1}{2}\frac{1}{\sqrt{n}}$, where n is the sample size. This is also known as the *standard error of the sample proportion,* and we encountered it initially in Chapter 5.

All we have to do is to impose this information onto the Empirical Rule graphic, and we have an explicit representation of the sampling distribution of the sample proportion. See the figure to the right.

A lot of useful knowledge emerges immediately. Just make sure you are not confused by the notation. Along the horizontal axis we have potential values of the sample proportion, which we introduced in Chapter 5 as $\hat{p}$. Reading directly off the graphic, we can say:

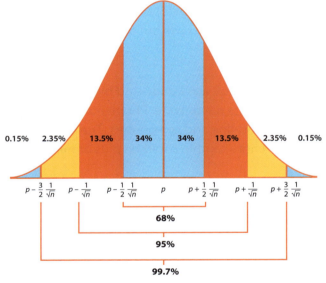

- 68% of all $\hat{p}$ values based on a simple random sample of size n will fall between $p - \frac{1}{2}\frac{1}{\sqrt{n}}$ and $p + \frac{1}{2}\frac{1}{\sqrt{n}}$.
- 95% will fall between the values of $p - \frac{1}{\sqrt{n}}$ and $p + \frac{1}{\sqrt{n}}$.
- Finally, 99.7% of all $\hat{p}$ values based on a simple random sample of size n will fall between the values of $p - \frac{3}{2}\frac{1}{\sqrt{n}}$ and $p + \frac{3}{2}\frac{1}{\sqrt{n}}$.

This may be a little hard to appreciate at first glance, but these are powerfully useful statements. For example, if you know that 95% of all $\hat{p}$ values based on a simple random sample of size n will fall between the values of $p - \frac{1}{\sqrt{n}}$ and $p + \frac{1}{\sqrt{n}}$, then you know that for 95% of all samples of size n from this population, the sample proportions produced from those samples will be no further than $\frac{1}{\sqrt{n}}$ from the parameter p. That's a very concrete, albeit probabilistic, statement that links your sample statistic to the location of the parameter.

Example: Gun Laws

Consider, again, the CBS News/*New York Times* poll concerning gun laws we first saw in Chapter 5.

Title: Poll: Majority of Americans Back Stricter Gun Laws

Authors: Sarah Dutton, Jennifer De Pinto, Anthony Salvanto, Fred Backus, and Leigh Ann Caldwell

Source: CBS News January 17, 2013.

http://www.cbsnews.com/8301-250_162-57564597/poll-majority-of-americans-back-stricter-gun-laws/

As the president outlined sweeping new proposals aimed to reduce gun violence, a new CBS News/*New York Times* poll found that Americans back the central components of the president's proposals, including background checks, a national gun sale database, limits on high capacity magazines and a ban on semi-automatic weapons. Asked if they generally back stricter gun laws, more than half of respondents—54%—support stricter gun laws … That is a jump from April—before the Newtown and Aurora shootings—when only 39% backed stricter gun laws but about the same as ten years ago.

…

This poll was conducted by telephone from January 11–15, 2013, among 1,110 adults nationwide. Phone numbers were dialed from samples of both standard land-line and cell phones. The error due to sampling for results based on the entire sample could be plus or minus three percentage points.

The parameter of interest, *p,* is the actual proportion (or percentage, just be consistent) of all adult Americans who would have said they support stricter gun laws, had they all been asked. They weren't all asked, of course, only 1110 were. Of these 1110 we are told that 54% said they supported stricter gun laws. By this point we know quite well that another well-chosen probability sample of size 1110 wouldn't yield the 54%. But what do we know? From what we have just learned, we know that 95% of all samples of size 1110 will yield a sample percentage that is no further than 3% ($\frac{1}{\sqrt{1110}} = 0.03$) away from the population percentage $p \times 100\%$.

Think about this! You don't know *p* and you can't know *p*. There are close to 250,000,000 adult Americans in the U.S.! But just based on a probability sample of size 1110, you can say that there is a 95% chance that such a sample will produce a sample percentage that is no further than 3% (to the left or right) from the true percentage who would have said they support stricter gun laws, had they all been asked.

Example: And More Gun Laws

Consider, again, the CBS News/*New York Times* poll concerning gun laws just discussed in the example above. Before the poll was even taken, what could you say about the chances of a probability sample of size 1110 producing a sample proportion who say they support stricter gun laws that was somewhere between 1.5% below or 4.5% above the true proportion in the population who would have agreed, had they all been asked?

This is a simple Empirical Rule question nested in the context of a sampling distribution. We know that the standard deviation of the sampling distribution of the sample proportion,

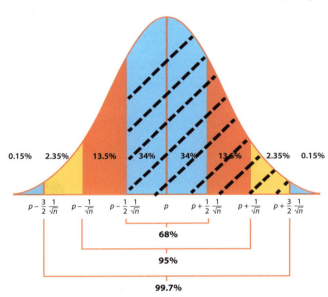

based on samples of size 1110, is $\sigma = \frac{1}{2}\frac{1}{\sqrt{1110}} = 0.015$, or 1.5%. It follows that the question is asking about the likelihood of seeing a sample proportion, $\hat{p}$, that is between one standard deviation below the parameter and three above.

The area is shaded in the figure below: 34% + 34% + 13.5% + 2.35% = 83.85%. So there are about 84 chances in 100 that a sample of size 1110 will produce a $\hat{p}$ that is in the specified range. Key to being able to answer such a difficult-sounding question was that your sample had to be a probability sample.

Sampling Distribution of the Sample Mean ———————

Results very similar to what we have just described apply equally to the sampling distribution of the sample mean. Let's collect those relevant facts here.

The sampling distribution of the sample mean is:

■ **Bell-shaped;**

■ **Has mean, μ,** the population average of the quantity of interest;

■ **Has a standard deviation we will estimate as** $\sigma_{\bar{x}} = \frac{\sigma}{\sqrt{n}}$, where n is the sample size, and **σ** is the standard deviation of all measurements in the population. This is also known as the *standard error of the sample mean,* which we encountered initially in Chapter 5. Since we realistically never know σ, we typically just estimate it with s, the standard deviation of all measurements in the sample.

Once again, all we have to do now is impose this information on what we know from the Empirical Rule to produce the following explicit representation of the sampling distribution of the sample mean.

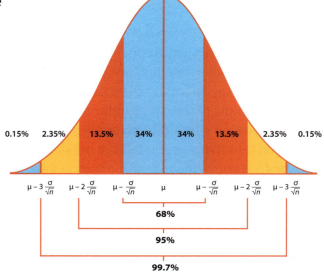

As before, we can immediately see the following:

- 68% of all $\bar{x}$ values based on a simple random sample of size n will fall between the values of $\mu - \frac{\sigma}{\sqrt{n}}$ and $\mu + \frac{\sigma}{\sqrt{n}}$.
- 95% will fall between $\mu - 2\frac{\sigma}{\sqrt{n}}$ and $\mu + 2\frac{\sigma}{\sqrt{n}}$.
- Finally, 99.7% of all $\bar{x}$ values based on a simple random sample of size n will fall between the values of $\mu - 3\frac{\sigma}{\sqrt{n}}$ and $\mu + 3\frac{\sigma}{\sqrt{n}}$.

Example: Student Debt

In 2018 the nonprofit organization *Summer and Student Debt Crisis* planned to survey 7095 adults with student loans in 2018 so they could estimate the average outstanding student debt in the population of 900,000 from which the sample was to be taken. Of course, we don't know the population standard deviation, but for now, let's assume that σ is fairly estimated as $25,000. What are the chances that the mean debt of the 7095 that will be sampled is no further than $594 from the actual population mean debt?

The graphic below, describing the sampling distribution of the sample mean, answers this for us quite easily. Since we are assuming $\sigma = 25000$, we know

$$2 \times \frac{\sigma}{\sqrt{7095}} = 2 \times \frac{25000}{\sqrt{7095}} = 593.6.$$

Therefore, we are being asked to specify the chance of seeing a sample mean $\bar{x}$ that is within two (sampling distribution) standard deviations of the population mean μ. It follows from the graphic that 95% of all sample means from this population, based on probability samples of size 7095, will fall no further than about $594 from the actual population mean.

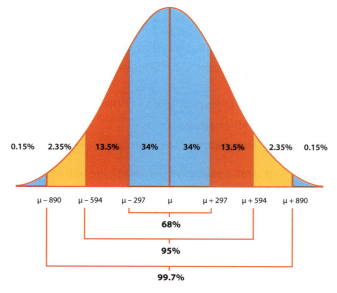

So Where DO Confidence Intervals Come From?

The discussion above has already answered this question for us, but it is quite possibly still a little opaque and hidden in the notation. Let's make the connection explicit. We are going to focus only on the sampling distribution of the sample proportion, but it will be easy to see that the exact same reasoning applies to the sampling distribution of the sample mean.

Remember that the horizontal axis of the sampling distribution (shown below) describes where the sample proportions $\hat{p}$ appear. We know, for example, that 95% of all such sample proportions will be no further than $\frac{1}{\sqrt{n}}$ from the parameter p. It follows that we know there is a 95% chance, given a sample proportion, that if we move $\frac{1}{\sqrt{n}}$ to the right of $\hat{p}$, and $\frac{1}{\sqrt{n}}$ to the left of $\hat{p}$, then that interval will be wide enough to cover p. See the graphic.

The finicky interpretation of a 95% confidence interval, first seen in Chapter 5, becomes crystal clear: there is a 95% chance that a probability sample of size n will produce a sample proportion, such that the interval $\left(\hat{p} - \frac{1}{\sqrt{n}}, \hat{p} + \frac{1}{\sqrt{n}}\right)$ will be wide enough to capture (cover) the parameter p.

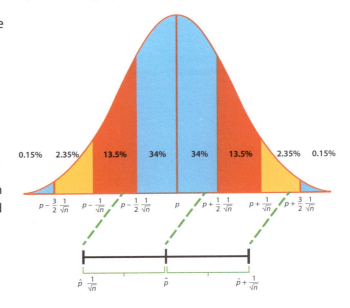

The same is true, of course, for other levels of confidence exhibited by the Empirical Rule. In fact, it is easy to see that we are not restricted to just the three levels shown in the graphic. Recall Table 5.2 recorded the confidence coefficients we need to use with the margin error to form confidence intervals. The MOE had the form:

$$\text{MOE} = \frac{z^*}{2}\left[\frac{1}{\sqrt{n}}\right].$$

As we were reminded above, we called $\frac{1}{2}\left[\frac{1}{\sqrt{n}}\right]$ the standard error in Chapter 5, and we learned in this chapter that this is just another name for the standard deviation of the sampling distribution of the sample proportion. It follows that z^* is just *counting* the number of those particular standard deviations to the left and right of the sample proportion that are needed to capture the parameter with the corresponding confidence. For example, a 95% confidence level corresponds to a z^* of 2, and we are thus instructed to move 2 standard errors to the left and right of the sample proportion in order to form our confidence interval. That is, we use the interval

$$\left(\hat{p} - \frac{2}{2}\frac{1}{\sqrt{n}}, \hat{p} + \frac{2}{2}\frac{1}{\sqrt{n}}\right) = \left(\hat{p} - \frac{1}{\sqrt{n}}, \hat{p} + \frac{1}{\sqrt{n}}\right).$$

How could we use the same reasoning to produce a 70% confidence interval? We know from Table 5.2 that z^* has to be 1.04. It follows that $\frac{z^*}{2} = \frac{1.04}{2} = 0.52$. Therefore, 70% of all probability samples of size n will produce sample proportions that are no more than $(0.52) \times \left[\frac{1}{\sqrt{n}}\right]$ from the population parameter, p. It follows that the formula for a 70% confidence interval for a sample

proportion based on a simple random sample of size n is $\left(\hat{p} - (0.52)\frac{1}{\sqrt{n}}, \hat{p} + (0.52)\frac{1}{\sqrt{n}}\right)$.

So this is where confidence intervals originate: from sampling distributions that inherit their probabilistic (confidence) properties from the probability inherent in the original sampling. No probability in the sampling? Then no sampling distribution with nice properties and no confidence interval. We can summarize what we have learned in the graphic below.

If the sample that is taken is a **probability sample** then…

The **sampling distribution** of the sample statistic will be
- bell-shaped;
- peak above the parameter;
- and have a predictable standard error.

Bell-shaped distributions admit predictable probability statements, thanks to the Empirical Rule.

Therefore, we can assess **how likely it is that a sample of size n will produce a statistic** that is no further than a given distance from the parameter.

It is really amazing that elementary statistical inference can make such far-reaching statements. It is worth repeating that at this point it should be abundantly clear why we *have* to start with an actual probability sample, and not just a sample that we heuristically deem fair. The probability that we ultimately need to make confidence statements is *inherited* from this probabilistic start. If you start with a convenience sample of some kind or another, then there is no such inheritance and as a result, no legitimate claim to the formulas for margin of error.

TAKE-HOME POINTS

- Sampling variability is predictable for common statistics computed from probability samples.
- The Empirical Rule infuses bell-shaped sampling distributions with simple probability.
- Common MOE and confidence interval formulas critically depend on the original samples having been probability samples.

Chapter 6 Exercises

Reading Check

Short Answer: Please provide brief, concise answers to the following questions.

1. What is a sampling distribution?

2. What characteristics will a sampling distribution have for probability samples?

3. What is the standard error of the sample mean, based on a simple random sample?

4. In the sports marketing example, about what percentage of all test scores would be expected to fall between 58 and 76?

5. Specify an interval to the right of the population proportion p where you can expect about 47.5% of all the sample proportions to fall.

Beyond the Numbers

1. Connecting Relative Area to Probability

The likelihood of college students playing corn hole is high, but have you ever thought about how likely it is that you will actually get the bag in the hole? If we take some liberties both with the game and with the specifics of the underlying probability, we can gain some understanding about the natural relationship between relative area and common-sense probability! The key will be to not overthink the problem.

a. Suppose you aren't very good at corn hole. In fact, suppose the best you can do is to know that your throw will at least hit some random place on the board. If the surface of an official corn hole board looks like the diagram shown, what are the chances that it will hit the hole? Think of the bag as a very small item, and remember you don't really aim. Rather, your toss just lands at a random place on the board.

   i. About 2.5 chances in 10 since 28.27/1152 = 0.0245

   ii. About 41 chances in 100 since 1152/28.27 is 40.75

   iii. About 2.5 chances in 100, since 28.27/1152 = 0.0245

   iv. About 4.1 chances in 100 since 1152/28.27 is 40.75

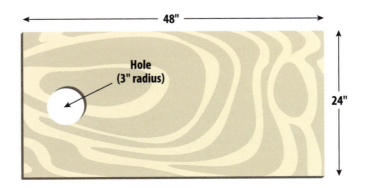

Now suppose the corn hole board has a funny bell-shape, with the hole delineated by the two vertical line segments. Granted, this would be the height of nerd tailgating. Suppose also that the entire bell-shaped sheet is 8 ft², just like a regulation corn hole board surface. We are still only guaranteed that our throw will land someplace randomly on the board.

b. Suppose you toss the bean bag at random toward the board, just as in a. Which of the two bell boards, A or B, would be better to play on and why?

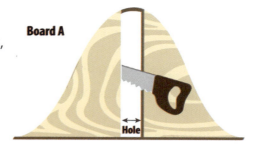

Board A

   i. Board A since there is a smaller chance of the bag going in the hole

   ii. Board A since there is a greater chance of the bag going in the hole

   iii. Board B since there is a smaller chance of the bag going in the hole

   iv. Board B since there is a greater chance of the bag going in the hole

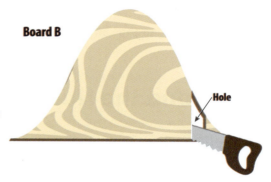

Board B

c. If you are just throwing the bean bag at the bell board at random, then how would you figure out the chances of the bag going into the hole in Board A?

   i. 8 divided by the area of the entire board

   ii. The area of the hole in A divided by the area of the hole in B

   iii. The area of the board divided by the area of the hole

   iv. The area of the hole, divided by the area of the entire board

2. Face in Class Books

In a 2012 *Washington Post* article entitled "Is College Too Easy? As Study Time Falls, Debate Rises," Daniel de Vise reports that "over the past half-century, the [average] amount of time college students actually study—read, write, and otherwise prepare for class—has dwindled from 24 hours a week to about 15 …." No standard deviation is given, but let's assume that standard deviation is 2.5 hours.

a. Suppose a college student is selected at random. Use the empirical rule to estimate how likely it is that this student studies between 10 and 17.5 hours per week.

   i. About 50% of the time

   ii. About 81.5 chances in 100

   iii. About 13.5 chances in 100

   iv. About 2.5 chances in 100

b. Suppose a college student is selected at random. Use the empirical rule to estimate how likely it is that this student studies between 17.5 and 20 hours per week.

   i. About 50% of the time

   ii. About 81.5 chances in 100

   iii. About 13.5 chances in 100

   iv. About 2.5 chances in 100

c. Suppose a college student is selected at random. Use the empirical rule to estimate how likely it is that this student studies more than 20 hours per week.

   i. About 50% of the time

   ii. About 81.5 chances in 100

   iii. About 13.5 chances in 100

   iv. About 2.5 chances in 100

3. Class on Facebook?

A 2012 study measured "the efficacy of social networking systems as instructional tools." The study surveyed 186 students about the use of social networking systems as an active part of the semester class structure. One question asked and answered by 181 of the 186 students, along with the results received, is shown below.

Question from the study:
There are no specific benefits that make Facebook a better forum for class discussions and announcements than a learning management system like Blackboard. Do you agree or disagree?

Make sure you can read the table. If you are having trouble interpreting the data, ask your instructor to explain it. The mean of these 181 answers is 3.15, and the standard deviation is 1.05.

TABLE 6.2 Study Results[1]

Response	Number of Subjects Choosing This Response
1 – Strongly Disagree	9
2 – Disagree	43
3 – Neutral/Undecided	59
4 – Agree	52
5 – Strongly Agree	18

1   Buzzetto-More, N. "Social Networking in Undergraduate Education,"*Interdisciplinary Journal of Information, Knowledge, and Management Volume 7, 2012.*

a. Use the empirical rule to estimate how likely it is that an answer to this question will be in the interval 2.10 to 4.20. What was the actual percentage of answers in this interval?

   i. Estimated is 81.5% and actual is 79.20%

   ii. Estimated is 68% and actual is 61.33%

   iii. Estimated is 95% and actual is 94.28%

   iv. Estimated is 16% and actual is 9.9%

b. Use the empirical rule to estimate how likely it is that an answer to this question will be above 4.20. What was the actual percentage of answers in this interval?

   i. Estimated is 81.5% and actual is 79.20%

   ii. Estimated is 68% and actual is 61.33%

   iii. Estimated is 95% and actual is 94.28%

   iv. Estimated is 16% and actual is 9.9%

c. This example is somewhat atypical since only five outcomes are possible, and the empirical rule doesn't strictly apply. It still provides useful estimates, however. Graph the distribution of these 181 answers and show that it is, indeed, bell-shaped.

d. Confirm that the mean is 3.15 and the standard deviation is 1.05, as claimed. You are encouraged to use computer software (such as Microsoft Excel or Apple Numbers) as directed by your instructor. Provide detailed instructions as to how you accomplished this task.

4. Denying the Pass

Coach Linguini is trying to develop the defense of his new center Mo Biggs. Mo is an amazing talent with a truly unbelievable 12-foot wingspan (distance from tip of left fingers to tip of right fingers with arms outstretched). Mo's team is playing an opponent known to pass right into the middle of the lane. (See vertical red dotted line in graphic.) Coach wants Mo to play defense by shuffling his feet along the blue dotted line tangent to the free throw circle. Coach knows that, depending on what the opposing offense is doing, Mo might have to move to the left or the right, but he feels Mo can deny a pass into the center of the lane—even if only the smallest tip of one of Mo's hands is in the center of the lane.

a. How far to the left of the center line can Mo move the center of his body and still be able to deny the pass to the lane?

   i. 18′

   ii. 12′

   iii. 6′

   iv. ±3′

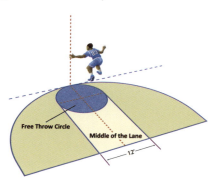

Team data suggest that Mo will play within six feet on either side of the center line 68% of the time. He will play six-to-twelve feet to the left of the center line 13.5% of the time, and six-to-twelve feet to the right of the center line 13.5% of the time. Finally, he will only play in the lanes twelve-to-eighteen feet (left or right of center line) 2.35% of the time, respectively. (See diagram.)

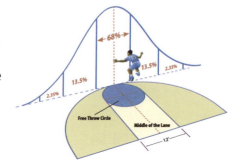

b. About what percentage of the time will Mo be in position to deflect a pass that is thrown dead center into the middle of the lane (right through the vertical red line)? Assume all passes thrown by the opposition will be at approximately chest level for Mo.

   i. About 68% of the time

   ii. About ±34% of the time

   iii. About 95% of the time

   iv. About 81.5% of the time

5. Making the Pass

Off the court, Mo has his own problems trying to pass statistics class. In particular, he is trying to understand confidence intervals for proportions a little better. The sampling distribution of the sample proportion is shown again here. Think of the horizontal axis in the figure as where the sample proportion $\hat{p}$ moves about from sample to sample, much the way Mo's feet do on defense. The diagram just tells us the chance of $\hat{p}$ being in certain places along the x-axis. The very center of the diagram is where the unknown parameter p resides.

a. Refer to the diagram to the right. Suppose a sample of size n produced a $\hat{p}$ that lands right on $p - \frac{1}{2}\frac{1}{\sqrt{n}}$. How far away from p is $\hat{p}$?

   i. It would be about $p + \frac{1}{2}\frac{1}{\sqrt{n}}$ to the left.

   ii. It would be about $\frac{1}{2}\frac{1}{\sqrt{n}}$ to the right.

   iii. It would be about $\frac{1}{2}\frac{1}{\sqrt{n}}$ to the left.

   iv. It would be about $p - \frac{1}{2}\frac{1}{\sqrt{n}}$ to the left.

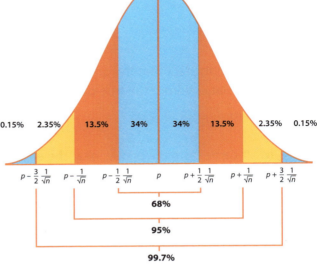

b. Think of width of the interval $\hat{p} \pm k$ as the "wingspan" of $\hat{p}$. What does k have to be if you want the interval to overlap the parameter p 95% of the time?

   i. $\hat{p} \pm \frac{1}{\sqrt{n}}$

   ii. $\frac{1}{2}\frac{1}{\sqrt{n}}$

   iii. $\hat{p} \pm \frac{1}{2}\frac{1}{\sqrt{n}}$

   iv. $\frac{1}{\sqrt{n}}$

c. Revisit what you know about the sampling distribution of the sample proportion, provided the initial samples are probability samples. Discuss how critical or not those known characteristics were for answers to the above questions to be possible. A complete and coherent answer is expected.

6. Sizing up a Poll

At the beginning of the 2011–2012 NBA season, the players were locked in a labor dispute, and the fans had become disgruntled. Read the following excerpt from a Poll Position media report that appeared during that time period.

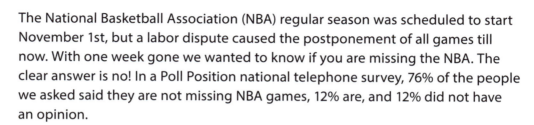

Title: 76%: We Do Not Miss the NBA

Author: Larry Register

Source: https://bleacherreport.com/articles/930316-nba-lockout-76-percent-of-those-polled-do-not-miss-professional-basketball

> The National Basketball Association (NBA) regular season was scheduled to start November 1st, but a labor dispute caused the postponement of all games till now. With one week gone we wanted to know if you are missing the NBA. The clear answer is no! In a Poll Position national telephone survey, 76% of the people we asked said they are not missing NBA games, 12% are, and 12% did not have an opinion.
>
> Poll Position's scientific telephone survey of 1,179 registered voters nationwide was conducted November 6, 2011, and has a margin of error of ±3%.

a. Explain where the margin of error of ±3% came from?

i. From $\left[\frac{76}{2}\right]/12 \approx 3$

ii. From the fact that the only 3% of those surveyed answered outside of two standard deviations from the true parameter

iii. From $\frac{1}{\sqrt{1179}} \approx 0.03$

iv. It came from the article, near the end.

b. Polling organizations don't typically just come up with a reasonable sounding sample size, then take the sample, find the statistic, and construct their confidence intervals. Most often, they reverse engineer the problem—starting with a margin of error and some associated confidence level that they can live with, and then working backwards to the size of the sample needed to achieve that outcome.

In the above poll about the NBA lockout of 2011, what size sample would Poll Position have needed to take in order to construct a 99.9% confidence interval with a width of 0.02 or less?

i.   About 27,060

ii. About 9,181

iii. About 13,530

iv. About 11,000, more or less

Projects

1. Rolling Bells

This project is flexible enough to be done alone, in small groups in class, or outside of class. Your instructor will clarify how he wants this done. If done in class, he may use dice, spinners, Legos, magnetic strips, or objects completely other than those described below. The point of the exercise will be unchanged.

Part 1: Generate the Statistics

a. Roll a standard six-sided die 25 times and record the percentage of times an even face appears. Repeat this 50 times. Record your 50 sample percentages in a table similar to the one below.

b. Describe the population you are sampling.

c. Will these samples be equivalent to simple random samples? Explain.

d. In this simulation you know the parameter in the population. What is it?

TABLE 6.3

Trial	% Even	Trial	% Even	Trial	% Even	Trial	% Even	Trial	% Even
1		11		21		31		41	
2		12		22		32		42	
3		13		23		33		43	
4		14		24		34		44	
5		15		25		35		45	
6		16		26		36		46	
7		17		27		37		47	
8		18		28		38		48	
9		19		29		39		49	
10		20		30		40		50	

Part 2: Create the Sampling Distribution

Once you have calculated your percentages of even faces, create a sampling distribution of the 50 statistics. Unless your instructor asks you to do otherwise, use software (an online histogram applet such as http://www.shodor.org/interactivate/activities/Histogram/ is particularly convenient) to construct the sampling distribution of the 50 percentages. Make sure your plot has customized labels and title. Fix the *x*-minimum to be 0, and choose an interval width of size 6.

Part 3: Connect to Probability

a. What is the standard deviation of the sampling distribution of the sample proportion in this simulation?

b. What proportion of all of your actual sample statistics are within two standard deviations of the parameter?

c. How does your answer in b. compare to what the Empirical Rule (repeated to the right) predicts will be the case?

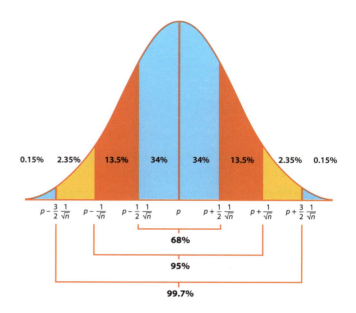

Part 4: The Deliverable

Unless your instructor tells you otherwise, all questions should be answered in a typed document. Number the parts and the questions within the parts. The table and histogram must be generated by software (not hand drawn) and included in the actual document (not appended in a separate document or spreadsheet). Use complete sentences, correct spelling, and make sure your explanations are cogent.

2. Repetition Confidence

This project is flexible enough to be done alone, in small groups in class, or outside of class. Your instructor will clarify how she wants this done. If done in class, she may use dice, spinners, Popsicle sticks, magnetic strips, or objects completely other than those described below. The point of the exercise will be unchanged.

Part 1: Generate the Data

a. Roll a standard six-sided die 75 times and record the percentage of times a 1, 2, 3, or 4 lands face up. Repeat this 60 times. Record your 60 sample percentages in a table similar to the one below. Use the columns with headings "Sample %."

b. Describe the population you are sampling.

c. Will these samples be equivalent to simple random samples? Explain.

d. In this simulation you know the parameter in the population. What is it?

TABLE 6.4

Sample %	Sample +/–12%	Sample %	Sample +/–12%	Sample %	Sample +/–12%	Sample %	Sample +/–12%	Sample %	Sample +/–12%

Part 2: Form the Confidence Intervals

a. Since these are samples of size $n = 75$, the 95% margin of error associated with each sample statistic is 0.12. Explain why this is?

b. For each of your sample statistics, compute a 95% confidence interval and record alongside that statistic, and under the columns with headings "Sample +/– 12%." For example, if you generate a sample statistic of 0.64 (64%), you would record (64 – 12, 64 + 12), that is, (52, 76) beside the 64% in your table.

Part 3: Connect to Probability

a. What percentage of all the confidence intervals you computed contain the parameter?

b. How is the answer to the question in a. connected to the precise interpretation of what kind of "confidence" a confidence interval provides? Explain carefully.

Part 4: The Deliverable

Unless your instructor tells you otherwise, all questions should be answered in a typed document. Number the parts and the questions within the parts. The table must be generated by software (e.g. Word, and not hand drawn) and included in the actual document (not appended in a separate document or spreadsheet). Use complete sentences, correct spelling, and make sure your explanations are cogent.

3. Reality Video

Introduction

This module explained to us where confidence intervals come from. Now, we want to give you a chance to both show you know what you are doing and to help teach others.

The Assignment

Your instructor may have you do this as individuals or may form you into teams. We'll refer to "teams" below, although you may be a team of one depending on what your instructor wants. Each team is required to do the following:

a. Pick one or more topics from this chapter. Your instructor may want to approve them before you proceed.

b. Construct a three-minute video that teaches the chosen topic to the class. You must carefully and correctly use the language of the chapter in your presentation, and you are required to give an example that is NOT in the chapter.

c. Post the video on YouTube and send the link to your instructor in the manner he stipulates.

7

CHAPTER 7
Sampling: When Probability Isn't Enough

Introduction

The margin of error is a useful method for quantifying sampling variability, also called "random sampling error." Non-sampling errors pose challenges to sampling integrity that can't be addressed by the margin of error. We will use the following working definition of "non-sampling error."

Non-sampling error is an error caused by something other than the fact that a sample was selected, instead of the entire population. These errors can be present in censuses as well as samples.

Some typical non-sampling errors originate from the following:

- Coverage problems;
- Biased questionnaires or poorly constructed questionnaires;
- Interviewer bias, real or perceived;
- Subject nonresponse;
- False information provided by the subjects, intentional or accidental.

It is becoming recognized that non-sampling errors may be a bigger threat to survey integrity than sampling error. Sampling error was understandable in the sense that the probability inherited from the random act of sampling would guarantee a patterned behavior in statistics from sample to sample. There is no analogous mathematical way of understanding or quantifying non-sampling error. Rather, the best we can do is to be aware of the likely sources and try to mitigate their effects as much as possible. This chapter looks at two of the sources of non-sampling error listed above, as a means of increasing that awareness. Further sources are discussed in the exercises. It is important to keep this in mind:

The margin of error offers absolutely no quantification of non-sampling error and no direct or indirect assessment of its magnitude.

This is still widely misunderstood by those who try to interpret elementary statistical inference.

Biased or Poorly Constructed Questions

If a questionnaire has poorly constructed questions, then the responses that are collected may end up being misleading. Survey design is not a mathematical topic and, as such, not something statisticians are necessarily better at than anyone else. Still, we all should be aware of some of the more common mistakes made in question construction.

Double Barrel Question

E.g. "Do you plan to leave your job and look for another one during the coming year?"

Survey design experts call this a "double-barreled" question since it has two questions embedded in one. The problem is that it might be possible to agree with one of the questions but not both. If a respondent chooses the answer, "Yes," have they rationalized that this is the correct way to answer if they agree with at least one of the parts? Or perhaps the respondent thought they could only answer yes if they agreed with both parts.

The No Way Out Question

E.g. "Have you often, sometimes, hardly ever, or never felt badly because you were unfaithful to your wife?"

Believe it or not, this was a real question once asked in a Harris Poll. In response, 1% said often, 14% said sometimes or hardly ever, and 85% said they never felt bad because of this. Does this mean that 85% of the sample had cheated on their wives and not felt bad? Surely a large part of that 85% accounts for people who have never cheated on their wives, and thus choose the only category that seems to apply.

The Leading Question

E.g. "Should we increase taxes in order to get more housing and better schools, or should we keep them about the same?"

The problem here is that the question is leading the respondent. Even if you don't think taxes should be increased, few people would be against "more housing" and "better schools."

It will be hard to construct reliable inferences from data collected from these kinds of questions. As beneficiaries of survey data, and the attending inferences, we need to be alert to these kinds of unintentional biases that the questions may have caused. If we are charged with creating and implementing the survey, our task is even harder. We need to acquaint ourselves more broadly with what is known about survey construction. Often, a pilot (preliminary) sample is taken just so interviewees can offer feedback on what they might have found misleading or confusing. That information is then used to revise the questionnaire as needed before the official sample is taken.

The Initiative on Survey Methodology at Duke University offers the following tips for mitigating non-response associated with question wording (see https://dism.duke.edu/survey-help/help-topics/, last accessed May 2021). Detailed discussion of these items goes beyond the scope of this chapter. However, excellent examples are provided on the referenced site.

- Avoid double-barreled questions;

- Avoid erroneous assumptions;

- Clarify ambiguous and imprecise terms or break them down into several questions;

- Define terms very specifically when necessary;

- Avoid loaded, leading, emotional, or evocative language as it can bias responses;

- Avoid confusing technical or academic terms;

- Balance questions to make positive and negative responses "ok";

- Consider providing counterarguments in the question itself;

- Avoid complex sentences.

Subject Nonresponse

Even if you select a simple random sample just perfectly from your population, what are you supposed to do if only 40% of those selected choose to answer the questions you are asking? Before you say that just wouldn't happen, look at the graphic shown here (based on 2018 data). Only mail surveys and in-person surveys had response rates that were above 40%, with the overall average response rate being 33%!

So you can expect that even after a polling organization has chosen its sample in just the perfect probabilistic way, it is only going to end up with a relatively small proportion of subjects who will respond. There is no easy way to fix this. A naïve approach is to say, "well, if they need a sample of size 1000 and they expect only a 50% response rate, then they should have selected a sample of size 2000 originally."

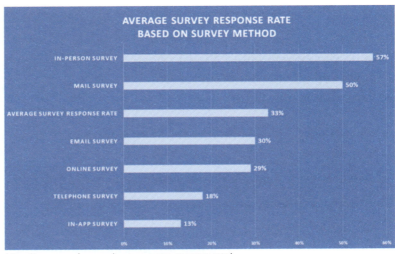

https://surveyanyplace.com/average-survey-response-rate/

While the arithmetic makes sense, the probability doesn't. That is, the real question is, would the sample of more or less 1000 that you ended up with be a simple random sample? The answer is 'no.' Oversampling in anticipation of a certain response rate fixes the actual response rate but not the underlying probability. We know quite well that it is the pristine probability associated with the original sample that produces an unencumbered mathematical path to margins of error and confidence intervals. Even if one were able to argue that a margin of error was still useful, though only approximately correct in some sense, it is impossible to know if the actual respondents differed in some systematic way from the original sample. Still, non-sampling errors are a messy business, and we will see below that oversampling is a common tool used by survey organizations.

There is no perfect way out of the non-response dilemma. It may well be that non-response errors are more damaging to the integrity of down-the-line inferences than is sampling variability. Let's look at a couple of sources of non-response errors and some of the ways experts try to mitigate the subsequent negative effect on the inference.

Questions Are Too Sensitive

Even if a question is worded perfectly, answers can be compromised if the question is sensitive. Surveys about drug use, abortion, what kind of sex you have and how often, how you spend your money, etc., are often not answered. Have a look at the table below that was published in the *Psychological Bulletin* in 2007.

You can read the table this way. Out of 1,000 people asked:

- About 81 would refuse to respond to a question about their total household income.

- Thirty would likely refuse to answer a question about the number of male sexual partners.

- Nearly nine would likely not answer a question about the age of their first sexual intercourse.

The Initiative on Survey Methodology at Duke University (referenced above), offers the following (paraphrased) advice for asking sensitive questions:

TABLE 7.1

Item	Percentage Nonresponse
Total household income	8.15
No. of lifetime male sexual partners	3.05
Received public assistance	2.22
No. of times had sex in past 4 weeks	1.37
Age of first sexual intercourse	0.87
Blood tested for HIV	0.65
Age of first menstrual period	0.39
Highest grade completed	0.04

- Always reassure respondents about their anonymity or confidentiality in the introduction to the survey;

- Avoid putting sensitive questions too early or too late in the survey;

- Make respondents feel comfortable telling the truth;

- Consider the mode in which the survey is deployed. Surveys administered in a manner that does not have a human interviewer may increase the respondent's willingness to answer a sensitive question.

Subjects Are Too Busy or Uninterested

As noted at the beginning of this chapter, questions don't have to be sensitive for there to be a substantial percentage of people sampled who will just not answer. They may be too busy, uninterested in the subject, or don't trust how the data will be used. What one has to worry about most is how non-respondents might differ systematically from others in the chosen sample. Do they have more sex or less sex? Do they make less money or make more but don't want to say? Once again, the Initiative on Survey Methodology at Duke offers the following (well-accepted) advice on how to increase response rates.

- Use call-backs and reminders. That is, sampled people should be contacted multiple times during the data collection period.

- Use refusal conversion techniques. In some cases, the survey will have staff employed who are skilled at contacting non-respondents and convincing them to answer.

- Use incentives. Offering an incentive is a way of creating a sense of obligation on the part of the sampled subject to respond.

- Employ oversampling. This was described earlier in this chapter.

We have paraphrased these to better fit the current discussion. Some of the suggestions are investigated in more detail in the exercises at the end of this chapter. We would like to think that any major poll that has produced information of consequence to us will have utilized one or more of these techniques to reduce non-response bias. In practice, however, much of the time these practices are too time consuming or expensive for polling organizations to use.

Consequences Are Real

So where does all this discussion on non-sampling errors and the challenge of getting a handle on their impact leave us? That is, non-sampling errors are clearly a problem and can compromise the integrity of even simple inferences. But just how bad is it? Different experts are going to have different opinions obviously. However, the Harris Poll has offered a very clear opinion. Have a look at the disclaimer that they now include with all their poll results.

> *All sample surveys and polls, whether or not they use probability sampling, are subject to multiple sources of error which are most often not possible to quantify or estimate, including sampling error, coverage error, error associated with nonresponse, error associated with question wording and response options, and post-survey weighting and adjustments. Therefore, Harris Interactive avoids the words "margin of error" as they are misleading.*

This is an astonishingly honest admission to be offered by a major polling organization! Harris won't report the MOE because they know the following:

- It might confuse people who think it addresses non-sampling as well as sampling errors;
- Common ways non-sampling error is addressed will compromise the theoretical integrity of the MOE;
- Non-sampling error might well be a much larger challenge than sampling variability to making good human inferences from the data they are reporting on.

Understanding the potential effects of non-sampling errors on our inferences is critical. Knowing how to potentially reduce some of those problems is useful and something we should expect of a reputable polling organization. Even so, the mathematical connections between the sampling activity and the margin of error will have been strained, and our ability to even get a handle on sampling variability—which we would like to think is predictable will have been compromised. How much it will have been compromised is typically not something that can be answered. The best advice perhaps is to just think of the margin of error that is reported in a complex, real-world survey as an approximate way of addressing sampling variability.

TAKE-HOME POINTS

- Non-sampling errors can substantially detract from the accuracy of survey data and are not addressed by the margin of error.
- While there are ways of mitigating non-sampling errors, these often come at the expense of the probability needed to ensure a pristine connection with the margin of error.

Chapter 7 Exercises

Reading Check

Short Answer: Please provide brief, concise answers to the following questions.

1. What kind of error does the margin of error address?

2. What is the definition of a non-sampling error?

3. What was the issue with the question "Have you often, sometimes, hardly ever, or never felt badly because you were unfaithful to your wife?" How could this question compromise the survey results?

4. What are some strategies for reducing non-response errors?

5. Give your own example of a survey question that would likely create a notable source of non-sampling error.

Beyond the Numbers

1. That's Getting Personal

Title: Playboy Survey Asks College Students about Sex Lives

Author: Marilyn King

Source: State News, September 17, 2009, http://www. statenews.com/index.php/article/2009/09/ playboy_survey_asks_college_students_about_sex_lives

Conducted at Playboy.com, the survey asked 5,000 U.S. university and college students about their sex lives. Of the respondents, 80% were male and 20% were female. One question asked the following:

> *"Are you in a nude picture on someone's camera phone?"*

Thirty-four percent said "yes."

a. Other samples of 5,000 people, asked the same question, would not produce a sample percentage of 34%. What is this kind of variability called?

    i. Non-sampling variability

    ii. Sampling variability

    iii. Margin of error variability

    iv. Non-response variability

b. If the sample had truly been a random sample of all university and college students, what percentage of women would you expect to have been in the sample?

    i. About 5%

    ii. About 34%

    iii. About 50%

    iv. About 66%

c. Name at least one error you'd expect this survey to suffer from, even if all 15.9 million college and university students in the United States had answered, not just 5,000. Discuss how this error might be addressed by a reputable polling organization.

2. Bully Folly

On October 4, 1994, the *San Francisco Examiner* went to press with the headline "1 in 4 Youths Abused, Survey Finds." The telephone survey of 2,000 children ages 10–16 asked in part: "In the past year have you been slapped, kicked, punched, hit, or threatened with an object by an adult, sibling, or another child?"

Give at least one non-sampling error that makes this headline deceptive. Suggest a better headline using this question. Then keep the headline that is reported and suggest one or more questions that would better inform that headline.

3. Healthcare and Harris Reform

Title: Many Americans Still Confused about New Healthcare Reform Law and Its Provisions

Source: Harris Interactive, July 29, 2010, http://www. prnewswire.com/news-releases/many-americans-still-confused-about-new-healthcare-reform-law-and-its-provisions-99541539.html

Harris Interactive is a huge polling organization. As noted in this chapter, it took the unprecedented step of including a disclaimer at the end of its surveys. This article below offers an example of what a typical Harris poll now says.

> Not sure what's in—and not in—the new healthcare legislation signed into law by President Barack Obama in March? You're not alone. More than 2,100 adults were given a list of 18 reform items and asked to identify what's included and what's not included in the law. Only four items were correctly identified by the majority of those polled.

Most (58%) know that the reform package will prohibit insurers from denying coverage to people because they are already sick; 55% know the law permits children to stay on their parents' insurance plan until age 26; and 52% realize that people who don't have insurance will be subject to financial penalties. Additionally, half are aware that employers with more than 50 employees will have to offer their workers affordable insurance. …

Methodology This survey was conducted online within the United States from July 15 to 19, 2010, among 2,104 adults (aged 18 and over). Figures for age, sex, race/ethnicity, education, region, and household income were weighted where necessary to bring them into line with their actual proportions in the population. Propensity score weighting was also used to adjust for respondents' propensity to be online.

All sample surveys and polls, whether or not they use probability sampling, are subject to multiple sources of error which are most often not possible to quantify or estimate, including sampling error, coverage error, error associated with nonresponse, error associated with question wording and response options, and post-survey weighting and adjustments. Therefore, Harris Interactive avoids the words "margin of error" as they are misleading. All that can be calculated are different possible sampling errors with different probabilities for pure, unweighted, random samples with 100% response rates. These are only theoretical because no published polls come close to this ideal.

Respondents for this survey were selected from among those who have agreed to participate in Harris Interactive surveys. The data have been weighted to reflect the composition of the adult population. Because the sample is based on those who agreed to participate in the Harris Interactive panel, no estimates of theoretical sampling error can be calculated.

Full data available at https://theharrispoll.com. The results of this Harris Poll may not be used in advertising, marketing, or promotion without the prior written permission of Harris Interactive.

These statements conform to the principles of disclosure of the National Council on Public Polls.

In November 2013, Harris Interactive was purchased by Nielsen and now publishes as The Harris Poll. The same disclaimer about the margin of error is still being used.

a. What is the ratio of a 95% margin of error to an 80% margin of error for the true proportion of all Americans who believe that the reform package will prohibit insurers from denying coverage to people?

   i.   0.014

   ii.   0.050

   iii.   0.020

   iv.   1.563

b. What is meant by the following statement that appears in the **Methodology** section?

"Therefore, Harris Interactive avoids the words 'margin of error' as they are misleading."

   i. There are so many different kinds of non-sampling error that go into a survey, only one of which the MOE addresses.

   ii. There are so many different kinds of error that go into a survey, only one being non-sampling error, which is what the MOE addresses.

   iii. There are so many different kinds of error that go into a survey, only one being random sampling error, which is what the MOE addresses.

   iv. There are so many different kinds of error that go into a survey, only one being biased sampling error, which is what the MOE addresses.

c. What is meant by the following statement that appears in the **Methodology** section?

"Because the sample is based on those who agreed to participate in the Harris Interactive panel, no estimates of theoretical sampling error can be calculated."

   i. Even if there were no non-sampling errors, the MOE is suspect because the data are based on a sample from those who agreed to participate.

   ii. Even if there were significant sampling error, the MOE would be suspect because of all the errors due to how the health-reform questions were designed.

   iii. In the complete absence of non-sampling error, the MOE would have been appropriate for this survey.

   iv. In the complete absence of sampling errors, the MOE would still have been useful to have as a way of summarizing selection bias.

d. Which type of error does the Harris Poll seem to be claiming is the most difficult to get a handle on? Support your answer.

4. Survey Says!

The MOE helps us get a handle on sampling variability. Non-sampling variability is a lot tougher to tame. Give detailed answers to the following questions. You may need to do some research on your own.

a. Give a plausible example of a type of survey that might have substantial coverage error, and explain why. Note that coverage error results from not allowing all members of the survey population to have an equal or nonzero chance of being sampled. What can you do to limit coverage error?

b. Give a plausible example of what might cause non-response error. Note that non-response error results from not being able to interview people who would be eligible to take the survey. In your answer, say something about how you might be reassured that non-responses did not create any substantial error.

c. Perhaps the most pervasive and vexing form of "non-sampling" error in surveys is question order. As you consider the following scenario, note that measurement errors result when surveys do not survey what they intended to measure.

Imagine that a restaurant owner was conducting a customer satisfaction survey. With no prior survey background, he creates a short survey with questions ordered like this:

1. Please rate the temperature of your entrée.

2. Please rate the taste of your food.

3. Please rate the menu selection here.

4. Please rate the courtesy of your server.

5. Please rate the service you received.

6. Please rate your overall experience at this restaurant.

What might be unwise about the order of these questions?

d. Offer some suggestions on how you might rearrange the survey above. If desired, consult additional information about question order at https://www.pewresearch.org/methods/u-s-survey-research/questionnaire-design/.

5. Pay to Play

Title: Efficacy of Incentives in Increasing Response Rates

Authors: Fahimi, M., Whitmore, R.W., Chromy, J.R., Siegel, P.H., & Cahalan, M.J. (2006).

Source: Presented at Second International Conference on Telephone Survey Methodology, Miami, FL, https://www.rti.org/sites/default/files/resources/TSM2006_Fahimi-efficacy_paper.pdf

An internet search on the effects of incentives on increasing response rates will reveal a lot of research on the topic. One particular study conducted at the Research Triangle Institute in Raleigh-Durham, North Carolina, is particularly interesting. In Phase I of the study, 1,197 subjects were contacted and asked to complete a survey. However, those subjects were randomly split into three subgroups. Subjects in the first group were not offered any monetary reward for early completion of the survey. Subjects in the second were offered $20 for early completion, and those in the third group were offered $30 for early completion. Here is what the researchers found with respect to the number of subjects who responded to the survey:

TABLE 7.2

Incentive Group (Early Response)	Number of Respondents	Number of Non-Respondents
Group 1 ($0)	66	336
Group 2 ($20)	120	271
Group 3 ($30)	138	266
Total	324	873

a. What two fractions produced from the table would be used to support or refute a claim that incentives matter for early completion of a survey?

    i. 66/336 and 258/537

    ii. 66/402 and 258/795

    iii. 402/873 and 795/873

    iv. 336/402 and 537/795

b. What two fractions produced from the table would be used to support or refute a claim that a high incentive is more effective than a low incentive at achieving early completion of a survey?

    i. 138/266 and 120/271

    ii. 404/873 and 391/873

    iii. 404/471 and 391/471

    iv. 138/404 and 120/391

6. No Pay Replay

There were 873 non-respondents to the survey introduced in Problem 5. In Phase II of the study, these 873 subjects were contacted a second time and asked to complete the survey. No additional incentives were offered, and the offers of incentives for early completion had already expired. Here is what the researchers found in Table 7.3:

a. What percentage produced from the table would be used to support or refute a claim that no-incentive follow-up requests are effective at increasing response rates?

    i. 32%

    ii. 51%

    iii. 48%

    iv. 17%

TABLE 7.3

Incentive Group (Early Response)	Number of Respondents	Number of Non-Respondents
Group 1 ($0)	109	227
Group 2 ($20)	91	180
Group 3 ($30)	96	170
Total	296	577

b. Use data from the table to support or refute a claim that the effectiveness of no-incentive follow-up requests depends both on whether an incentive for early completion had originally been offered and whether that incentive was large or small.

7. Some Pay Saves Day

There were 577 non-respondents still remaining after the Phase II attempt. In Phase III, these 577 were contacted again and asked to complete the survey. This time, however, the remaining non-respondents were randomly divided into two groups: one would receive no compensation for completion, while the other would receive $30 for completion. Here is what the researchers found:

TABLE 7.4

Incentive Group (Early Response)	Number of Respondents	Number of Non-Respondents
NF1 ($0)	98	190
NF2 ($30)	135	154

a. What pair of fractions from the table would be used to support or refute a claim that incentives matter when conducting follow-ups for survey completion?

   i.   98/233 and 190/344

   ii. 135/233 and 154/344

   iii. 98/190 and 135/154

   iv. 135/289 and 98/288

b. At the end of Phase III, 233 people responded. All of these 233 had been contacted two other times over the course of the experiment. Explain why this could create confounding if complex statistical methods are not used to mitigate.

8. No Pay for Your Say

This question assumes you were assigned and completed questions 5–7. Take a step back and look at the results of those three questions. Use what you have learned to describe how you would incentivize (or not) a survey that you wanted to administer. Your answer should include comments about incentives for early completion as well as incentives for follow-up completion. Remember, you are not being asked to design an experiment. That's what the authors above did. Rather, you are being asked to use what you have learned from their study to decide whether you would want to attach an incentive plan to your survey, and if so, what it should look like.

9. Knotty Not

Title: Poll on Doubt of Holocaust Is Corrected

Authors: Michael R. Kagay

Source: *https://www.nytimes.com/1994/07/08/us/poll-on-doubt-of-holocaust-is-corrected.html*

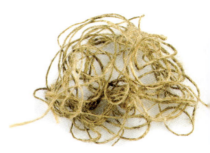

Just how important is question wording in a survey? A mixed-up 1992 Roper poll makes the answer clear. While this happened a long time ago, it remains one of the most sobering examples of how seriously we have to take non-sampling errors. In the article referenced above, a fatal flaw in the original poll is discussed, and the poll's first results are compared with the results of a corrected follow-up poll. Find the article, read it, and then answer the following questions.

a.   What was the exact wording of the question asked in the original 1992 poll, and what percentage of those surveyed responded that it was possible that the Nazi extermination of the Jews never happened?

b.   What was the problem with the original question, and how was that problem corrected in the follow-up poll?

10. When You Say It *That* Way

Title: Question Wording

Authors: Pew Research Center

Source: http://www.people-press.org/methodology/questionnaire-design/question-wording/

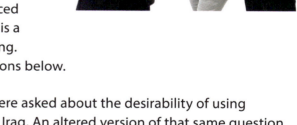

The Pew Research Center is one of the most prolific polling organizations operating today. The referenced article, provided by the Pew Research Center itself, is a short tutorial on the importance of question wording. Find the article, read it, and then answer the questions below.

a.   In the primary example provided, Americans were asked about the desirability of using military action to end Saddam Hussein's rule in Iraq. An altered version of that same question added the clause "even if it meant that U.S. forces might suffer thousands of casualties." What was the percentage drop in affirming answers?

    i.   25%

    ii.   35%

    iii.   15%

    iv.   The article just says the percentage changed but doesn't say by how much.

b. The article lists several tips for survey construction. Which of the following four statements are the two "important things to consider in crafting survey questions" that the article mentions?

c. "Ask questions that are clear and specific and that each respondent will be able to answer."

d. "Try to limit the number of response options in an effort to elicit more truthful answers."

e. "If possible, make sure an interviewer is present, in an attempt to reduce social desirability bias."

f. "Ask only one question at a time."

    i. A and B

    ii. A and D

    iii. B and C

    iv. B and D

g. Should marijuana be legalized in your state? Suppose you are charged with constructing a survey to collect public opinion on this controversial question. Construct two versions of the same survey question. The first version should be free of wording bias. The second version should exhibit a gentle, but clear push toward legalization.

Projects

1. Does Race Still Matter?

Introduction

This chapter focused on how complex it can be to produce quality data from surveys, even with careful probability sampling.

The Assignment

For this project you are asked to learn more about stereotype threat and how that phenomenon affects data quality from surveys.

Title: Stereotype Threat and Race of Interviewer Effects in a Survey on Political Knowledge

Authors: Darrin Davis and Brian Silver

Source: *American Journal of Political Science*, Vol. 47, No. January 2003, pp 33-45. http://www.researchgate.net/publication/228828389_Stereotype_threat_and_race_of_interviewer_effects_in_a_survey_on_political_knowledge

The study referenced above was published in 2003, almost 35 years after an earlier study suggested that race was a big factor in non-sampling survey errors. In the original 1968 study, interviewers in Detroit asked black residents, "Do you personally feel that you trust most white people, some white people, or none at all?" When the interviewer was white, 35%

answered "most." When the interviewer was black, only 7% answered "most." Is society beyond this problem now? Access the article above and read it. You may want to ignore the more complicated statistical summaries, but you should read the non-technical sections of the paper in detail.

As you read the paper, focus on answering the questions listed below. Organize these answers into a typed, professional paper that is presented in your own words. The paper should read like a report to your supervisor, not like a bulleted list of answers to the questions. It is up to you to organize the paper around these questions and then to construct transitions among the different topics.

i.   Describe social desirability bias and explain how it affects the accuracy of survey data.

ii. Explain the concept of stereotype threat in the context of gathering survey data.

iii. Explain how the 2003 experiment was designed to eliminate social desirability bias as a cause for any treatment differences that would be observed.

iv. The authors asked seven political questions to two groups of subjects. Describe what distinguished the two treatments assigned to the groups.

v.   The authors also collected racial information on the interviewers and the interviewees. Explain how this was done.

vi. A primary deliverable of this study was the evaluation of the effect of the (perceived) race of the interviewer on the political knowledge of blacks and whites. What did the study find?

vii. What do you think this study says about the racial landscape of 2003 compared to the racial landscape of 1968?

2. Reality Video

Introduction

This chapter reminded us of some basic statistical constructs, and how their computation, use, or display can affect our human inferences. Now, we want to give you a chance to both show you know what you are doing and to help teach others.

The Assignment

Your instructor may have you do this as individuals or may form you into teams. We'll refer to "teams" below, although you may be a team of one depending on what your instructor wants. Each team is required to do the following:

a. Pick one or more topics from this chapter. Your instructor may want to approve them before you proceed.

b. Construct a three-minute video that teaches the chosen topic to the class. You must carefully and correctly use the language of the chapter in your presentation, and you are required to give an example that is NOT in the chapter.

c. Post the video on YouTube and send the link to your instructor in the manner he stipulates.

8

CHAPTER 8
The Language of Decision Making

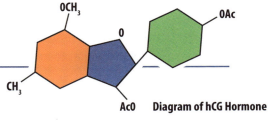

Diagram of hCG Hormone

Introduction

There aren't any statistics available on exactly how many home pregnancy tests (HPTs) are used each year, but it is safe to say that they are used in abundance. An HPT typically uses a urine sample and works by using chemistry to bond the glycoprotein hCG to an antibody and an indicator. hCG is typically presented just after fertilization.

The goal of this screening test is to let the user know if she is likely pregnant. This is called a "positive" outcome in the language of screening tests. If the test is unable to suggest the user is pregnant, that is called a "negative" outcome. Don't attach social meanings to positive and negative. When a test such as the HPT thinks it has found what it was designed to find, it is common to call that a positive outcome.

All screening tests—whether they are screening for depression, concussion, bowel cancer, whatever—are faced with delivering the same kind of dichotomous decision. Either the test indicates you likely have what the test is designed to look for (a "positive" or "yes" result), or the test indicates that you don't seem to have what the test is designed to look for (a "negative" or "no" result). And all such tests have some risk of the same two possible mistakes:

> The test can say you do have what the test is looking for, when you really don't. The test has wrongly come back positive. This is called a **false positive.** The rate at which this is expected to happen for the test in question is called the **false positive rate** for that test.

> Or the test can say you don't have what the test is looking for, when you really do. The test has wrongly come back negative. This is called a **false negative.** The rate at which this is expected to happen for the test in question is called the **false negative rate** for that test.

Evaluating Screening Tests

Every screening test has to be evaluated to see how often it is susceptible to false positives and how often it is susceptible to false negatives. To compute the false negative rate and false positive rate, we need two types of information:

- The classification of each subject by the screening test as either a positive or a negative.
- The true classification of the subject. That is, whether each subject was, in reality, a positive or negative.

The latter information is usually available from some more extensive, perhaps more invasive or expensive testing, often called the "gold standard." The expense or inconvenience of the gold standard is why screening tests are developed. Nevertheless, to evaluate the efficacy of a new screening test, or a screening test being used in a new way, this information has to be available.

For example, while an HPT provides an inexpensive screen for pregnancy, a more definite (and expensive) option is to just go to the doctor for a blood test. Likewise, there are concussion screenings that are important to do on the sideline of a sporting event, but MRIs could provide definitive answers if available and affordable.

We will spend more time learning about how to compute false positive rates and false negative rates later. But, first, let's become acquainted with more language common to screening tests. If you have a medical screening test performed in a clinic, your doctor may not use the terms false positive rate or false negative rate when she talks to you about the reliability of the procedure. Instead, she may refer to the sensitivity and specificity of the test.

Specificity is the chance that a screening test will correctly identify a true negative outcome as a negative. Numerically, specificity is 1 – the false positive rate.

Sensitivity is the chance that a screening test will correctly identify a true positive outcome as a positive. Numerically, sensitivity is 1 – the false negative rate.

Example: Accuracy of HPTs

Let's look back at HPTs for a minute. How accurate are they, at least according to their box labels and instructional inserts? The claim made by one brand of HPT is shown here.

Clearblue Digital Pregnancy Test

- No test is more accurate—over 99% accurate from the day of your expected period
- Results 5 days sooner
- Unmistakably clear results in words
- Easy to use with its unique hourglass symbol to show you the test is working
- Fast—results in just 3 minutes
- Easy-to-use 1-step test

Notice that the label says the test is 99% accurate. There are two questions worth asking at this point if you ever plan to have any encounter with an HPT. Is this reference to accuracy a reference to sensitivity, specificity, or something else entirely? And, secondly, how was this assessment done? It turns out that the 99% means that 99% of the time the test will return a positive when a non-pregnant woman's urine is mixed with commercial hCG. Too much information, perhaps but that is how it is done. Is this a claim about sensitivity or specificity?

Remember, the hCG is the indicator of pregnancy, which is what the test is designed to look for. Since the urine being tested has been mixed with hCG, it is a proxy for urine from a pregnant woman. Hence, the claim is that 99% of the time the test will correctly identify a positive as a positive. So this is the laboratory claim about the sensitivity of the test. It follows that the false negative rate of this HPT is 1% or 0.01.

That's quite good. Actually, the sensitivity of most HPTs, tested in lab-controlled situations, is around 99%, but that may be deceptively high. Those tests were not performed using real women who were trying to use the tests in actual urine streams. Turns out that is a problem. A 1998 study in the *Archives of Family Medicine* found that the actual false negative rates, using real women in real situations, tended to be higher than 1%, sometimes much higher. See Table 8.1 below on the left.

What about false positive rates and specificity? For an HPT, a false positive is often thought of as the error that is less socially problematic. Hence, kits don't typically say anything about specificity. However, the study just mentioned also recorded the false positive rates for several different HPT kits in practice, and the results are shown in Table 8.2 on the right. Notice they varied widely by brand and could be either quite high or essentially 0.[1]

TABLE 8.1 [1]

Kit Type	% Time Kit Didn't Say Subject Was Pregnant, When She Actually Was
Ova II	47.4
Predictor	3
Answer	22
Daisy 2	18
e.p.t.	18
e.p.t. plus	10
Advance	14
Fact	0
Answer 2	0
First Response	7.1
Advance	8.8

TABLE 8.2 [1]

Kit Type	% Time Kit Said Subject Was Pregnant, When She Wasn't
Ova II	48.1
Predictor	4
Answer	36
Daisy 2	36
e.p.t.	25
e.p.t. plus	8
Advance	9
Fact	6.5
Answer 2	6.5
First Response	0
Advance	0

How well does an HPT perform in practice? The data from the two tables above were combined to allow for a head-to-head comparison of several different brands. Keep in mind this study is from 1998 and current brands may produce different results. As of 2014, this was the only extensive study available.

1   Data from study in *Arch Fam Med* Sept/Oct 1998. A meta-analysis by Bastian, et al.

There are several interesting things to note.

- First, one of the kits (OVA II) was just bad, with a high false negative rate and a high false positive rate.

- While a few of the kits had similar false positive and false negative rates, by and large those that had good false positive rates had worse false negative rates, and vice versa.

- The tension between false positive rates and false negative rates is a common phenomenon for screening tests of all kinds. Ratcheting down the false negative rate typically—not always—ratchets up the false positive rate, and the reverse is true as well.

TABLE 8.3 [2]

Kit Type	FNR	FPR
Ova II	47.4	48.1
Predictor	3	4
Answer	22	36
Daisy 2	18	36
e.p.t.	18	25
e.p.t. plus	10	8
Advance	14	9
Fact	0	6.5
Answer 2	0	6.5
First Response	7.1	0
Advance	8.8	0

So what is the best test of the ones shown? That depends on what you are looking for. The one with the best balance is probably Predictor with a 3% and 4% false negative and false positive rate, respectively. But there are two with an estimated 0% false negative rate, and if that is the error that one wants to keep the lowest, then those two would be the most desirable.

Practicing the Computations

The actual computation of sensitivity and specificity is pretty easy, provided you are comfortable with fractions. This section will use several examples to help you understand how the computations are done.

Example: ImPACT Test for Concussions

Let's start with the ImPACT test. ImPACT stands for "Immediate Post-Concussion Assessment and Cognitive Testing" and is designed to be a screening test for concussion that is somewhere between an on-field diagnosis and an MRI. It takes about twenty-five minutes to complete, and it measures attention span, working memory, sustained and selective attention time, response variability, non-verbal problem solving, and reaction time. Subjects are given a total score for each of these categories.

Just how well does ImPACT work as a screening test? To answer this question, recall we need two things. We need data on how the test categorized athletes. That is, we need a study population so we can know how many athletes ImPACT said were concussed and how many ImPACT said were not concussed. Secondly, we need some gold standard results, such as would be provided by an MRI, so that we can know whether the athlete really was concussed or not.

---

2   Data from study in *Arch Fam Med* Sept/Oct 1998. A meta-analysis by Bastian, et al.

Researchers publishing in the Archives of Clinical Neuropsychology had just that information for 138 athletes. See the table below.

TABLE 8.4 Rule: Beyond a certain level on the ImPACT scale, you are said to be concussed.[3]

ImPACT Prediction	Actual Status		Totals
	Athlete was, in reality, not concussed	Athlete was, in reality, concussed	
ImPACT said "not concussed"	59	13	72
ImPACT said "concussed"	7	59	66
Totals	66	72	138

Notice that of those 138 athletes, 66 really were not concussed and 72 were. In a somewhat confusing coincidence, notice that ImPACT said that 72 were not concussed and 66 were. Of the 66 that were truly not concussed (true negatives), ImPACT said seven were. So ImPACT had seven instances of a false positive, and, hence, the estimated false positive rate is 7 out of 66 or 0.11: 11% if you prefer percentages. Similarly, of the 72 who were truly concussed according to an MRI (true positives), ImPACT said 13 were not. So the false negative rate is 13 out of 72 or 0.18 or 18%.

It follows that the estimated specificity for ImPACT is 100% minus 11% or 89%. And the estimated sensitivity is 100% minus 18% or 82%.

Example: Beck Depression Inventory Test

The Beck Depression Inventory (BDI) is still commonly used, especially with students, to screen for depression. The Inventory consists of 21 questions about how the subject has been feeling in the last week. For example, one question is designed to have you assess how sad you feel, ranging from "I do not feel sad" (which gets scored a 0) to "I am so sad or unhappy that I can't stand it" (which gets scores a 3). All questions are scored similarly, so the subject's total score can range from 0 to 3×21 or 63. High scores are indications of depression; different cutoffs for quantifying "high" are used for different purposes.

The table below has the results for 95 patients who took the BDI. The gold standard, or actual status of each patient (depressed or not depressed), was determined by a long clinical interview as required by the Diagnostic and Statistical Manual of Mental Disorders IV. A cutoff of 10 was used for determining if a screened subject was, in fact, likely to be clinically depressed. This is still a common cutoff to be used.

---

3   "Sensitivity and specificity of the ImPACT Test Battery for concussor in athletes." *Archives of Clinical Neuropsychology,* 2005.

TABLE 8.5 **Rule: If your total score on the Beck Inventory is 10 or greater, then you will be categorized as depressed.[4]**

Beck Inventory Indication	Actual Status as determined by Diagnostic and Statistical Manual of Mental Disorders IV		Totals
	Patient was, in reality, not depressed	Patient was, in reality, depressed	
Beck said "not depressed"	66	5	71
Beck said "depressed"	12	12	24
Totals	78	17	95

Notice that of the 95 patients participating, 78 were not truly clinically depressed, but 17 were. The Beck Inventory said that 71 of those 95 were not depressed and 24 were. More importantly, of the 78 who really were not depressed, Beck said 12 were. So the BDI has an estimated false positive rate of 12 over 78 or 14%. Similarly, of the 17 who really were depressed, Beck said five were not. So the BDI has an estimated false negative rate of 5 out of 17 or 29%.

It follows that the sensitivity is estimated at 71% and the specificity at 85%. Both are good, but not spectacular. As you can see, the computation of false positive and false negative rates is really very straightforward, especially when the data are already arrayed in a table for us.

Example: Field Sobriety Tests

There are three common testing procedures that are implemented during a field sobriety test (FST) at a sobriety checkpoint:

- A test for visual nystagmus (HGN);
- The one-legged stand (OLS);
- The walk and turn (WAT).

The visual nystagmus test is typically implemented as follows. The detained driver is told to follow (with his eyes) a pen or a finger to the left and to the right. The officer administering the test is going to be looking for a lack of smooth pursuit (to left or right), onset of eye jerking (nystagmus) before 45 degrees (to left or right), and presence of sustained nystagmus when the driver looks all the way to the left and to the right. An infraction is recorded for any one of these six clues that the officer will check for.

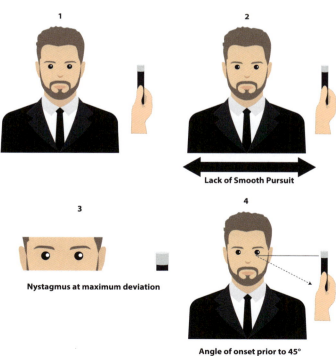

Lack of Smooth Pursuit

Nystagmus at maximum deviation

Angle of onset prior to 45°

---

4   "Sensitivity and specificity of Depression Questionnaires in a College-Age Sample." *Journal of Genetic Psychology, 2008,* 169(3), 281–288.

The one-legged stand requires the driver to stand in a particular way, then to lift a leg about 6 inches off the ground with the foot pointed out, not down. Arms have to be at the side and the driver is required to count 1001, 1002, … until told to stop. An infraction will be recorded every time the driver sways, has to use arms to balance, hops, or puts his foot down.

Finally, the walk and turn requires the driver to look down as his feet, arms to his side, and take (usually) 9 heel-to-toe steps along a straight line, turn and do the same back. Infractions are recorded if the driver can't keep his balance, uses arms for balance, steps off line, etc.

Therefore, the data that are recorded for a full field sobriety test are infraction counts for each of the three challenges. A well-known national database has archived the FST data from 296 subjects who participated in the National Highway and Transportation Safety Agency's 1998 San Diego field sobriety test reliability study. Of the 296 subjects from the San Diego study, 267 had blood alcohol content of 0.04% or greater. This was considered legally intoxicated for the purposes of that study. So the data set has 267 who were, in truth, drunk, and 29 who were not.

The way in which these kinds of counts are used to flag (or fail to flag) someone as likely intoxicated varies a great deal. For purposes of this example, we are going to create something called a Total Field Sobriety Test score that is just the sum of the number of infractions in each of the three observed categories. Also, we will assume that the detaining officer will use the rule that a Total Field Sobriety score of 4 or greater will be an indication of possible intoxication.

TABLE 8.6

Case	HGN	OLS	WAT	Total FST	Actual BAC
229	0	0	1	1	0
254	0	0	1	1	0.02
66		1		1	0.067
142	2	0	0	2	0.005
217	2	0	0	2	0.03
191	2			2	0.034
182	2	0	0	2	0.038
109	2			2	0.04
259	2	0	0	2	0.04
199	0	1	1	2	0.048
13	2	0	0	2	0.05
2	2			2	0.06
67	2		1	3	0.022
145	0	3	0	3	0.03
53	2	0	1	3	0.032
15	2	0	1	3	0.04
287	2	0	1	3	0.04
89	2	1	0	2	0.05
123	0	1	2	3	0.05
258	2	1	0	3	0.053
35	2	0	2	3	0
11	2	1	1	4	0.01
247	4			4	0.016
6	2	0	2	4	0.02
294	2	0	2	4	0.02
231	2	0	2	4	0.03
74	2	1	1	4	Actual
214	2	1	1	4	0.04
58	4			4	0.05
14	2	1	1	4	0.058
34	4	0	0	4	0.058
12	2	0	2	4	0.06
211	2	1	1	4	0.06
232	4	0	0	4	0.06
293	2	2	0	4	0.06
130	2	1	1	4	0.07
297	4			4	0.08
271	2	0	2	4	0.1
119		2	2	4	0.121

A portion of the San Diego Field data set is shown in Table 8.6. The Total Field Sobriety Test score was added, the data were ordered by that value, and all the subjects who had a score less than or equal to 4 are shown. For example Case Number 220 got a 0 on the HGN, a 0 on the OLS and a 1 on the WAT. So his Total FST is 0 + 0 + 1 or 1, and that is what is shown in the Total FST column.

Recall that a score of 3 or less would be categorized as "sober" by the rule we adopted for this illustration. How many of those are in the entire data set? Just count the number of rows in the portion of the data set shown that correspond to a 3 or below on the Total FST. There are 20 of those. (See red circle below).

Of those 20 that the FST judged to be sober, how many really were sober? That is, of those 20, how many had an Actual BAC (Blood Alcohol Content) less than 0.04? Once again, count. For example, Case 229 was judged sober by the FST (Total FST of 1) and really was sober according to the BAC (had a 0). There were 9 of those 20 who were really sober. Nine of the subjects that the FST said were sober actually were sober. (See blue circle below).

We can now fill out the table shown below, using what we know about the totals, the results of the counting we just did, and subtraction.

And the table is complete. Now we can compute the estimated false negative and false positive rates for these important data.

TABLE 8.7 **Using Total FST ≥ 4 to Designate "Drunk"**

FST Decision	Actual BAC		Totals
	< 0.04% (Sober)	≥ 0.04% (Drunk)	
Sober	9	11	20
Drunk	20	256	276
Totals	29	267	296

Given

- Of the 29 subjects who really were sober, the FST said 20 were drunk. So the false positive rate is 20 out of 29, or 69%. Hence, the specificity is 100% – 69%, or 31%

- Likewise, of the 267 subjects who really were drunk, 11 were classified as sober by the FST. Hence, the false negative rate is 11 out of 267 or 4%. It follows that the sensitivity is 96%

Typical of a field sobriety test, the sensitivity is very good and the specificity is not so good. The test is excellent at catching drunk drivers, but the price paid is that it also falsely accuses a rather large percentage of sober drivers. You will not be surprised to know that these differences are often at the heart of arguments made by opposing attorneys in drunk driving court cases that are based on field sobriety test results.

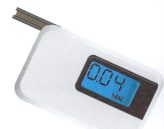

So what if we change the rule that triggers the FST to categorize a participant as drunk? For example, what if instead of a 4 or above being taken to mean "intoxicated" one took 2 or above? This will make it a lot easier to catch all the drunks. It will also catch even more people who are truly sober. So the FPR will surely go up, and the FNR will go down. It is, therefore, no surprise that how well a screening test performs with respect to sensitivity and specificity will be directly related to the cutoff that is used to identify a "positive" outcome.

A Little More Language

Sensitivity is the chance a screening test will come back positive, *if*, in fact, *the subject has the condition* being screened for. We could turn this around and ask what are the chances the subject has the condition, *if*, in fact, *the screening test came back positive?* A similar inversion of specificity leads to the following additional two ways the efficacy of a screening test can be evaluated.

> The **positive predictive value (PPV)** of a screening test is the chance that the subject really has the condition, given the screening test produced a positive outcome.

> The **negative predictive value (NPV)** of a screening test is the chance that the subject really does not have the condition, given the screening test produced a negative outcome.

PPV and NPV might well be of more interest to you personally than sensitivity and specificity when you receive the results from a screening test.

Example: ImPACT Test for Concussions (Revisited)

Recall the study introduced above on ImPACT.

There were 66 athletes who were judged to be concussed by ImPACT. Of those 66, 59 were actually concussed. So the positive predictive value is 59/66 = 0.89. Similarly, there were 72 athletes who were judged not to be concussed by ImPact. Of those 72, 59 were actually not concussed. So the negative predictive value is 59/72 = 0.82.

TABLE 8.8 **Rule: Beyond a certain level on the ImPACT scale, you are said to be concussed.**

ImPACT Prediction	Actual Status		Totals
	Athlete was, in reality, not concussed	Athlete was, in reality, concussed	
ImPACT said "not concussed"	59	13	72
ImPACT said "concussed"	7	59	66
Totals	66	72	138

Bayes' Rule

As was just noticed, the positive predictive value is just the reverse of sensitivity in that what was conditioned on was flipped. This is similar for specificity and the negative predictive value. It is useful to have notation for helping make this critical difference more apparent and to fully understand why we would never expect sensitivity and PPV (or specificity and NPV) to be the same.

Notation

The notation P(A|B) is read as "the probability of A, given B, has occurred." So the "|" symbol is read as "given." Formally, A and B are called *events* and P(A|B) is a *conditional* probability. Likewise, P(A) is read as "the probability of A." It is not conditioned on anything else having occurred. Hence, it is called an *unconditional* probability.

> Bayes' Rule is a useful way of relating conditional and unconditional probabilities. According to this rule, for any two events A and B, we have:
>
> $$P(A|B) = \frac{P(B|A) \times P(A)}{P(B)}$$

Let's use "T+" to denote the event "the screening test concludes that the condition (disease, pregnancy, etc.) is present." Likewise "T−" is notation for the event "the screening test concludes that the condition is not present." "CP" denotes that a condition is actually present, while "CA" denotes that a condition is actually absent. We can recast definitions we have already encountered in this new notation.

- $P(T+|CP)$ is the sensitivity of a test.
- $P(T-|CA)$ is the specificity of a test.
- $P(CP|T+)$ is the positive predictive value of a test.
- $P(CA|T-)$ is the negative predictive value of a test.
- $P(CP)$ is the prevalence of the condition in the population.

Immediately we can use Bayes' Rule to relate sensitivity to positive predictive value:

$$P(T+|CP) = \frac{P(CP|T+) \times P(T+)}{P(CP)}; \quad \text{that is:} \quad \text{Sensitivity} = \frac{PPV \times P(T+)}{\text{Prevelance}} = PPV \times \left[\frac{P(T+)}{\text{Prevelance}}\right]$$

It becomes quite clear, then, that sensitivity and positive predictive value are not at all the same. Bayes' Rule shows us that, in fact, sensitivity is PPV times the ratio of $P(T+)$ (which we didn't give a separate name) and prevalence.

Example: Confirming Bayes' Rule

For the ImPACT screening we have the following:

- Sensitivity = 59/72
- PPV = 59/66
- Prevalence = 72/138
- $P(T+)$ = 66/138

It follows that:

$$\frac{59}{72} = \frac{\frac{59}{66} \times \frac{66}{138}}{\frac{72}{138}} = \frac{\frac{59}{138}}{\frac{72}{138}} = \frac{59}{138} \times \frac{138}{72} = \frac{59}{72}$$

and Bayes' Rule is confirmed! Of course, we didn't really need to confirm it. It is always true! While Bayes' Rule does relate quantities like sensitivity and PPV, the larger point is that it makes crystal clear that the probabilities (or rates) $P(A|B)$, $P(B|A)$, $P(A)$, and $P(B)$ should not be expected to be the same for any events A and B.

TAKE-HOME POINTS

- The false positive rate and false negative rate are common numerical assessments of efficacy of a screening test.

- Simple fractions can be used to compute sensitivity, specificity, positive predictive value, and negative predictive value when both the test results and the truth are arrayed in a 2×2 table.

- There is no reason to think that assorted conditional probabilities are the same as each other or are the same as related unconditional probabilities. Bayes' Rule makes this abundantly clear.

Chapter 8 Exercises

Reading Check

Short Answer: Please provide brief, concise answers to the following questions.

1. What are the two ways in which a screening test can be wrong?

2. What is a false positive rate?

3. What is sensitivity in terms of the false negative rate?

4. Give two reasons why the "99% accuracy" claims listed on home pregnancy tests might be misleading.

5. Regarding the ImPACT example, how many athletes studied were not in fact concussed?

6. Regarding the ImPACT example, how many athletes studied did the screening test find to be concussed?

7. Regarding the Beck Inventory example, in what percentage of cases did the screening test make the right decision?

8. Regarding the Field Sobriety Test example, what was the sensitivity when a cutoff of 4 was used?

9. In what sense are positive predictive value and sensitivity similar?

10. What is the larger purpose of Bayes' Rule being introduced in this chapter?

Beyond the Numbers

1. Grading the FOB

The Fecal Occult Blood (FOB) test is used to screen people for bowel cancer. As expected, if it comes up positive, then you are told you likely have bowel cancer. If it comes up negative, you are told you are likely okay. So the outcome is dichotomous: either a "thumbs up" or a "thumbs down." A study involving 203 people was designed to assess how well the FOB works. Two things were recorded for each participant in the study: whether the FOB said they had bowel cancer and whether they really did (as determined by an endoscopy). The results are in Table 8.9.

a. How do you know from the table that there were 203 people in the study?

b. How many times did the FOB make the *right* decision?

   i. 19 times

   ii. 3 times

   iii. 200 times

   iv. 184 times

TABLE 8.9 Performance of the FOB

FOB Test	Patients with Bowel Cancer (as confirmed on endoscopy)	
	Positive	Negative
Positive	2	18
Negative	1	182

c. What percentage of the time did the FOB make the *wrong* decision?

    i. About 9% of the time

    ii. About 1% of the time

    iii. About 49% of the time

    iv. About 41% of the time

d. What might be the consequences of the FOB screening test saying "positive" when the patient really didn't have bowel cancer?

    i. This would potentially create unnecessary anxiety for the patient.

    ii. There are no immediate consequences since not having bowel cancer is a very positive outcome.

    iii. A patient with a potentially fatal disease might be deprived of the quickest post-screening intervention possible.

    iv. The patient would now be much more likely to also have a positive outcome on the gold standard treatment, thereby prolonging an unnecessary stay in the health care system.

e. How often did the error of saying "positive" when the patient really didn't have bowel cancer occur?

    i. 18 times

    ii. 2 times

    iii. 1 time

    iv. 182 times

f. What might be the consequences of the FOB saying "negative" when the patient really did have bowel cancer?

    i. This would potentially create unnecessary anxiety for the patient.

    ii. There are no immediate consequences since not having bowel cancer is a very positive outcome.

    iii. A patient with a potentially fatal disease might be deprived of the quickest post-screening intervention possible.

    iv. The patient would now be much more likely to also have a positive outcome on the gold standard treatment, thereby prolonging an unnecessary stay in the health care system.

g. How often did the error saying "negative" when the patient really did have bowel cancer occur?

    i.   18 times

    ii. 2 times

    iii. 1 time

    iv. 182 times

h. Looking at your answers to parts e. and g., is the FOB a good test or not? Defend your answer.

2. More on the FST

We were introduced to the NHTSA's 1998 San Diego field sobriety test validation study in this chapter. A small part of the study's data set is presented in the table below. It contains all 29 of the legally sober drivers in the study. Your instructor will explain what is in the data table and give an example of how to read it if you are unsure. See the appendix for a table of the full data set or go online at www.statconcepts.com/datasets where you can download the full data as an Excel file.

TABLE 8.10

HGN	OLS	WAT	Actual BAC	HGN	OLS	WAT	Actual BAC
0	0	1	0.000	4	1	0	0.020
2	0	2	0.000	2	—	1	0.022
2	5	2	0.000	2	3	1	0.027
2	0	0	0.005	6	2	2	0.028
2	1	1	0.010	0	3	0	0.030
2	1	2	0.010	2	0	0	0.030
4	—	—	0.016	2	0	2	0.030
2	2	2	0.017	2	1	2	0.030
4	2	4	0.017	—	2	3	0.030
0	0	1	0.020	2	0	1	0.032
2	0	2	0.020	2	—	—	0.034
2	1	2	0.020	3	3	3	0.037
2	3	3	0.020	2	0	0	0.038
2	0	2	0.020	2	3	1	0.039
2	1	2	0.020				

All other participants had a BAC of .04% or above, and thus were legally drunk.

a. Recall that this study assumed that a BAC of 0.04% or above means that a person is legally drunk. There were 296 participants in the study, and part of the table is already filled out for the participants who were legally drunk. Use the rule that any score in any of the three categories that is a "2" or higher means the roadside test tagged the participant as drunk. What are the values of the missing entries in Table 8.11?

   i.   A = 3; B = 2; C = 29; D = 29; E = 291;

   ii. A = 3; B = 26; C = 29; D = 5; E = 291;

   iii. A = 27; B = 2; C = 29; D = 29; E = 267;

   iv. A = 2; B = 27; C = 29; D = 4; E = 292;

TABLE 8.11

SFST Decision	Real BAC Results		Totals
	< 0.04%	≥ 0.04%	
Sober	A	2	D
Drunk	B	265	E
Totals	C	267	296

b. Suppose we change the rule to say that any score in any of the three categories that is a "3" or higher means the roadside test tagged the participant as drunk. This is what the table of counts will look like with this new rule in Table 8.12.

   According to this new table, what is the specificity of the FST with this new rule?

TABLE 8.12 Rule: Any score in any of the three categories that is a "3" or higher will be taken as indication you are drunk.

FST Decision	Real BAC Results		Totals
	< 0.04 (Sober)	≥ 0.04 (Drunk)	
Sober	18	22	40
Drunk	11	245	256
Totals	29	267	296

   i.   245/267
   ii. 18/29
   iii. 245/256
   iv. 18/40

c. According to this new table, what is the positive predictive value of the FST with this new rule?

   i.   245/267
   ii. 18/29
   iii. 245/256
   iv. 18/40

d. According to this new table, what is the negative predictive value of the FST with this new rule?

   i.   245/267
   ii. 18/29
   iii. 245/256
   iv. 18/40

e. According to this new table, what is the sensitivity of the FST with this new rule?

    i.   245/267

    ii. 18/29

    iii. 245/256

    iv. 18/40

f. Look at your answers to the above four questions. Is the FST with the "≥ 3 rule" a good test or not? Explain.

3. Bending the Rules

We were introduced to the NHTSA's 1998 San Diego field sobriety test validation study in this chapter. You will need access to the full data set for this question. See the appendix for a table of the full data set or go online at www.statconcepts.com/datasets where you can conveniently download the data as an Excel file.

a. Suppose we say that a Total FST (that is, a sum of one's HGN, OLS, and WAT scores) of 2 or more means a person is drunk. Then the 296 subjects in the study would be categorized as shown in Table 8.13.

For this rule, the false positive rate (FPR) is 27/29 = 0.931 (93.1%), and the sensitivity is 266/267 = 0.996 (99.6%). Different rules have different FPRs and sensitivities. That is, if the rule is "You are drunk if your total FST is greater than or equal to 4," the FPR and sensitivity will be different from those we just calculated. In fact, these data were collected in an example in the chapter (FPR was 69% and Sensitivity 96%).

Change the rule as needed and find the remaining six entries of the Table 8.14 (rounded to two decimal places).

TABLE 8.13 Using Total FST ≥ 2 to Designate "Drunk"

| FST Decision | Actual BAC | | Totals |
	< 0.04% (Sober)	≥ 0.04% (Drunk)	
Sober	2	1	3
Drunk	27	266	293
Totals	29	267	296

TABLE 8.14 Rule: You are drunk if your total FST score is greater than or equal to k.

k	FPR (x)	Sensitivity (y)
1	A	D
2	0.931	0.996
3	B	E
4	0.690	0.960
5	C	F

    i.   A = 0.48; B = 0.91; C = 1.00; D = 0.98; E = 0.35; F = 0.69;

    ii. A = 1.00; B = 0.79; C = 0.48; D = 1.00; E = 0.98; F = 0.91;

    iii. A = 1.00; B = 0.79; C = 1.00; D = 0.98; E = 0.48; F = 0.91;

    iv. A = 0.48; B = 0.91; C = 0.98; D = 0.35; E = 0.69; F = 0.79;

b. Based on the values you produced for the table in part a., what would be your choice for the best rule? Defend your answer.

c. Patients with suspected hypothyroidism were screened by Goldstein and Mushlin (*J. Gen. Intern. Med.* 1987;2:20–24) using thyroxine levels, often abbreviated as T4. Thyroxine is a hormone secreted into the bloodstream by the thyroid. The authors looked at three cutoff values for thyroxine level: 5 or less; 7 or less; 9 or less. With these three cutoffs they found the following:

TABLE 8.15

Cutoff (Rule)	Sensitivity	Specificity
≤ 5 suggests hypothyroidism	0.56	0.99
≤ 7 suggests hypothyroidism	0.78	0.81
≤ 9 suggests hypothyroidism	0.91	0.42

A plot of sensitivities (along a *y*-axis) versus FPRs (along an *x*-axis) for different rules is called a receiver operating characteristic curve (ROC). An ROC is a convenient way of deciding what cutoff rule is best for a particular screening test. Which of the following plots shown below is the correct ROC plot for the hypothyroid study?

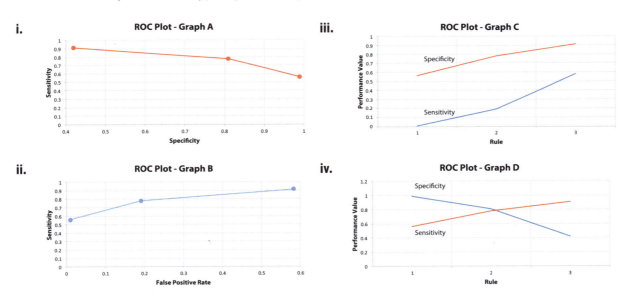

i. ROC Plot - Graph A

ii. ROC Plot - Graph B

iii. ROC Plot - Graph C

iv. ROC Plot - Graph D

4. Stairway Stumble

Title: Ruling a Diagnosis In or Out with "SpPin" and "SnNOut": A Note of Caution

Authors: D. Pewsner, M. Battaglia, C. Minder, A. Marx, H. Bucher, and M. Egger

Source: *BMJ*, Volume 329, July 2004, pp. 209 ff.

The authors of this study describe a situation whereby a physician attending to the patients of a colleague on vacation encountered a *"40-year-old teacher who had sprained her right ankle. Returning from a conference, she had stumbled while walking down the stairs with a heavy bag. Examination revealed a moderately swollen lateral right ankle."*

The attending physician screened the woman using the Ottawa ankle rules screening test, and upon finding no tenderness of the bone, ruled out a fracture without requiring an x-ray. His reasoning was based on the "SnNOut" guidelines, which many physicians use today. According to these guidelines, high sensitivity **(Sn)** and a negative test result **(N)** implies that probability the condition will be present, given the test came back negative, will be low and, hence, a real positive can safely be ruled out **(Out).** Is it reasonable to assume that sensitivity can stand on its own this way? This exercise takes a deeper look at this question.

Let "T+" denote the event of a screening test saying that a condition is present. Likewise "T−" denotes a screening test saying that a condition is absent. CP denotes that a condition is actually present, whereas CA denotes that a condition is actually absent. Diagnostic data for the Ottawa Ankle Test are collected in Table 8.16.

TABLE 8.16

a. What is the sensitivity of the test (rounded to two decimal places)?

Predicted by the Ottawa Ankle Test	Truth Regarding Fracture		Totals
	CA	CP	
T−	51	5	56
T+	277	88	365
Totals	328	93	421

    i. 0.95

    ii. 0.09

    iii. 0.16

    iv. 0.81

b. What is the chance the condition will be present, given the test came back negative, for the Ottawa Ankle Test (rounded to two decimal places)?

    i. 0.95

    ii. 0.09

    iii. 0.16

    iv. 0.81

c. Do the two values you found in parts a. and b. support the SnNOut idea? Why or why not?

Now let's change the data and suppose the screen test produced the results shown in Table 8.17.

TABLE 8.17 Ankle Test Results

d. What is the sensitivity of the test based on these new screening results (rounded to two decimal places)?

Predicted by the Ottawa Ankle Test	Truth Regarding Fracture		Totals
	Not Fractured	Fractured	
Not Fractured	5	21	26
Fractured	26	369	395
Totals	31	390	421

i. 0.95

ii. 0.08

iii. 0.16

iv. 0.81

e. What is the chance the condition will be present, given the test comes back negative, for the Ottawa Ankle Test based on these new screening results (rounded to two decimal places)?

i. 0.95

ii. 0.08

iii. 0.16

iv. 0.81

f. The prevalence of a condition is just the probability that the condition will be present in the population studied. What is the prevalence based on each of the two sets of screening data?

i. 0.22 based on the first; 0.93 based on the second

ii. 0.93 (rounded) in both cases

iii. 0.22 (rounded) in both cases

iv. 0.93 based on the first; 0.22 based on the second

g. Use Bayes' Rule and what you found out about prevalence to argue that SnNOut guidelines are ill-advised in some cases. Be specific and detailed in your answer.

5. CAGE Practice

Title: Ruling a Diagnosis In or Out with "SpPin" and "SnNOut": A Note of Caution

Authors: D. Pewsner, M. Battaglia, C. Minder, A. Marx, H. Bucher, and M. Egger

Source: *BMJ*, Volume 329, July 2004, pp. 209 ff.

The authors of this study describe a situation whereby a physician attending to the patients of a colleague on vacation encountered a *"40-year-old teacher who had sprained her right ankle. Returning from a conference, she had stumbled while walking down the stairs with a heavy bag. Examination revealed a moderately swollen lateral right ankle. The patient was able to walk but was clearly in pain. Her breath smelt of alcohol."*

After ruling out an ankle fracture, the attending physician administered a screening test known as the CAGE test, a test used to screen for possible alcohol problems. This examination consists of four questions. Patient agreement with two or more of these questions suggests a problem with alcohol. Here are the data for one assessment of the overall quality of this test:

TABLE 8.18 **Alcohol Test Results**

Predicted by CAGE Test	Truth Regarding Problem		Totals
	No Alcohol Problem	Alcohol Problem	
No Alcohol Problem	400	57	457
Alcohol Problem	1	60	61
Totals	401	117	518

a. What is the sensitivity of the test based on these screening results (rounded to three decimal places)?

   i.   0.998

   ii. 0.226

   iii. 0.513

   iv. 0.888

b. What is the specificity for the test based on these screening results (rounded to three decimal places)?

   i.   0.998

   ii. 0.226

   iii. 0.513

   iv. 0.888

c. *Overall Accuracy* is defined as Overall Accuracy = (Prevalence) × Sensitivity + (1 – Prevalence) × Specificity. The prevalence of a condition is just the probability that the condition will be present in the population studied. What is the overall accuracy for the CAGE test based on the screening results above (answer rounded to three decimal places)?

   i.   0.998

   ii. 0.226

   iii. 0.513

   iv. 0.888

d. Does the overall accuracy computation seem reasonable in this problem? Explain why you say it does or does not.

6. Relevant Prevalence

Title: The Use of "Overall Accuracy" to Evaluate the Validity of Screening or Diagnostic Tests

Authors: A. Alberg, J. Park, B. Hager,, M. Brock, and M. Diener-West

Source: *J Gen Intern Med* 2004; 19:460–465.

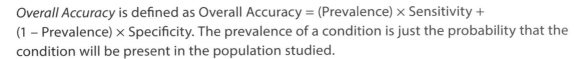

Overall Accuracy is defined as Overall Accuracy = (Prevalence) × Sensitivity + (1 − Prevalence) × Specificity. The prevalence of a condition is just the probability that the condition will be present in the population studied.

The authors of this study discuss the pitfalls of using overall accuracy as a summary measure of how well a test is doing. The following table regarding a screening test for liver cancer was recreated from summary numbers presented in this article.

a. What is the sensitivity of the test based on these screening results (rounded to three decimal places)?

i. 0.949

ii. 0.050

iii. 0.400

iv. 0.922

b. What is the specificity for the test based on these screening results (rounded to three decimal places)?

i. 0.949

ii. 0.050

iii. 0.400

iv. 0.922

TABLE 8.19 Cancer Test Results

Predicted by Test	Truth Regarding Cancer		Totals
	No Liver Cancer	Liver Cancer	
No Liver Cancer	543	18	561
Liver Cancer	29	12	41
Totals	572	30	602

c. What is the overall accuracy for this cancer screening test based on the screening results above (answer rounded to three decimal places)?

i. 0.949

ii. 0.050

iii. 0.400

iv. 0.922

d. Overall accuracy is not a completely reasonable concept in this application? Explain why it is not. What is causing the problem? Be specific.

7. Will Count for Sex

Overall Accuracy is defined as Overall Accuracy = (Prevalence) × Sensitivity + (1 − Prevalence) × Specificity. The prevalence of a condition is just the probability that the condition will be present in the population studied.

The use of overall accuracy as a measure of a screening test's quality can have some unexpected side effects. Take gender identification, for example. Suppose we have a test that is designed to identify gender simply by counting the number of letters in a person's name. Too many letters and the test identifies the person as a female. Too few and the test identifies the person as a male.

Absurd? Surely. But the important question before us is whether the test has good overall accuracy.

These are real data from a class of statistics students at the University of Kentucky. There were 10 males and 58 females included in the sample. The test was designed to identify a positive outcome (female) if there were 9 or more letters in a person's name. Here are the results:

a. What is the sensitivity of the test based on these screening results (rounded to three decimal places)?

 i. 0.853

 ii. 0.000

 iii. 0.809

 iv. 0.948

b. What is the specificity for the test based on these screening results (rounded to three decimal places)?

 i. 0.853

 ii. 0.000

 iii. 0.809

 iv. 0.948

TABLE 8.20 Gender Test Results

Predicted by Test	Truth Regarding Gender		Totals
	Male	Female	
Male	0	3	3
Female	10	55	65
Totals	10	58	68

c. What is the overall accuracy for this gender screening test based on the screening results above (answer rounded to three decimal places)?

 i. 0.853

 ii. 0.000

 iii. 0.809

 iv. 0.948

d. Is this test for gender "accurate"? Overall accuracy is not a completely reasonable concept in this application? Explain why it is not. What is causing the problem? Be specific.

8. Let the Algebra Speak!

This exercise takes a strictly algebraic look at the concept of the "overall accuracy" of a screening test. Recall the following:

Overall Accuracy = (Prevalence) × Sensitivity + (1 – Prevalence) × Specificity

The prevalence of a condition is just the probability that the condition will be present in the population studied. Look closely at this simple equation for overall accuracy, and answer the following two questions.

a. The table below lists some generic values for sensitivity, specificity, and prevalence. Match each triplet of sensitivity, prevalence, and specificity values with the letter that best describes the resulting overall accuracy value from the list alongside the table. Choose your answer from the list of possible answers shown here.

   i. I = B; II = E; III = C; IV = A; V = D

   ii. I = C; II = A; III = E; IV = D; V = B

   iii. I = E; II = A; III = D; IV = B; V = C

   iv. I = B; II = D; III = E; IV = A; V = C

TABLE 8.21

Sensitivity	Prevalence	Specificity	Overall Accuracy (Record Letter of Best Match from List on Right)
High	High	Low	i.
Low	Low	High	ii.
Any Value between 0 and 1	Around ½	Any Value between 0 and 1	iii.
Low	Low	Low	iv.
Low	High	Low	v.

A. High, similar to Specificity

B. Low, similar to Sensitivity

C. High, similar to Sensitivity

D. Low, Similar to Specificity

E. Roughly the average of Sensitivity and Specificity

b. It's your first job out of college. Your boss has just asked you to use *overall accuracy* to order several different screening tests for a condition that is important—but rare—in the field you work in. Based on what you have learned from this activity, explain to your boss why that is not a good idea.

9. Rarely Wrong

Have a look at the Bayes' Rule equation again. Write out the equation with the positive predictive value on the left hand side. Assuming it makes sense to think of both sensitivity and P(T+) as not changing with prevalence*, what happens to the positive predictive value as the prevalence in the population increases? What are the implications of what you have found? Explain.

*It turns out that these are not completely reasonable assumptions. But by making them, you can use Bayes' rule to discover something interesting about the relationship between PPV and prevalence. What you discover can be shown to be true even without these assumptions.

10. Confirming Bayes' Rule One More Time

The Beck Depression Inventory (BDI) was introduced in this chapter. The following table has the results for 95 patients who took the BDI. The gold standard, or actual status of each patient (depressed or not depressed), was determined by a long clinical interview as required by the Diagnostic and Statistical Manual of Mental Disorders IV. A cutoff of 10 was used for determining if a screened subject was, in fact, likely to be clinically depressed. This is still a common cutoff to be used.

TABLE 8.22 **Rule: If your total score on the Beck Inventory is 10 or greater then you will be categorized as depressed.**[5]

Beck Inventory Indication	Actual Status as determined by Diagnostic and Statistical Manual of Mental Disorders IV		Totals
	Patient was, in reality, not depressed	Patient was, in reality, depressed	
Beck said "not depressed"	66	5	71
Beck said "depressed"	12	12	24
Totals	78	17	95

Recall, the specificity for this test (based on these screening data) was found to be 66/78.

a. What is the negative predictive value?

    i. 71/95

    ii. 66/71

    iii. 78/95

    iv. 17/95

---

5 "Sensitivity and specificity of Depression Questionnaires in a College-Age Sample." *Journal of Genetic Psychology,* 2008, 169(3), 281–288.

b. What is the probability the test comes back negative?

    i. 71/95

    ii. 66/71

    iii. 78/95

    iv. 17/95

c. What is the probability the condition is absent in the population?

    i. 71/95

    ii. 66/71

    iii. 78/95

    iv. 17/95

d. Write out Bayes' Rule for $P(CA|T-)$ – the negative predictive value of the test. Verify Bayes' Rule for the NPV. Make sure you leave all numbers in fractional form. Do NOT convert to decimals.

Projects

1. Statistical Wooziness

Introduction

This activity is suitable for in-class or for groups to do outside of class provided data from each group are made available for pooling in Part II of the activity. Inexpensive versions of real beer goggles can be simulated by smearing clear glue on cheap safety glasses or swim goggles, or by covering either with a cellophane wrap that is smeared with petroleum jelly. The real point of this exercise is to have a chance to generate your own field sobriety test data. This can be done in a variety of ways.

The Assignment

Part I

If you are stopped for suspicion of driving while intoxicated, you may be asked to perform a One-Leg Stand as part of a field sobriety test (FST). In the One-Leg Stand test, the suspect is instructed to stand with one foot approximately six inches off the ground and count seconds aloud as "one thousand one, one thousand two, etc." until told to put the foot down. The officer times the subject for 30 seconds and looks for four indicators of impairment: swaying while balancing, using arms to balance, hopping to maintain balance, and putting the foot down (all the way or anywhere below the six-inch mark).

Form groups of 6 or less (or as directed by your instructor). Each group will be assigned the task of performing a One-Leg Stand test simulated as either:

▸ Sober—no beer goggles, beer goggles below the legal intoxication limit, or unaltered swim goggles, or

▸ Intoxicated—beer goggles above the legal intoxication limit, or altered swim goggles.

Take turns giving the One-Leg Stand test to each person in your group. Make sure it is a divided attention test. In particular, the person being tested *must* count aloud for 30 seconds and hold her foot six inches off the ground for the entire 30 seconds.

Count the number of times *any* of the above four impairment indicators occur. That total score is the classmate's One-Leg Stand score. Make sure everyone in the group is both tested and involved in the testing of others. Record the scores in Table 8.23.

TABLE 8.23 Your Group's One-Leg Stand Scores

Group Member	Simulated Drunk or Simulated Sober (Circle One)	
	Tally of Impairment Indicators	**Total of Impairment Indicators**
1		
2		
3		
4		
5		
6		

Part II

Suppose the Statistical State Police (SSP) decided that you are going to be considered "drunk" if your One-Leg Stand score is 5 or greater. As long as all data from all groups are available, the following table can be filled. Of course, all groups should have the same table entries if it is filled out correctly. You simply need to know if a group was being simulated as drunk or sober, and how many infractions each participant had.

TABLE 8.24 The Class's One-Leg Stand Results

Outcome of Field Sobriety Test	Actual Condition: Sober	Actual Condition: Drunk	Totals
Sober (OLS < 5)			
Drunk (OLS ≥ 5)			
Totals			

Final Product

Your job is to write a short (typed, professional) paper that is presented in your own words. The paper should read like a report to your supervisor, not like a bulleted list of answers to the questions. Here is what it should include:

i.   Both of the tables above (completed of course)

ii.   The sensitivity, specificity, positive predictive value, and negative predictive value of the test using the rule chosen (your instructor may ask for a different cutoff; follow her advice).

iii. An assessment of how well the test is performing as a screening test for intoxication. You must provide this assessment in the context of the quantities you provided in ii.

2. Party Test

Introduction

This activity is suitable for in-class, for homework, or for groups to do outside of class provided data from each group are made available for pooling in Part II of the activity.

The Assignment

Part I

Andrew Gelman is a professor of statistics and political science at Columbia University. He is also an expert on how and why people vote. After the 2012 Presidential election, Gelman weighed in on the old debate of who is richer, Democrats or Republicans. Gelman argued that, "both the Democrats and the Republicans are 'the party of the rich.' But Republicans more so than Democrats." You can find his full set of comments in the article referenced below.

Title: Richer People Continue to Vote Republican

Author: Andrew Gelman

Source: Campaigns and Elections, Political Economy (Comment), November, 14, 2012. https://statmodeling.stat.columbia.edu/2012/11/14/richer-people-continue-to-vote-republican/ used the graph shown below to support his statement. Carefully explain how the graph supports Gelman's point.

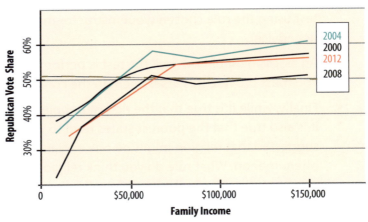

Richer Voters Continue to Lean Republican (data from exit polls)

Part II

Discretionary spending can be defined as the amount of money you use for non-essential items; money you spend as you see fit. Let's make up a screening rule that categorizes anyone who allocates $50.00 or more per week for discretionary spending as a Republican. Let's see how well this screening test works.

Your instructor may lead this activity in class, or you may be assigned to do it as homework. In either case, you are required to cross-classify *at least* 40 individuals into the following table. To do so, you must collect two pieces of information on each individual: level of discretionary spending and party affiliation. You should probably collect both anonymously.

Once the data are collected, fill out the table shown here.

TABLE 8.25 Party Affiliation Results

Predicted Affiliation Based on Discretionary Spending Amounts $50.00 or more → Republican	Actual Party Affiliation		Totals
	Republican	Non-Republican	
Republican			
Non-Republican			
Totals			

Final Product

Your job is to write a short (typed, professional) paper that is presented in your own words. The paper should read like a report to your supervisor, not like a bulleted list of answers to the questions. Here is what it should include:

i.   An introduction that addresses the Gelman graph and your comments from Part I.

ii. The completed table from Part II.

iii. The sensitivity, specificity, positive predictive value, and negative predictive value of the test using the rule chosen (your instructor may ask for a cutoff different than $50; follow her advice).

iv. An assessment of how well the test is performing as a screening test for party affiliation. You must provide this assessment in the context of the quantities you provided in iii.

v.   Finally, while it is true that the richest people tend to vote Republican (Gelman graph), it is also true that the poorest states tend to vote Republican. Some authors have suggested that religion might be one explanation for this seemingly contradictory phenomenon. What role do you think religion might play? Work your answer to this question into your final paper as a reflection.

3. A Test for Gender

Introduction—What's In This for You?

- You will further develop your skill with sensitivity and specificity computations.

- You will be exposed to the vague concept of "accuracy," which is what is used to actually judge how worthwhile field sobriety tests (FSTs) are Accuracy is a weighted average of sensitivity and specificity.

- You will experience—not just hear about—the argument behind some objections to FSTs.

- You will experience—not just hear about—some quirky behavior that can take place when validating a test in an environment that is overrun with "positives."

- You will have a chance to show that you can follow a fairly complex set of instructions.

What Do I Turn In?

Follow your instructor's lead.

Context and Definition of Gender Test

Let's suppose the NTHSA (National Traffic, Health, and Safety Administration) has decided it needs to develop a scientific roadside test that will let police officers determine the gender of drivers they encounter at traffic stops. Here is the test they give you:

Test: You will base your test on the number of letters in a driver's first and last names. Using your test, officers will determine a driver's gender simply by counting the number of letters in the driver's name. Too many letters and officers will decide the driver is a female. Too few and they will decide the driver is a male.

Now don't laugh too loudly or be too skeptical. One just needs to show a test like this has good accuracy, as defined below. That will be enough in some parts of the government to suggest that the test has been validated—that it is a good test, if you will. Read on.

Data for Tasks 1 and 2

To look into how well this test performs, we have taken the names of everyone from a past class (just like this one) and we have counted the letters in their names and identified them as male or female. That list is available in Table 8.33 at the end of this project description.

Task 1—The Initial Validation Attempt

This job will require some concentration. Please notice that there were 58 females in this past class and 41 males. You are going to set certain cutoffs for name length, and compute the specificity and sensitivity. Then, you will vary the cutoffs. For example, if you follow the rule "Designate Gender as Female if Name Length Is > 6," your results will be as follows:

TABLE 8.26 Using Total Name Length > 6 to Designate "Female"

Test Decision	Actual Gender		Totals
	Female	Male	
Female	58	41	99
Male	0	0	0
Totals	58	41	99
	Sensitivity = (58/58)*100% = 100%	Specificity = (0/41)*100% = 0%	

Accuracy = (58/99)*Sensitivity + (41/99)*Specificity = (58/99) = 0.59

Do this same thing for each of the tables shown here. Fill in the whole table and compute the accuracy, just like in the example

TABLE 8.27 Using Total Name Length > 9 to Designate "Female"

Test Decision	Actual Gender		Totals
	Female	Male	
Female			
Male			
Totals	58	41	99
	Sensitivity =	Specificity =	

Accuracy = (58/99)*Sensitivity + (41/99)*Specificity =

TABLE 8.28 Using Total Name Length > 12 to Designate "Female"

Test Decision	Actual Gender		Totals
	Female	Male	
Female			
Male			
Totals	58	41	99
	Sensitivity =	Specificity =	

Accuracy = (58/99)*Sensitivity + (41/99)*Specificity =

Task 2—Reflection on the Initial Validation Attempt

Based on accuracy, is the test you've been given a good one or not? Higher accuracy makes for a better test in the minds of some. Be very clear.

Data for Tasks 3 and 4

These are the same data as Table 8.33 except now **only the first 10 males are included.** Please notice that the students on that list have just been referred to as "Student 1," "Student 2," and so on. There are now 10 males and 58 females.

TABLE 8.29 Reduced Data Set for Tasks 3 and 4

Student	Length of Name	Student	Length of Name
Male 1	21	Male 6	15
Male 2	12	Male 7	13
Male 3	12	Male 8	12
Male 4	10	Male 9	15
Male 5	14	Male 10	13

Task 3—The Initial Validation Attempt on the Subsetted Data

Just as before, you are going to set certain cutoffs for name length and compute the specificity and sensitivity. Then, you will vary the cutoffs. For example, if you follow the rule "Designate Gender as Female if Name Length Is > 6," your results will be the same as below.

TABLE 8.30 Using Total Name Length > 6 to Designate "Female"

Test Decision	Actual Gender		Totals
	Female	**Male**	
Female	58	10	**68**
Male	0	0	**0**
Totals	**58**	**10**	**68**
	Sensitivity = (58/58)*100% = 100%	Specificity = (0/10) * 100% = 0%	
Accuracy = (58/68)*Sensitivity + (10/68)*Specificity = (58/68) = 0.85			

Now do this for the two tables below. Fill in the whole table and compute the accuracy, just like in the example.

TABLE 8.31 Using Total Name Length > 9 to Designate "Female"

Test Decision	Actual Gender		Totals
	Female	**Male**	
Female			
Male			
Totals	58	10	68
	Sensitivity =	Specificity =	
Accuracy = (58/68)*Sensitivity + (10/68)*Specificity =			

TABLE 8.32 Using Total Name Length > 12 to Designate "Female"

Test Decision	Actual Gender		Totals
	Female	Male	
Female			
Male			
Totals	58	10	68
	Sensitivity =	Specificity =	
Accuracy = (58/68)*Sensitivity + (10/68)*Specificity =			

Task 4—Reflection on This Initial Validation Attempt

Is your test performing better or worse on the subsetted data? Explain why.

Task 5—Putting This All Together

Opponents to field sobriety test validation studies argue that by keeping the cutoff really low (like our "> 6" rule) and by having the targeted population dominate the test data (drunks for FST, but females in this exercise), the concept of accuracy becomes deceptive. Does your study support that argument or refute it? Explain.

TABLE 8.33

Student	Gender	Length of Name	Student	Gender	Length of Name	Student	Gender	Length of Name
Female 1	F	10	Female 36	F	13	Male 13	M	16
Female 2	F	13	Female 37	F	14	Male 14	M	10
Female 3	F	12	Female 38	F	14	Male 15	M	11
Female 4	F	12	Female 39	F	14	Male 16	M	12
Female 5	F	13	Female 40	F	13	Male 17	M	10
Female 6	F	15	Female 41	F	10	Male 18	M	10
Female 7	F	13	Female 42	F	13	Male 19	M	12
Female 8	F	12	Female 43	F	12	Male 20	M	14
Female 9	F	11	Female 44	F	18	Male 21	M	11
Female 10	F	11	Female 45	F	13	Male 22	M	12
Female 11	F	14	Female 46	F	14	Male 23	M	19
Female 12	F	22	Female 47	F	15	Male 24	M	19
Female 13	F	11	Female 48	F	10	Male 25	M	15
Female 14	F	15	Female 49	F	13	Male 26	M	11
Female 15	F	13	Female 50	F	14	Male 27	M	12
Female 16	F	11	Female 51	F	12	Male 28	M	11
Female 17	F	9	Female 52	F	17	Male 29	M	14
Female 18	F	11	Female 53	F	11	Male 30	M	11
Female 19	F	13	Female 54	F	14	Male 31	M	15
Female 20	F	17	Female 55	F	11	Male 32	M	14
Female 21	F	13	Female 56	F	9	Male 33	M	9
Female 22	F	12	Female 57	F	13	Male 34	M	13
Female 23	F	13	Female 58	F	14	Male 35	M	16
Female 24	F	17	Male 1	M	21	Male 36	M	12
Female 25	F	9	Male 2	M	12	Male 37	M	9
Female 26	F	12	Male 3	M	12	Male 38	M	9
Female 27	F	13	Male 4	M	10	Male 39	M	16
Female 28	F	13	Male 5	M	14	Male 40	M	14
Female 29	F	14	Male 6	M	15	Male 41	M	12
Female 30	F	13	Male 7	M	13			
Female 31	F	11	Male 8	M	12			
Female 32	F	14	Male 9	M	15			
Female 33	F	11	Male 10	M	13			
Female 34	F	15	Male 11	M	12			
Female 35	F	13	Male 12	M	12			

4. Reality Video

Introduction

This chapter introduced us to sensitivity and specificity as the language of decision making for screening tests.

The Assignment

Your instructor may have you do this as individuals or may form you into teams. We'll refer to "teams" below, although you may be a team of one depending on what your instructor wants. Each team is required to do the following:

a.   Pick one or more topics from this chapter. Your instructor may want to approve them before you proceed.

b.   Construct a three-minute video that teaches the chosen topic to the class. You must carefully and correctly use the language of the chapter in your presentation, and you are required to give an example that is NOT in the chapter.

c.   Post the video on YouTube and send the link to your instructor in the manner he stipulates.

9 CHAPTER 9
Hypothesis Testing: Concepts and Consumption

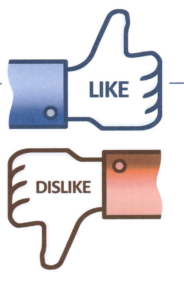

Introduction

Chapter 8 introduced us to screening tests. Screening tests ultimately have to facilitate a choice between two outcomes. A positive outcome means the test has uncovered evidence of what it was designed to find. For instance, if you score high enough on a field sobriety test, then the outcome will suggest you are intoxicated. A negative outcome simply means that the test did not uncover adequate evidence of what the test was looking for. If you have a low score on a field sobriety test, then it will likely be assumed you are sober, whether you are or not.

Experiments facilitate a choice similar to screening tests. We can think of a positive outcome as when the experiment has produced adequate evidence that the treatment is effective. A negative outcome is when the experiment has failed to produce enough evidence to say the treatment is effective. What constitutes adequate evidence is the crux of statistical hypothesis testing, and as we will see, that discussion has a significant overlap with the language used to assess the validity of a screening test. We will use this similarity to make the abstract language of hypothesis testing more accessible.

Let's start by looking at an example that has been in the news for some time, surfacing first in 2009. The drug Flibanserin was originally being studied as a potential anti-depressant, but was shown to be inadequate for that purpose. What researchers noticed during those clinical trials was that a number of women who participated in the trials indicated that their sex drive had increased. This prompted the manufacturer to develop a completely new clinical trial wherein 1,378 premenopausal women satisfying the experimental protocol were randomly assigned to one of two groups. One group took 100 mg of Flibanerin, and the other took a placebo. All participants were required to keep a daily journal about their sex drive.

In what sense is this experiment like a screening test? First of all, the experiment is designed to produce evidence that will allow an informed choice between one of two possible outcomes:

■ Either one concludes that there is enough evidence to say Flibanserin is better than a placebo for increasing sex drive (a positive outcome); or

■ One has to conclude that there is not enough evidence to say Flibanserin is better than a placebo for increasing sex drive (a negative outcome).

Secondly, just like a typical screening test, this choice will have to be made based on the evidence at hand, and the choice could always be the wrong one.

■ The experimental data could suggest the Flibanserin is effective, when in reality it is not.

■ The experimental data could fail to suggest Flibanserin is effective, when in reality it is.

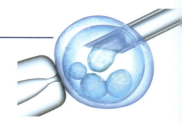

Finally, just as error rates were used to quantify the risk in using a screening test to facilitate a decision, so will similar quantities in hypothesis testing. In particular, a criterion will be set for the maximum allowable risk of saying Flibanserin is effective, when it really is not.

■ The experimental data have to be convincing enough that Flibanserin is actually effective, that this is unlikely to be the wrong decision to make.

We will revisit the Flibanserin example at different times because its journey to eventually FDA approval is a near-perfect example of the language and limitations of hypothesis testing.

To be clear, hypothesis testing is not just the purview of experimentation, medical or otherwise. But it is there that we can most clearly illustrate the relevant language and concepts, so we will mostly stay within that context as we proceed.

A Paradigm for Hypothesis Testing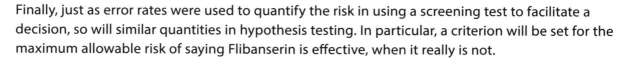

For medical experiments, we can view the generic choice being made as one of concluding "Treatment is Not Effective" and "Treatment Is Effective." The details will depend on the actual experiment, whether two treatments were being compared, if a treatment was being compared to a placebo, etc.

Hypothesis testing is the statistical paradigm for making and justifying the choice between saying a treatment is effective and failing to be able to say the treatment is effective.

Unlike the screening tests we saw in previous chapters, hypothesis testing is not a questionnaire, or a physical examination, or a blood test. Rather, it is a collection of mathematical steps that take the data the experiment has generated, and in the presence of a check on the risk of being wrong, suggests whether it is apropos to side with the positive outcome.

The decisions made in hypothesis testing come with the same risks as, say, the decision made by a home pregnancy test, or the Beck Depression Inventory that screens for clinical depression. A test of hypothesis may produce insufficient evidence that the treatment is effective, when unbeknownst to the experimenters, it is. Likewise, the test may suggest that the data produced by the experiment are sufficient to say that the treatment is effective, when, in fact, it may not be effective. Hence, just like common screening tests, a test of hypothesis may be susceptible to false negatives and false positives.

TABLE 9.1

Conclude from Experimental Results	Truth	
	Treatment Really Is Ineffective	Treatment Really Is Effective
Treatment Ineffective	True Negative	False Negative
Treatment Effective	False Positive	True Positive

We are well aware at this point that both sensitivity and specificity help us understand the risks in using a particular screening test. For example, we found in Chapter 8 that the sensitivity of the ImPACT test for concussions was about 0.82, while the specificity was approximately 0.89. That means, of course, that if we use this test on an athlete and find it comes back positive, then there are 11 chances in 100 (100–89) that this was a false positive. That is, if we conclude the athlete is concussed based on the screen, then there are 11 chances in 100 that this is the wrong decision. If your job is to apply this test to screen for possible concussions, then you have to be aware that your assessments need to be interpreted in the presence of this risk, and the risk of a false negative.

In hypothesis testing there are concepts of risk that are very similar to that of false positive rate and false negative rate. A false negative would correspond to the decision to not conclude the treatment was effective, when you should have concluded it was. While sensitivity is very important in hypothesis testing, it is a complicated concept in that context, corresponding somewhat to what statisticians call the "power" of the testing procedure. Luckily, many commonly used statistical tests reported in the media possess good sensitivity. Now this isn't true for all types of hypothesis testing, but a deeper discussion in that direction would be beyond the scope of this reading. In simple situations you can think of power as just $1 - \beta$, where β is the Type II error rate (see definition on the next page).

It turns out that the specificity of the testing procedure is what drives the practical decision of choosing between "Treatment Is Not Effective" and "Treatment Is Effective." That is, in choosing between those two possible conclusions, the risk of saying the treatment is effective, when in fact it is not, is the hurdle that has to be cleared. At the expense of introducing some addition notation, we will be able to make this discussion more precise.

There is a well-used, special notation for the choice being made in hypothesis testing.

> The **null hypothesis**, often denoted by H_0, is the hypothesis that the treatment is not effective.

> The **alternative hypothesis**, often denoted by H_A, is the hypothesis that the treatment is effective.

In hypothesis testing, a false positive is when the results from a hypothesis test suggest H_A is true when H_0 really is. Colloquially, this is a "wrong" positive. In statistical science this has its own name.

> A **Type I error** is when data suggest H_A is true, but H_A really isn't. The Type I error rate is the risk of this happening. The Type I error rate is typically denoted by α.

Similarly, a false negative is when the results from a hypothesis test suggest H_A is not true, when it really is. Colloquially, this is a "wrong" negative.

> A **Type II error** is when data fail to suggest H_A is true, but H_A really is. The Type II error rate is the risk of this happening. The Type II error rate is typically denoted by β.

An acceptable Type I error rate should be set before even collecting the experimental data. The Type I error rate forms a hurdle that the data must clear before there is any support for saying the alternative hypothesis is likely true. It is important for that hurdle to be set before any judgment is formed (consciously or unconsciously) about the ability of the data to meet that challenge.

It is common to call the preset false positive rate "alpha" or the "alpha level." If the data collected from the experiment allow for the rejection of H_0 in the presence of this alpha level (typically taken to be 0.05), then the results are said to be "statistically significant."

> Hypothesis testing results are **statistically significant** if it is possible to conclude that H_A is likely true, in the presence of the pre-determined alpha level that provides a check on the risk involved in this conclusion.

We've been talking about testing a hypothesis, but how do we actually do it? How do we decide on whether we can reject H_0 at the predetermined alpha level? What are the steps involved? We won't actually delve into the details until a later chapter, and we don't actually need the details to be intelligent consumers of the results of statistical hypothesis testing. What we do need to know is the following. Once the data from an experiment are collected, the subsequent analysis often produces something called a **p-value**. This is a particularly complex probability to understand, but not necessarily to compute. More importantly, once it is computed, the logic of deciding it is safe to accept H_A or not is clear. What one does is to simply compare the p-value to the alpha level. If the p-value is smaller than the alpha level, then it is safe to go with H_A. That is, the data have met the risk of accepting H_A when in fact H_A is not true, a risk quantified by the alpha level. If the p-value is not smaller than the alpha level, then it is not safe to say H_A is true.

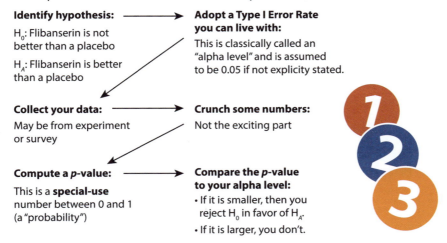

Identify hypothesis:

H_0: Flibanserin is not better than a placebo

H_A: Flibanserin is better than a placebo

Adopt a Type I Error Rate you can live with:

This is classically called an "alpha level" and is assumed to be 0.05 if not explicity stated.

Collect your data:

May be from experiment or survey

Crunch some numbers:

Not the exciting part

Compute a p-value:

This is a **special-use** number between 0 and 1 (a "probability")

Compare the p-value to your alpha level:

• If it is smaller, then you reject H_0 in favor of H_A.

• If it is larger, you don't.

> A **p-value** associated with a statistical test of hypothesis is a special use number (a probability, in fact) between 0 and 1. If it is smaller than the preset alpha level, then it is safe to accept H_A as likely true. Else it is too risky.

We will have more to say about what a p-value really is, how it is often misinterpreted, and eventually how to compute it for some particularly simple hypotheses. But first, let's practice applying what we have learned about the basic precepts of statistical hypothesis testing.

Reading with a Statistical Eye

Example: Multivitamins and Cancer

The following *New York Times* article on multivitamins appeared on October 17th, 2012. The article addresses a clinical trial that stretched for more than a decade. Male doctors were followed during that period of time, and those taking a daily multivitamin were compared to those who took a placebo (dummy pill).

Title: Multivitamin Use Linked to Lowered Cancer Risk

Author: Roni Caryn Rabin

Source: *New York Times*

> After a series of conflicting reports about whether vitamin pills can stave off chronic disease, researchers announced on Wednesday that a large clinical trial of nearly 15,000 older male doctors followed for more than a decade found that those taking a daily multivitamin experienced 8% fewer cancers than the subjects taking dummy pills.

The findings were to be presented Wednesday at an American Association for Cancer Research conference on cancer prevention in Anaheim, Calif., and the paper was published online in the *Journal of the American Medical Association*.

The reduction in total cancers was small but statistically significant, said the study's lead author, Dr. J. Michael Gaziano, a cardiologist at Brigham and Women's Hospital and the VA Boston Healthcare System. While the main reason to take a multivitamin is to prevent nutritional deficiencies, Dr. Gaziano said, "It certainly appears there is a modest reduction in the risk of cancer from a typical multivitamin."

What did the researchers find? You can see from the article that those taking the multivitamins "experienced eight percent fewer cancers ...." But it is the first line of the last paragraph that invokes a statistical badging of sorts. "The reduction in total cancers was small but statistically significant ...." With what we have learned so far in this chapter, we can say the following:

- The null and alternative hypotheses seem to be

  - H_0: Multivitamins are no more effective than a placebo at reducing cancer risk.
  - H_A: Multivitamins are more effective than a placebo at reducing cancer risk.

- The results were statistically significant, so

  - H_A was accepted.
  - It must have been that a *p*-value was computed and found to be smaller than a presumed alpha level of 0.05.
  - Therefore, the risk of siding with H_A, when H_A was really not true, was assessed to be less than 5 chances in 100. This was presumed to be a tolerable risk to take.

Example: Subway versus McDonalds

The excerpt shown below is from a 2013 article that appeared in the *Chicago Tribune*. You should be able to find the entire article by just searching on the title. This study was not an experiment but an observational study of adolescents who went to McDonald's and Subway. Researchers collected data on what those adolescents actually ate at each of the restaurants and kept track of the total number of calories consumed. What did they find? Look at the last paragraph. They bought an average of 1,038 calories at McDonald's and 955 at Subway. But that calorie difference was "not statistically significant."

Title: Teens Ate 'Too Many Calories' at Subway and McDonald's, Study Says

Author: Mary MacVean

Source: *Chicago Tribune*

Adolescents who went to a McDonald's and Subway in Los Angeles bought about the same number of calories at each, despite Subway's reputation as a healthier place to eat, researchers said.

The menus are not the point, lead researcher Dr. Lenard Lesser of the Palo Alto Medical Foundation Research Institute said by phone. "Our study was not based on what people have the ability to pick, our study was based on what adolescents actually selected in a real-world setting."

The adolescents bought an average of 1,038 calories at McDonald's and 955 calories at Subway. The calorie difference was not statistically significant, the researchers said. Their work was published Monday in the *Journal of Adolescent Health*.

- The null and alternative hypotheses seem to be

  - H_0: The number of calories consumed at Subway is the same as at McDonald's.
  - H_A: The number of calories consumed at Subway is less than at McDonald's.

- The results were NOT statistically significant, so

  - H_A could not be accepted.
  - It must have been that a *p*-value was computed and found to be larger than a presumed alpha level of 0.05.
  - Therefore, the risk of siding with H_A, when H_A was really not true, was assessed to be more than 5 chances in 100. This was presumed to be an intolerable risk to take, so the alternative was not accepted.

Example: Mental Illness and Obesity

The *New York Times* study shown here—again you can almost surely find the full article if you search on the title—addresses weight loss in people with serious mental illness. While we don't have an abundance of information from the article, we do see in the last paragraph that 24 well-designed studies of weight loss programs for the mentally ill were scrutinized. We are told that "most achieved statistically significant weight loss, but very few achieved 'clinically significant' weight loss."

Title: A Battle Plan to Lose Weight

Author: Catherine Saint Louis

Source: *New York Times*

People with serious mental illnesses, like schizophrenia, bipolar disorder, or major depression, are at least 50% more likely to be overweight or obese than the general population. They die earlier, too, with the primary cause heart disease.

Yet diet and exercise usually take a back seat to the treatment of their illnesses. The drugs used, anti-depressants and antipsychotics, can increase appetite and weight.

"Treatment contributes to the problem of obesity," said Dr. Thomas R. Insel, the director of the National Institute of Mental Health. "Not every drug does, but that has made the problem of obesity greater in the last decade."

It has been a difficult issue for mental health experts. A 2012 review of health promotion programs for those with serious mental illness by Dartmouth researchers concluded that of 24 well-designed studies, most achieved statistically significant weight loss, but very few achieved "clinically significant weight loss."

For any of the programs that achieved statistically significant weight loss, we know the following:

- The null and alternative hypotheses would be something like

  ▸ H_0: On average, there was no change in weight as a result of the program.

  ▸ H_A: On average, there was a reduction in weight as a result of the program.

- If the results were statistically significant, we can say

  ▸ H_A could be accepted.

  ▸ It must have been that a p-value was computed and found to be smaller than a presumed alpha level of 0.05.

  ▸ Therefore, the risk of siding with H_A, when H_A was really not true, was assessed to be less than 5 chances in 100. This was presumed to be a tolerable risk to take, so the alternative was accepted.

This article has an interesting nuance. The researchers also concluded that few of the studies found any *practically significant* weight loss. That is to say, the change in weight, probably on average, for patients in a given study was big enough to be statistically significant, but not big enough to be practically significant. More will be said on this important distinction later.

Example: Presidential Payments

The following *Tribune* article has a look at a study that compared the pay of presidents at various universities with measures of prestige for those universities. The findings are in the last paragraph. "[N]o statistically significant relationship was observed between academic quality and presidential pay."

Title: How to Tell If College Presidents Are Overpaid

Author: Richar Veddar

Source: *Chicago Tribune*

The Chronicle of Higher Education tells us the median salary of public university presidents rose 4.7% in 2011–12 to more than $440,000 a year. This increase vastly outpaced the rate of inflation, as well as the earnings of the typical worker in the U.S. economy. Perhaps, most relevant for this community, it also surpassed the compensations growth for university professors.

There appears to be neither rhyme nor reason for vast differences in presidential pay. David R. Hopkins, the president of Wright State University—an unremarkable commuter school ranked rather poorly in major-magazine rankings—makes far more than the presidents of the much larger, and vastly more prestigious, University of California at Berkeley, University of North Carolina at Chapel Hill, or the University of Wisconsin.

My associate Daniel Garrett analyzed the relationship between presidential compensation and academic performance for 145 schools, using the Forbes magazine rankings of best colleges. (Full disclosure: My Center for College Affordability and Productivity compiles those rankings for Forbes.) Adjusting for enrollment differences, no statistically significant relationship was observed between academic quality and presidential pay.

- The null and alternative hypotheses would be something like

  - H_0: There is no relationship between presidential pay and performance of a university.
  - H_A: There is a relationship between presidential pay and performance of a university.

- Since the results were NOT statistically significant, we can say

  - H_A could not be accepted.
  - It must have been that a *p*-value was computed and found to be larger than a presumed alpha level of 0.05.
  - Therefore, the risk of siding with H_A, when H_A was really not true, was assessed to be greater than 5 chances in 100. This was presumed to be an intolerable risk to take, so the alternative could not be accepted.

Example: Basketball Streaking

You might want to search on the title of this *New York Times* article or take a minute to read the article more thoroughly before continuing. The so-called "hot hand" has been debated in sports since at least 1985. Coaches and players have long believed that players go on "streaks" and get "hot" with the tendency to continue playing very well once they start a streak of outstanding plays. This was first challenged in 1985 when researchers studied performance records from two pro and one university basketball team finding that players "statistically were not more likely to hit a second basket after sinking a first." More recently, other researchers have accessed much larger amounts of data and have concluded that "basketball players experienced statistically significant and recognizable hot periods over an entire game or two …."

Title: Are 'Hot Hands' in Sports a Real Thing?

Author: Gretchen Reynolds

Source: *New York Times*

Winning streaks in sports may be more than just magical thinking, several new studies suggest.

Whether you call them winning streaks, "hot hands," or being "in the zone," most sports fans believe that players, and teams, tend to go on tears …. But our faith in hot hands is challenged by a rich and well-regarded body of science over the past 30 years, much of it focused on basketball, that tells us our belief is mostly fallacious. In one of the first and best-known of these studies, published in 1985, scientists parsed records from the Philadelphia 76ers, the Boston Celtics, and the Cornell University varsity squad and concluded that players statistically were not more likely to hit a second basket after sinking a first.

In the most wide-ranging of the new studies, Gur Yaari, a computational biologist at Yale, and his colleagues fathered enormous amounts of data about an entire season's worth of free throw shooting in the NBA and 50,000 games bowled in the Professional Bowlers Association. In these big sets of data, which were far larger than those used in, for instance, the 1985 basketball study, success did slightly increase the chances of subsequent success—though generally over a longer time frame than the next shot. Basketball players experienced statistically significant and recognizable hot periods over an entire game or two, during which they would hit more free throws than random chance would suggest.

- The null and alternative hypotheses would be something like

  ▸ H_0: Scoring streaks last no longer than would be expected by chance.

  ▸ H_A: Scoring streaks last longer than would be expected by chance.

- Since the results were found to be statistically significant in this article, we can say

  ▸ H_A could be accepted.

  ▸ It must have been that a *p*-value was computed and found to be smaller than a presumed alpha level of 0.05.

  ▸ Therefore, the risk of siding with H_A, when H_A was really not true, was assessed to be smaller than 5 chances in 100. This was presumed to be a tolerable risk to take, so the alternative was accepted.

Notice that the article is explicit about how much more data this study had access to than previous studies. More data means more information, but more data can also confuse us about what it really means for results to be statistically significant. All things being equal, if you have enough data then you can likely tag even the tiniest actual differences as statistically significant differences. That's important to keep in mind. One of the exercises at the end of this chapter explores that connection in more depth.

The Elusive and Confusing *p*-value

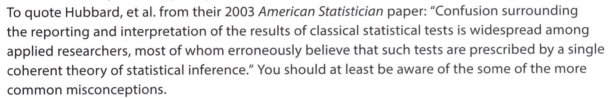

If you follow the examples in the previous section, then you have acquired the primary skill that was intended. However, the simplified, almost algorithmic, way the material was presented belies the conceptual complexity of statistical hypothesis testing. To quote Hubbard, et al. from their 2003 *American Statistician* paper: "Confusion surrounding the reporting and interpretation of the results of classical statistical tests is widespread among applied researchers, most of whom erroneously believe that such tests are prescribed by a single coherent theory of statistical inference." You should at least be aware of the some of the more common misconceptions.

- Foremost, there is confusion over what a *p*-value really is. It is surprisingly common to conflate the *p*-value with the alpha level.

- Second, many practitioners treat statistical science as if it was a single, unified science, or at least a science born of a single source. Neither is true. There are different theories— mathematical, but also philosophical—about what it means to test a hypothesis and, in practice, ideas are mixed from several origins and used as if they can co-exist when sometimes they really can't.

We are going to avoid getting into too much of the philosophy. However, everyone who uses a *p*-value, whether they computed it or someone else did, as part of how they digest hypothesis testing, should be aware of what it really is, and what it is not. Let's get the technical definition over with first.

> A ***p*-value** is the probability of seeing data that are as inconsistent with H_0, or more so, than the data you have on hand for the test, assuming the null is actually true.

This is a hard idea. The smaller the *p*-value, the smaller the chances of seeing data more inconsistent with the null hypothesis than the data you have, assuming the null is true. For example, suppose you have the following null and alternative hypotheses:

- H_0: No more than 80% of Americans favor the legalization of marijuana.

- H_A: More than 80% of Americans favor the legalization of marijuana.

Let's assume we have the results from a sample poll of 100 Americans that produced the statistic "89% of the sample favor the legalization of marijuana." If we assume the null is true, then the poll results are inconsistent (contradictory to) that presumption. A *p*-value quantifies that inconsistency. In this case, the *p*-value can be computed to be 0.012. You can think of this as saying there are only 12 chances in 1000 that, assuming the true proportion of Americans who think marijuana should be legalized is no more than 0.80, you would ever see data more inconsistent with that null than the 0.89 you observed from your sample. In the inverted logic of inference, this might be taken to suggest that the data and the null are unlikely to coexist. The data are real and the null is just a proposition. So it is suggested that null is unlikely to be true.

You have to be careful though. Having a tiny *p*-value does not reduce your risk of wrongly going with H_A when H_A is really not true. That's the job of the alpha level. Users and consumers of hypothesis testing, even some statisticians, continually conflate the two. They are fundamentally different. Perhaps the most common mistake is to say that a *p*-value of 0.001 means the results are somehow "more significant" than if the *p*-value had been 0.04.

That's a very common misinterpretation you want to avoid making. Part of the reason for this confusion is that there are different statistical theories about how inference should be practiced. In truth, using a conglomeration of a null, an alternative, an alpha level, and a *p*-value forces the merger of at least two perspectives on inference that are, at a basic level, not compatible. Be that as it may, this is exactly how inference is most often practiced and is sure to be the underlying paradigm behind hypothesis tests you are going to read about in the media. It may not even be such a big deal if we focus only on the decision being made about whether

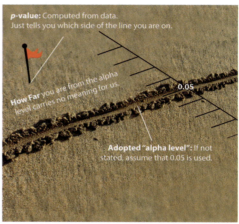

it is safe to go with H_A or not. It's when people start to articulate the risks involved in that decision that the seams between the different philosophical approaches become apparent.

For more discussion on this topic, please see "Confusion over Measures of Evidence (*p*'s) versus Errors (*a*'s) in Classical Statistical Testing," by Raymond Hubbard, M.J. Bayarri, Kenneth N. Berk, and Matthew A. Carlton. *American Statistician*, Vol 57, No.3, 2003. pp. 171–182.

TAKE-HOME POINTS

- Statistical hypothesis testing can be thought of as a screening test that chooses between a null hypothesis and an alternative hypothesis. This choice is based on comparing a *p*-value to a preset Type I error rate (a "false positive rate").

- If the *p*-value is smaller than the Type I error rate, then H_A can be accepted. Else it cannot be accepted.

- A *p*-value is best thought of as a special-use probability that identifies which side of the alpha level cutoff your data end up on. A *p*-value is not the same as an alpha level, and the two should not be confused.

Chapter 9 Exercises

Reading Check

Short Answer: Please provide brief, concise answers to the following questions.

1. What is a false positive rate in the context of hypothesis testing?

2. What is the goal of hypothesis testing?

3. What is a Type I error, and how is it related to an "alpha level?"

4. What does it mean to say that the results of a hypothesis test are statistically significant?

5. How is *p*-value used in testing a hypothesis?

6. Recall the study that compared calories consumed at McDonald's to calories consumed at Subway. What had to be true about the size of the *p*-value, had it been reported, and why do you know that?

7. Regarding the vitamin study, what can you assume the preset probability of rejecting H_0 is if H_0 is in fact true?

8. Some of the obesity weight-loss studies' results were not statistically significant, so H_A was not accepted. Explain why this does not mean that H_0 can automatically be accepted. Hint: What does a Type I error address?

9. In the vitamin study, the results were said to be "small, but statistically significant." Give an intuitive reason as to why you might not be surprised that a trial involving more than 15,000 subjects would produce this kind of conclusion.

10. Why is a *p*-value of 0.001 not "more significant" than a *p*-value of 0.05?

Beyond the Numbers

1. Pumpkin Powered Prostates

Pumpkins and Prostate Health

Title: Pumpkin Seed Oil May Be a Halloween Treat

Author: Elena Conis

Source: *Los Angeles Times,* October 25, 2010, https://www.latimes.com/archives/la-xpm-2010-oct-25-la-he-nutrition-lab-pumpkin-20101025-story.html

According to the article in the *Los Angeles Times* by Elena Conis, for centuries pumpkin seeds have been a home remedy used to control or increase the frequency of urination in adults, children, and livestock. Knowing this about pumpkin seeds has prompted researchers to explore the possibility of a link between eating the seeds and better prostate health.

German researchers have been very involved in exploring this possible connection. The article summarized one study, the results of which were published in a German journal in 2000. According to the *Los Angeles Times*, the study:

> … randomly selected among about 500 men to take either 1,000 milligrams of pumpkin seed oil extract or a placebo every day for 12 months. Symptoms improved in 65% of the men who took the oil, which the researchers interpreted as a promising (and statistically significant) result, even though symptoms also improved in 54% of the men who took the placebo.

a. What is the alternative hypothesis that is being tested?

i. Pumpkin seed oil is no better at improving urinary symptoms than a placebo is.

ii. Taking pumpkin seed oil will improve symptoms in males by around 11%.

iii. 65% of men who take pumpkin seed oil will like it better than a placebo.

iv. Pumpkin seed oil is better at improving urinary symptoms than a placebo is.

b. Was the alternative accepted or not? How do you know?

i. H_A was accepted because the results were not statistically significant.

ii. H_A was accepted because the results were statistically significant.

iii. H_A was not accepted because the results were not statistically significant.

iv. H_A was not accepted because the results were statistically significant.

c. Can we be certain that pumpkin seed oil is effective?

i. No, because the results were not statistically significant.

ii. Yes, because the results were statistically significant.

iii. Yes, because the p-value must have been less than the pre-set alpha level.

iv. No, because there is always a chance a Type I error occurred.

2. Stutter Stopper?

Drugs for Stutterers

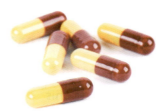

Title: Drug for Stutterers Shows Promise: Indevus Says Pill Reduced Incidents for Most in 1st Trial

Author: Stephen Heuser

Source: *Boston Globe*, May 25, 2006, http://archive.boston.com/business/globe/articles/2006/05/25/drug_for_stutterers_shows_promise//

The following is an extract from a *Boston Globe* article on stuttering:

A potential pill to treat stuttering took a step forward yesterday when Indevus Pharmaceuticals Inc. of Lexington said its experimental drug reduced stuttering in a majority of patients in its first clinical trial. The 132-patient trial is the largest human test ever conducted on a drug for stuttering, according to the company.

…

The Indevus drug, called pagoclone, was given to 88 patients in escalating doses, with the rest of the trial subjects receiving a placebo. The patients were then tracked using several widely accepted measures of stuttering.

…

On a third rating scale, based on doctors' impressions, the pagoclone patients scored a "numerically superior rating" to the placebo group, but the finding did not reach statistical significance.

a. What is the null hypothesis that is being tested?

i. Pagoclone is no better at improving stuttering symptoms than a placebo is.

ii. Pagoclone is no better at improving stuttering symptoms than Indevus.

iii. 88 out of 132 patients who take Pagoclone will experience improvements with their stuttering symptoms.

iv. Pagoclone is better at improving stuttering symptoms than a placebo is.

b. Was the alternative accepted or not? How do you know?

i. H_A was accepted because the results were not statistically significant.

ii. H_A was accepted because the results were statistically significant.

iii. H_A was not accepted because the results were not statistically significant.

iv. H_A was not accepted because the results were statistically significant.

c. Can we be certain that Pagoclone is ineffective?

i. No, because the results were not statistically significant.

ii. Yes, because there is always a chance a Type II error occurred.

iii. Yes, because the p-value must have been greater than the pre-set alpha level.

iv. No, because there is always a chance a Type II error occurred.

3. Statistics and Libido

Hypoactive Sexual Desire Disorder— Is There a Cure for Low Libido?

While researching a drug treatment for depression, German drugmaker Boehringer Ingelheim (BI) GmbH, claimed to have stumbled upon a libido-enhancing treatment for women. The drug, flibanserin, apparently did not lift women's moods, but during the study for depression, participants taking flibanserin reported higher sexual appetites on measures of well-being.

This prompted the company to conduct clinical trials specifically investigating flibanserin's usefulness in treating hypoactive sexual desire disorder (that is, low libido). BI ran three separate studies, and each study lasted 24 weeks.

Altogether, the studies included more than 5,000 female participants in the United States and Europe. The average age of study participants was 35, and most were married or in stable monogamous relationships of an average length of more than 10 years.

During the six months of the study, participants received either flibanserin or a placebo. Every day of the study, all participants were asked to record information about their sexually satisfying events as well as rate their desire levels. Changes from baseline (pre-study) were measured and compared. Participants on flibanserin reported an increase in the number of satisfying sexual events and a decrease in distress related to sexual situations.

a. An experiment like the one described above ultimately has to make a choice between two possible outcomes. What are those outcomes here?

   i. A choice has to be made between Flibanserin being no better than a placebo and Flibanserin being better than a placebo.

   ii. A choice has to be made between Flibanserin creating a notable change from baseline libido and Flibanserin not creating a notable change from baseline libido.

   iii. A choice has to be made between Flibanserin being better for women under 35 than for women over 35 years of age.

   iv. A choice has to be made between Flibanserin not being approved by FDA and Flibanserin being approved by FDA.

b. What is a false negative in the context of the Flibanserin testing?

   i. Since "Flibanserin is not effective" is the "negative" response, a false negative would be data from the study suggesting that the Flibanserin is not effective when it really is.

   ii. A false negative is when Flibanserin really is effective regardless of what the study data say.

   iii. Since "Flibanserin is effective" is the "negative" response, a false negative would be data from the study suggesting that Flibanserin is effective when it really is not.

   iv. A false negative is when Flibanserin really is not effective regardless of what the study data say.

c. What is a false positive in the context of the Flibanserin testing?

   i. A false positive is when Flibanserin really is effective regardless of what the study data say.

   ii. Since "Flibanserin is effective" is the "positive" response, a false positive would be data from the study suggesting that Flibanserin is effective when it is really is not.

   iii. Since "Flibanserin is not effective" is the "positive" response, a false positive would be data from the study suggesting that Flibanserin is not effective when it really is.

   iv. A false positive is when Flibanserin really is not effective regardless of what the study data say.

4. Rocky Biloba

Drug Effectiveness

Title: Ginkgo Biloba Ineffective against Dementia, Researchers Find

Author: Roni Caryn Rabin

Source: *New York Times*, November 18, 2008, http://www.nytimes.com/2008/11/19/health/research/18gingko.html

The following is an extract from a *New York Times* article on ginkgo biloba:

> The largest and longest independent clinical trial to assess ginkgo biloba's ability to prevent memory loss has found that the supplement does not prevent or delay dementia or Alzheimer's disease, researchers are reporting.

> The study is the first trial large enough to accurately assess the plant extract's effect on the incidence of dementia, experts said, and the results dashed hopes that it is an effective preventative. In fact, there were more cases of dementia among participants who were taking ginkgo biloba than among those who were taking a placebo, though the difference was not statistically significant.

"We were disappointed," said Dr. Steven T. DeKosky, dean of the School of Medicine at the University of Virginia and the principal investigator. "We were hopeful this would work."

a. An experiment like the one described above ultimately has to make a choice between two possible outcomes. What are those outcomes here?

    i. A choice has to be made between ginko biloba being no better than a placebo and ginko biloba being better than a placebo.

    ii. A choice has to be made between ginko biloba creating a notable change from baseline cognitive ability and ginko biloba not creating a notable change from baseline cognitive ability.

    iii. A choice has to be made between ginko biloba being better for Alzheimer's patients than for patients with other forms of dementia.

    iv. A choice has to be made between ginko biloba not being approved by the University of Virginia and ginko biloba being approved by the University of Virginia.

b. Which outcome was chosen? How is this related to the Type I Error Rate (false positive rate)?

    i. Since the results were NOT statistically significant, we know the p-value computed must have been smaller than a Type II error rate (our analogy to an FPR) that we assume was taken to be 0.05. This means that H_A was not chosen, so not enough evidence to say ginko biloba was effective.

    ii. Since the results were NOT statistically significant, we know the p-value computed must have been bigger than a Type II error rate (our analogy to an FPR) that we assume was taken to be 0.05. This means that H_A was chosen, so enough evidence to say ginko biloba was effective.

    iii. Since the results were NOT statistically significant, we know the p-value computed must have been bigger than a Type I error rate (our analogy to an FPR) that we assume was taken to be 0.05. This means that H_A was not chosen, so not enough evidence to say ginko biloba was effective.

    iv. Since the results were NOT statistically significant, we know the p-value computed must have been smaller than a Type I error rate (our analogy to an FPR) that we assume was taken to be 0.05. This means that H_A was chosen, so enough evidence to say ginko biloba was effective.

5. Unofficially Dangerous?

Drug Side Effects

Title: Supreme Court Rules against Zicam Maker

Author: Adam Liptak

Source: *New York Times*, March 23, 2011, http://www.nytimes.com/2011/03/23/ health/23bizcourt.html

The following is an excerpt from a *New York Times* article on Zicam:

> The Supreme Court unanimously ruled on Tuesday that investors suing a drug company for securities fraud may rely on its failure to disclose scattered reports of adverse affects [sic] from an over-the-counter cold remedy that fell short of statistical significance.
>
> The case involved Zicam, a nasal spray and gel made by Matrixx Initiatives and sold as a homeopathic medicine. From 1999 to 2004, the plaintiffs said, the company received reports that the products might have caused some users to lose their sense of smell, a condition called anosmia.
>
> Matrixx did not disclose the reports, and in 2003, the company said it was "poised for growth" and had "very strong momentum" though, by the plaintiffs' calculations, Zicam accounted for about 70% of its sales.
>
> …
>
> In the case before the justices, Matrixx Initiatives Inc. v. Siracusano, No. 09-1156, lawyers for Matrixx argued that it should not have been required to disclose small numbers of unreliable reports, which were the only ones available in 2004, they said. They added that the company should face liability for securities fraud only if the reports had been … statistically significant.

a. This is an unusual situation. What were the two outcomes being compared when the words "statistically significant" were used in the article?

  i. It appears that Matrixx was testing the alternative that Zicam does not cause anosmia versus the null that Zicam does cause anosmia.

  ii. It appears that Matrixx was testing the null that Zicam does not cause anosmia versus the alternative that Zicam does cause anosmia.

  iii. It appears that Matrixx was testing whether Zicam was an alternative to anosmia for sinus issues.

  iv. It appears that Zicam was testing the null that Matrixx does not cause anosmia versus the alternative that Matrixx does cause anosmia.

b. What was Matrixx's claim before the Supreme Court and how was that related to either a Type I or a Type II error rate?

   i. Matrixx was claiming that failure to disclose results that were not statistically significant should not have produced liability. The possibility that Zicam really did cause anosmia even though the results were not statistically significant is what the Type II error rate addresses.

   ii. Matrixx was claiming that failure to disclose results that were not statistically significant should not have produced liability. The possibility that Zicam really did cause anosmia even though the results were not statistically significant is what the Type I error rate addresses.

   iii. Matrixx was claiming that disclosing results that were statistically significant should not have produced liability. The possibility that Zicam really did not cause anosmia even though the results were statistically significant is what the Type II error rate addresses.

   iv. Matrixx was claiming that disclosing results that were statistically significant should not have produced liability. The possibility that Zicam really did not cause anosmia even though the results were statistically significant is what the Type I error rate addresses.

6. Prescription to Pass

Drug Safety

Title: Trial Intensifies Concerns about Safety of Vytorin

Author: Alex Berenson

Source: *New York Times*, July 22, 2008, http://www.nytimes.com/2008/07/22/business/22drug.html

The following is an extract from a *New York Times* article on Vytorin:

> In a clinical trial, the cholesterol-lowering drug Vytorin did not help people with heart-valve disease avoid further heart problems but did appear to increase their risk of cancer, scientists reported Monday.
>
> …
>
> Vytorin and Zetia, a companion drug, are prescribed each month to almost three million people worldwide and are among the world's top-selling medicines.
>
> …

In the Seas trial, which involved nearly 1,900 patients whose heart valves were partially blocked, participants were given either Vytorin or a placebo pill that contained no medicine. Scientists hoped that the trial would show that patients taking Vytorin would have a lower risk of needing valve replacement surgery or having heart failure. But the drug did not show those benefits.

"No significant difference was observed between the treatment groups for the combined primary endpoint," Dr. Terje Pedersen, the principal investigator for the study and a professor medicine at Ulleval University Hospital in Norway, said. The primary endpoint is the result that scientists hope to prove when they conduct a clinical trial.

However, patients taking Vytorin in the Seas trial did have a sharply higher risk of developing and dying from cancer. In the trial 102 patients taking Vytorin developed cancer, compared with 67 taking the placebo. Of those, 39 people taking Vytorin died from their cancer, compared with 23 taking the placebo.

The absolute numbers of cancer cases were relatively small. But they reached statistical significance, meaning the odds were less than 5% that they were the result of chance.

a. What alternative hypothesis was under scrutiny when the phrase "significant difference" was used?

i. That Vytorin was no better than a placebo with respect to risk of needing valve replacement or having heart failure.

ii. That Vytorin was better than a placebo with respect to risk of needing valve replacement or having heart failure.

iii. That Vytorin was no worse than a placebo with respect to the risk of developing cancer at some point after being taken.

iv. That Vytorin was worse than a placebo with respect to the risk of developing cancer at some point after being taken.

b. What null hypothesis was under scrutiny when the phrase "statistical significance" was used?

i. That Vytorin was no better than a placebo with respect to risk of needing valve replacement or having heart failure.

ii. That Vytorin was better than a placebo with respect to risk of needing valve replacement or having heart failure.

iii. That Vytorin was no worse than a placebo with respect to the risk of developing cancer at some point after being taken.

iv. That Vytorin was worse than a placebo with respect to the risk of developing cancer at some point after being taken.

7. Albatross Adaptability

Animal Sexuality?

Title: Can Animals Be Gay?

Author: Jon Mooallem

Source: *New York Times*, March 31, 2010, http://www.nytimes. com/2010/04/04/magazine/04animals-t. html?pagewanted5all

Evolutionarily speaking, reproduction is more than just the physical act of sex. Reproduction is only successful if offspring are able to reach sexual maturity themselves and further reproduce. So forming a zygote only gets you to the halfway mark. In the case of the albatross and other oviparous species where the young require parental care, maybe same-sex couples are more effective at raising the fledglings.

Comment by a reader:

Is there data comparing chick viability from same-sex versus male-female nests? The research cites ⅓ of the albatross pairs as same-sex, and I'd say that's a statistically significant percentage, so it can't be just some mutation in some of the birds. I'm very curious to learn whether this behavior is inherited or learned. Fascinating research!

We've learned that when phrases like "statistically significant" are being used, some comparison is being made. What, if anything, is being compared here in the reader's comment about "⅓ of the albatross pairs as same-sex" being statistically significant? Is the phrase "statistically significant" being used in a meaningful way? Why or why not?

8. Do I Have To?

Eat Your Veggies

Title: Eating Vegetables Doesn't Stop Cancer

Author: Tara Parker-Pope

Source: *New York Times*, April 8, 2010, http://well.blogs.nytimes. com/2010/04/08/eating-vegetables-doesnt-stop-cancer/

A recent *New York Times* article reported, "A major study tracking the eating habits of 478,000 Europeans suggests that consuming lots of fruits and vegetables has little if any effect on preventing cancer." The study, which was published in *The Journal of the National Cancer Institute*, "tracked 142,605 men and 335,873 women for an average of nearly nine years. Eating more vegetables was associated with a small but statistically significant reduction in cancer risk."

a.   What are the null (H_0) and alternative (H_A) hypotheses in the context of the Parker-Pope article?

b. What was the conclusion in their test? Did they reject H_0 or fail to reject H_0? How do you know?

c. Articulate what risk was involved, based on the decision made in the article, when the null (H_0) was rejected (or not). You should use the concept of false positive rate in your answer.

d. What does the article say about practical significance and statistical significance?

9. Dangerous Training

Crash Test

Title: Motorcycle Training Does Not Reduce Crash Risk, Study Says

Author: Cheryl Jensen

Source: *New York Times*, April 5, 2010, http://wheels.blogs.nytimes.com/2010/04/05/motorcycle-training-does-not-reduce-crash-risk-study-says/

Consider the following statements from a recent *New York Times* article discussing a recent study by the Highway Loss Data Institute regarding the effectiveness of training courses in improving motorcycle safety, and consider the questions below. The article says:

> "What is not so certain are the safety benefits of mandatory training programs for young drivers in some states. The study compared insurance claims in four states that require riders under 21 to take courses with states that do not. The study noted a 10% increase in crashes in states that required the courses."

> But that finding wasn't "statistically significant," Ms. McCartt [senior vice president for research at the Insurance Institute] said. That means the increase might or might not be real, although the institute found it worth noting. "It is important that it was going in the opposite direction of what people would expect," she said.

a. Identify the null (H_0) and the alternative (H_A) hypotheses in the context of this article.

b. Articulate what risk was involved, based on the decision made in the article, when the null (H_0) was rejected (or not). You should use the concept of Type I error in your answer.

10. Voucher Vouch

Voucher Controversy

Title: White House Ignores Evidence of How D.C. School Vouchers Work

Author: Editorial Board Opinion

Source: *Washington Post*, March 29, 2011, http://www.washingtonpost.com/opinions/ white-house-ignores-evidence-of-how-dc-school-vouchers-work/2011/03/29/AFFsnHyB_story.html

Consider the following statements from a recent *Washington Post* editorial discussing the strongly-worded dismissal of school vouchers. The article says:

> That dismissal might come as a surprise to Patrick J. Wolf, the principal investigator who helped conduct the rigorous studies of the D.C. Opportunity Scholarship Program and who has more than a decade of experience evaluating school choice programs.

> Here's what Mr. Wolf had to say about the program in Feb. 16 testimony to the Senate Committee on Homeland Security and Governmental Operations." In my opinion, by demonstrating statistically significant experimental impacts on boosting high school graduation rates and generating a wealth of evidence suggesting that students also benefited in reading achievement, the DC OSP has accomplished what few educational interventions can claim: It markedly improved important education outcomes for low-income inner-city students."

a. The phrase "statistically significant" is used in Mr. Wolf's testimony. What are the null (H_0) and alternative (H_A) hypotheses in the context of that testimony?

  i. H_0: Vouchers help graduation rates vs. H_A: No, they don't.

  ii. H_0: Vouchers help students get jobs vs. H_A: No, they don't.

  iii. H_0: Vouchers don't help students get jobs vs. H_A: Yes, they do.

  iv. H_0: Vouchers don't help graduation rates vs. H_A: Yes, they do.

b. What does the decision that was made have to do with a false positive rate?

  i. The FPR is essentially the same as a *p*-value. In this case, the FPR must have been smaller than 0.05.

  ii. The FPR is essentially the same as an assumed Type I error rate, typically taken to be 0.05. In this case, the *p*-value must have been bigger than 0.05.

  iii. The FPR is essentially the same as a *p*-value. In this case the FPR must have been bigger for the DC OSP than for non-voucher programs.

  iv. The FPR is essentially the same as an assumed Type I error rate, typically taken to be 0.05. In this case, the *p*-value must have been smaller than 0.05.

11. Legally Speaking

Title: Innocent versus Not Guilty: Jury Decision Based Entirely on Evidence

Author: Hugh Duvall

Source: http://www.defendingoregon.com/innocent-v-guilty/

Many attorneys have written about the difference between innocence and non-guilt. Most quickly turn to complex legalese. This article paints a particularly clear picture of the use, however:

> Juries never find defendants innocent. They cannot. Not only is it not their job, it is not within their power. They can only find them "not guilty." Once a person has been charged with having committed a crime, there is no mechanism by which that individual can prove his innocence. Yes, the law provides that the person is innocent unless proven guilty, but that is a legalism. It is not, nor could it be, a factual statement. The person, in fact, did or did not commit an offense. Each time a member of the media or other citizen states that William Kennedy Smith or one of the officers accused of beating Rodney King was found "innocent," they are not only incorrect, but are also ingraining within potential jurors a misconception about their role. They enhance the risk that enough jurors on a panel will retire into a jury room believing that it is their task to determine whether there is enough evidence to find a defendant innocent.

a.   The role of the prosecution is to establish guilt beyond a reasonable doubt. The defense has no such burden of proof. How does this affect the ability of a jury to find a defendant innocent?

b.   If "I" denotes the event "defendant is presumed innocent" and "E" denotes evidence presented by prosecution," then a juror's job is to evaluate $P(E \mid I)$—using notation we first encountered in Chapter 8. This probability can be ascertained to be big or small, depending on the case. Identify which of these (big or small) leads logically to a conclusion of guilty, and explain why the other can't logically lead to a conclusion of innocence.

c.   In this article, the author states that, "As a society, in administering the prosecution function, we must keep at the forefront of our mind that there is no way to reverse the implication of charging someone with a crime. Allowing ourselves to ignore the distinction between a jury's ability to find someone 'not guilty' and its inability to find someone 'innocent' works against this important interest." Explain in your own words what this means and describe why it is important in this discussion of "innocent" versus "not guilty."

d.   Is the distinction between accepting and failing to reject a null hypothesis a real distinction, or just semantics? Defend your answer based on what you have learned from your answers to parts a–c.

12. Alternative Evidence

Title: "It's Like ... You Know": The Use of Analogies and Heuristics in Teaching Introductory Statistical Methods

Author: Michael A. Martin, Australian National University

Source: *Journal of Statistics Education* Volume 11, Number 2 (2003). www.amstat.org/publications/jse/v11n2/martin.html

Analogies between the United States criminal justice system and hypothesis testing are very useful to the study of statistics. How so? They make it easier to understand the logic of an abstract task that is often seen as only having deductive relevance. In this article, Professor Michael Martin created a number of criminal trial analogies to explain statistics' most common (and most commonly confusing) concepts:

Criminal Trial Concepts

1. Defendant is innocent
2. Defendant is guilty
3. Verdict is to acquit
4. Verdict is to convict
5. Presumption of innocence
6. Conviction of an innocent person
7. Acquittal of a guilty person
8. Beyond reasonable doubt

Hypothesis Testing Concepts

1. Null hypothesis
2. Alternative hypothesis
3. Failure to reject the null hypothesis
4. Rejection of the null hypothesis
5. Assumption that the null hypothesis is true
6. Type I error
7. Type II error
8. Fixed (small) probability of Type I error

a. Take each pair of concepts in turn and explain how they are analogous. You must use the language of this chapter in your answers.

b. Is the distinction between accepting and failing to reject a null hypothesis a real distinction, or just semantics? Defend your answer based on what you have learned from explaining the analogies in part a.

13. Power and Beauty

The computation of power can be a complex endeavor, even for elementary forms of H_0 and H_A. However, we can gain valuable insights into the power of a statistical test by using a freely-available online tool to handle all of the complex computations for us.

Title: Power and Sample Size.com

Author: HyLown Consulting, LLC

Source: http://powerandsamplesize.com/Calculators/Test-1-Mean/1-Sample-Equality

Suppose you are interested in how people rate their looks on a 20 point scale, with a score of 0 meaning *unbearably ugly* and a score of 20 meaning *hopelessly gorgeous*. Let's suppose you want to test the following hypothesis:

- H_0: true population average rating is 10
- H_A: true population average rating is not 10

What we want to do is to answer the question:

"How likely is it that we will fail to reject H_0 if H_0 is truly false?"

The answer to this question leads us to the Type II error rate, from which power is easily computed. To come up with an answer, we first have to ask:

"What is the true average if H_0 is false?"

H_A only tells us, in this case, that the true average is something other than 10. We will begin by considering four possible averages that are different from 10: 10.5, 11, 11.5, and 12.

Access the applet at the web address listed in the **Source** line above. When the applet page loads:

- Select "Power"
- Enter the sample size (e.g. 50)
- Enter True Mean (e.g. 11)
- Enter Hypothesized Mean (always 10 for this problem)
- Enter the Standard Deviation as 5
- Leave the alpha level (Type I error rate) fixed at 5%
- Hit Calculate
- Record the Power value
- Do this for all True Mean and Sample Size combinations shown in the table below. One row is already done for you. Make sure you can confirm those entries and then finish finding the rest of the entries in the table. It is safest to reload the page between each calculation.

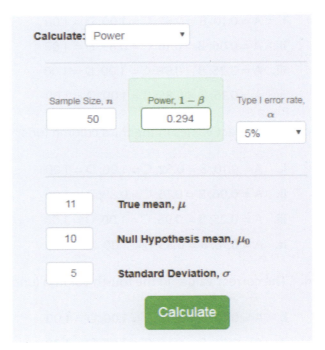

TABLE 9.2 Power Table

Sample Size	Possible Real Values of the Population Average (called True Mean on applet)			
	10.5	11	11.5	12
10	A	A	A	A
50	0.11	0.29	0.57	0.81
100	B	B	B	B
1,000	C	C	C	C
10,000	D	D	D	D

a. The correct values in the first column (under the True Mean of 10.5) are:

   i.   A = 0.10; B = 0.52; C = 1.00; D = 1.00

   ii. A = 0.06; B = 0.17; C = 0.89; D = 1.00

   iii. A = 0.25; B = 0.98; C = 1.00; D = 1.00

   iv. A = 0.16; B = 0.85; C = 1.00; D = 1.00

b. The correct values in the second column (under the True Mean of 11) are:

   i.   A = 0.10; B = 0.52; C = 1.00; D = 1.00

   ii. A = 0.06; B = 0.16; C = 0.89; D = 1.00

   iii. A = 0.25; B = 0.98; C = 1.00; D = 1.00

   iv. A = 0.16; B = 0.85; C = 1.00; D = 1.00

c. The correct values in the third column (under the True Mean of 11.5) are:

   i.   A = 0.10; B = 0.52; C = 1.00; D = 1.00

   ii. A = 0.06; B = 0.16; C = 0.89; D = 1.00

   iii. A = 0.25; B = 0.98; C = 1.00; D = 1.00

   iv. A = 0.16; B = 0.85; C = 1.00; D = 1.00

d. The correct values in the fourth column (under the True Mean of 12) are:

   i.   A = 0.10; B = 0.52; C = 1.00; D = 1.00

   ii. A = 0.06; B = 0.16; C = 0.89; D = 1.00

   iii. A = 0.25; B = 0.98; C = 1.00; D = 1.00

   iv. A = 0.16; B = 0.85; C = 1.00; D = 1.00

e. Look at the completed table. How does power change as sample size changes, regardless of the true average?

f.   Look at the completed table. How does power change as the possible true average changes, regardless of sample size?

g.   What would happen to power if you changed alpha? Investigate that question using the online calculator that you used to fill out the table above. Explain what you find. Think back to how sensitivity and specificity behaved in screening tests. How is this similar?

14. Error Rates and *p*-Values

In the classic David Bowie song "Space Oddity," Major Tom is admonished to take his protein pills and put his helmet on. That might be good advice here. This brief topic at hand concerns a concept that is too important to ignore, but too complex to present mathematically. Despite its importance, this concept is often misunderstood. It has even been said that "this misunderstanding ... is so deeply entrenched that it is not even seen as being a problem by the vast majority of researchers" (see "Confusion Over Measures of Evidence (*p*'s) Versus Errors (α's) in Classical Statistical Testing," by R. Hubbard and M.J. Bayarri. *The American Statistician,* August 2003, Vol. 57, No. 3.) Tighten your helmet straps. This exercise challenges your awareness in an oxygen-thin environment.

What is a Type I error, really?

If α is 0.05, then of all the experiments conducted at that level, at most 1 in 20 will result in the false rejection of the null hypothesis. Think of this as a prescription for behavior. If you use an alpha level of 0.05 for all your experiments, then over time you'll only end up wrongly rejecting your nulls 5% of the time. It is not a direct statement about any particular one of the individual experiments being reported. That's a challenging, but important distinction. However, it is similar in spirit to the interpretation of a confidence interval, which you learned about in another chapter.

What is a *p*-value then?

It is the probability of seeing data as, or more, inconsistent with the null hypothesis than what the experiment has just produced. A small *p*-value suggests the observed data would be a rare occurrence, under the assumption that the null is true.

So what's the problem?

There is a widespread tendency to misinterpret a *p*-value as an error rate. That is, it is a common mistake to conclude that a *p*-value of 0.05 means that there are only 5 chances in 100 that if you reject your null, you will do so wrongly. That is not true, of course. That conclusion is the purview of the alpha level, but as we just noted, that interpretation is an "over time" interpretation.

It is no surprise that people have wanted to know if they could say things more specific. For example, given a p-value of 0.03, is it possible to say how likely it is of wrongly rejecting the null, conditional on knowing you have that p-value? We know it is not 0.03. But what is it? One school of thought, not without its detractors, estimates the probability of wrongly rejecting the null given a computed p-value using the following formula:

$$\text{Conditional Error Rate} = \frac{1}{1 + \left[\frac{1}{-(2.718282)(p\text{-value})(\log(p\text{-value}))}\right]}$$

(where log is the *natural* log). So if we have a p-value of 0.03:

$$\text{Conditional Error Rate} = \frac{1}{1 + \left[\frac{1}{-(2.718282)(0.03)(\log(0.03))}\right]} = 0.222$$

According to this formula, a p-value of 0.03, corresponds to an estimated conditional error rate of 0.22! That is, *given* a p-value of 0.03 has been observed, there are almost 22 chances in 100 that if the null is rejected it will be a mistake. Make sure you can confirm the conditional error computation for a p-value of 0.03.

What do the conditional Type I errors look like for these different p-values? Using a software package (such as Microsoft Excel or Apple Numbers), fill out the rest of the entries in Table 9.3.

a. The correct values for the missing entries in the table are:

   i. A = 0.002; B = 0.018; C = 0.010; D = 0.289; E = 0.111; F = 0.385; G = 0.057

   ii. A = 0.010; B = 0.002; C = 0.018; D = 0.057; E = 0.289; F = 0.111; G = 0.385

   iii. A = 0.002; B = 0.010; C = 0.018; D = 0.057; E = 0.111; F = 0.289; G = 0.385

   iv. A = 0.010; B = 0.018; C = 0.002; D = 0.111; E = 0.385; F = 0.289; G = 0.057

TABLE 9.3

p-value	Conditional Error Rate
0.0001	A
0.0005	B
0.001	C
0.004	D
0.01	E
0.03	**0.222**
0.05	F
0.10	G

b. Look at the entries you found for the table. Suppose a practitioner misinterprets a p-value as the probability of wrongly rejecting the null. What does this table suggest the p-value would have to be before that error rate is approximately 5%?

c. This exercise is conceptually difficult and if misunderstood might serve to further conflate p-values and alpha levels, rather than better distinguish! Let's suppose you test a hypothesis with an alpha level of 0.05, and you find a p-value of 0.01. Explain what this exercise tells you about what both might really tell you about the risk of rejecting H_0 when H_0 is true.

Projects

1. Statistical Significance in the News

Introduction

Your job is to find an article in a major newspaper or popular magazine that uses the key phrase (or some slight variation thereof) "statistically significant" or "not statistically significant" to describe the results of a study or experiment. Your instructor may give you a list of articles to choose from. You may be required to use an article that is intended for general consumption, *not* a research article intended only for professor and scientist types.

The Assignment

You need to turn in a proofread, organized, typed response that includes *all* of the following. Your final product should be in paragraph form and not a list of bulleted items.

- A summary of the claim being made in the article; make sure you exhibit the article's use of the key phrase.

- The article's source information.

- A description of the null (H_0) and alternative (H_A) hypotheses that were tested. Make sure the description is in the context of the article.

- A discussion of the statistical risk that was involved in the decision made in the article. You should use the concept of a Type I error rate in your answer.

- A comparison of the notions of practical significance with statistical significance, in the context of the article.

2. Reality Video

Introduction

This chapter facilitated a transition from the language and ideas of screening tests to similar concepts in statistical hypothesis testing.

The Assignemnt

Your instructor may have you do this as individuals or may form you into teams. We'll refer to "teams" below, although you may be a team of one depending on what your instructor wants. Each team is required to do the following:

a. Pick one or more topics from this chapter. Your instructor may want to approve them before you proceed.

    b. Construct a three-minute video that teaches the chosen topic to the class. You must carefully and correctly use the language of the chapter in your presentation, and you are required to give an example that is NOT in the chapter.

    c. Post the video on YouTube and send the link to your instructor in the manner he stipulates.

10

CHAPTER 10
Hypothesis Testing: Computations

Introduction

There is virtually no end to the number of different hypotheses that can be tested with statistical science. Some of these hypotheses are quite complex, as are the tools to test them. If you are going to be a practicing professional statistician, then you are going to need to know a lot about these! This book is not designed to provide that kind of professional training. In this chapter, we are more interested in the reasoning that is common to almost all of these applications, even though the mechanics of the different implementations may vary widely.

Indeed, our goal is to learn how to test a limited number of simple hypotheses. And by "test" we mean we will learn how to actually complete the mathematical steps required to produce the p-value that is needed to make a decision about the viability of H_A. A command of those rudimentary mechanics will help us achieve our larger goal of being able to intelligently consume the byproducts of hypothesis testing.

From Words to Parameters

When we discussed hypotheses informally in Chapter 9, we typically used words. For example, if we are testing the usefulness of the failed depression drug Flibanserin as a libido aid, we discussed comparing a null of "Flibanserin is no better than a placebo" versus an alternative of "Flibanserin is better than a placebo."

> H_0: Flibanserin is no better than a placebo.

> H_A: Flibanserin is better than a placebo.

However, if we actually want to do the mathematics required to test this hypothesis, we have to structure the null and alternative more formally. In the Flibanserin case, for instance, the researchers might have the null say that the true average number of sexually satisfying events for all women who would use Flibanserin is the same as the true average number of sexually satisfying events for all women who might take a placebo instead.

And the alternative would then say something about how those means are not the same, perhaps that the mean response in the Flibanserin group is greater than the mean response in the placebo group. Something like this:

$$H_0 : \mu_{Flibanserin} \leq \mu_{Placebo}$$

$$H_A : \mu_{Flibanserin} > \mu_{Placebo}$$

where μ denotes the true average response. That is, if you think about a vaguely defined population of all women who would take Flibanserian, and then imagine adding up their recorded number of sexually satisfying events (the words used in the experiment) and dividing by the number of women in that population, you have $\mu_{Flibanserin}$. This is an unknown average associated with the full population of women using Flibanserin. The goal of hypothesis testing is to be able to make some decision about how that mean might compare to the analogous mean in a non-Flibanserin population of women, based only on the results of the experiment.

Keep in mind that the means being referred to are parameters. They describe much larger groups of subjects than those few that were recruited to participate in the experiment. But these overall averages are what we really want to know something about.

Example: Stress and Sleep

You probably won't be surprised to learn that stress affects the quality of your life. However, you may be surprised to learn that studies have suggested that stress affects college students' sleep more than alcohol, caffeine, or late-night electronics use, at least according to a 2009 study. One such study involved 1,125 college students, and for 68% of those in the sample, stress about school and life was their number one reason for not being able to sleep at night. This study is available on line in the *Journal of Adolescent Health* (Lund HG, et al. Sleep patterns and predictors of disturbed sleep in a large population of college students. *J Adolesc Health* online, 2009).

The 68% reported in the sleep study is a statistic, a sample percentage. That is, 68% of the 1,125 students (765 of them) listed stress as their primary reason for not being able to sleep. Let's suppose that a particular university plans to commit major funding for their own campus to the development of programs for reducing stress, once they can be confident that more that 65% of the student population is suffering from this problem. The business of hypothesis testing in this context might be to address whether it is safe to say that more than 65% of all college students feel this way, based on the 68% we saw in this one study.

So the hypothesis that needs to be tested would have the following form, where p is the true proportion of all colleges students who would have said stress is their primary sleep impediment, had they all been asked.

$$H_0 : p \leq 0.65$$

$$H_A : p > 0.65$$

In the next section we will learn the steps to test this hypothesis as well as how to test two variations.

When the Parameter Is a Proportion

Single-Sided with a Greater-than Alternative

In the example above, we just made up the 65%. In practice, this number emerges from the context. It is the crux of why one wants to even test a hypothesis. It may be the threshold beyond which public policy can be enacted, beyond which corporate funds can be committed, or beyond which a candidate wins the majority of the popular vote. Let's just use the notation p_0 as a placeholder for this value. The hypothesis of interest then has the generic form:

$$H_0 : p \leq p_0$$

$$H_A : p > p_0$$

The form of this hypothesis is said to be single-sided with a greater-than alternative. This just means the alternative hypothesis is making a claim about the parameter being bigger than some number, as opposed to smaller than or just not equal to some number.

We already know that to "test" this hypothesis (or any hypothesis) means the following:

- Start with a Type I error rate (alpha level) that is tolerable. This is usually taken to be 0.05;
- Collect the data associated with the study. For the sleep study the data were simple: 68% of the 1125 surveyed said that stress was the primary impediment to sleep for them. So the sample proportion who said it was the primary impediment is 0.68.
- **Compute a *p*-value** for the testing process;
- Compare the *p*-value to the alpha level. If it is greater, then it is not safe to assume H_A is true. If it is smaller, then it is safe to do so.

What we don't yet know is how to compute a *p*-value for this kind of hypothesis. The entire process can be broken down into three easy steps. Key to the process is the computation of something called the "standard score."

The **standard score** associated with a test of hypothesis about **a population proportion** is

$$z = \frac{\hat{p} - p_0}{\sqrt{\frac{p_0(1 - p_0)}{n}}}.$$

The anatomy of z is very straightforward:

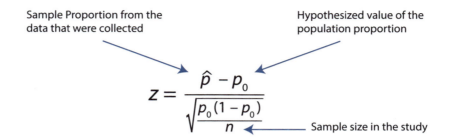

Sample Proportion from the data that were collected

Hypothesized value of the population proportion

$$z = \frac{\hat{p} - p_0}{\sqrt{\frac{p_0(1 - p_0)}{n}}}$$

Sample size in the study

We should point out that most introductory statistics books are going to take issue with this statistic being called a "z" statistic. Those objections would be completely legitimate. However, there is little to gain for our purposes if we split those mathematical hairs. Our goal is to simply get our hands on the computation of a p-value in some simple situations, solely because we want to further cement our understanding of what it means to test a hypothesis, and how that adds to our understanding of the issues that motivated the testing in the first place.

The three steps needed to compute a p-value for the test of the hypothesis shown above are really very simple:

1. Compute the associated standard score;

2. Take the standard score to a **standard score table** and look up the p-value;

3. Compare the p-value to the preset alpha level. If the p-value is smaller, then it is safe to conclude H_A is likely true. If the p-value is larger, then it is too risky to say that.

We will illustrate these steps when we revisit the stress example below. A copy of a standard score table like the one mentioned above is in the appendix of this book. It is also easy to just look up the p-value online using an applet. You just have to be careful. How the table (or applet) is arranged makes a difference.

Example: Stress and Sleep Revisited

In the stress example a sample proportion of $\hat{p} = 0.68$ students said school was a primary sleep impediment, out of $n = 1,125$ asked. Let's take a preset alpha level of 0.05 If we are testing:

$H_0 : p \leq 0.65$

$H_A : p > 0.65$

then p_0 is 0.65. We have all we need to compute the standard score associated with this hypothesis.

$$z = \frac{0.68 - 0.65}{\sqrt{\frac{(0.65)(1 - 0.65)}{1125}}} = \frac{0.03}{0.01422} = 2.11$$

To compute the *p*-value, we only have to take the value of *z* (2.11) to the *z* table shown on the right. Find the 2.1 down the left column and then the 0.01 across the top row. The *p*-value is 0.01743.

STANDARD SCORE TABLE

Example: If a standard score, *z*, is computed to be 1.73, then one would locate 1.73 in the table (see highlighting) and corresponding *p*-value would be 0.04182 ≈ 0.04. This would then have to be compared to the preset Type I error rate, often taken to be 0.05. This process only applies for simple hypotheses with a positive *z* score and a ">" in the alternative. Your instructor will show you how to use this table when *z* is negative and/or when the alternative is a "≠."

z	0	0.01	0.02	0.03	0.04	0.05	0.06	0.07	0.08	0.09
0	0.5	0.49601	0.49202	0.48803	0.48405	0.48006	0.47608	0.4721	0.46812	0.46414
0.1	0.46017	0.4562	0.45224	0.44828	0.44433	0.44038	0.43644	0.43251	0.42858	0.42465
0.2	0.42074	0.41683	0.41294	0.40905	0.40517	0.40129	0.39743	0.39358	0.38974	0.38591
0.3	0.38209	0.37828	0.37448	0.3707	0.36693	0.36317	0.35942	0.35569	0.35197	0.34827
0.4	0.34458	0.3409	0.33724	0.3336	0.32997	0.32636	0.32276	0.31918	0.31561	0.31207
0.5	0.30854	0.30503	0.30153	0.29806	0.2946	0.29116	0.28774	0.28434	0.28096	0.2776
0.6	0.27425	0.27093	0.26763	0.26435	0.26109	0.25785	0.25463	0.25143	0.24825	0.2451
0.7	0.24196	0.23885	0.23576	0.2327	0.22965	0.22663	0.22363	0.22065	0.2177	0.21476
0.8	0.21186	0.20897	0.20611	0.20327	0.20045	0.19766	0.19489	0.19215	0.18943	0.18673
0.9	0.18406	0.18141	0.17879	0.17619	0.17361	0.17106	0.16853	0.16602	0.16354	0.16109
1	0.15866	0.15625	0.15386	0.15151	0.14917	0.14686	0.14457	0.14231	0.14007	0.13786
1.1	0.13567	0.1335	0.13136	0.12924	0.12714	0.12507	0.12302	0.121	0.119	0.11702
1.2	0.11507	0.11314	0.11123	0.10935	0.10749	0.10565	0.10383	0.10204	0.10027	0.09853
1.3	0.0968	0.0951	0.09342	0.09176	0.09012	0.08851	0.08692	0.08534	0.08379	0.08226
1.4	0.08076	0.07927	0.0778	0.07636	0.07493	0.07353	0.07215	0.07078	0.06944	0.06811
1.5	0.06681	0.06552	0.06426	0.06301	0.06178	0.06057	0.05938	0.05821	0.05705	0.05592
1.6	0.0548	0.0537	0.05262	0.05155	0.0505	0.04947	0.04846	0.04746	0.04648	0.04551
1.7	0.04457	0.04363	0.04272	0.04182	0.04093	0.04006	0.0392	0.03836	0.03754	0.03673
1.8	0.03593	0.03515	0.03438	0.03362	0.03288	0.03216	0.03144	0.03074	0.03005	0.02938
1.9	0.02872	0.02807	0.02743	0.0268	0.02619	0.02559	0.025	0.02442	0.02385	0.0233
2	0.02275	0.02222	0.02169	0.02118	0.02068	0.02018	0.0197	0.01923	0.01876	0.01831
2.1	0.01786	0.01743	0.017	0.01659	0.01618	0.01578	0.01539	0.015	0.01463	0.01426
2.2	0.0139	0.01355	0.01321	0.01287	0.01255	0.01222	0.01191	0.0116	0.0113	0.01101
2.3	0.01072	0.01044	0.01017	0.0099	0.00964	0.00939	0.00914	0.00889	0.00866	0.00842
2.4	0.0082	0.00798	0.00776	0.00755	0.00734	0.00714	0.00695	0.00676	0.00657	0.00639
2.5	0.00621	0.00604	0.00587	0.0057	0.00554	0.00539	0.00523	0.00508	0.00494	0.0048
2.6	0.00466	0.00453	0.0044	0.00427	0.00415	0.00402	0.00391	0.00379	0.00368	0.00357
2.7	0.00347	0.00336	0.00326	0.00317	0.00307	0.00298	0.00289	0.0028	0.00272	0.00264
2.8	0.00256	0.00248	0.0024	0.00233	0.00226	0.00219	0.00212	0.00205	0.00199	0.00193
2.9	0.00187	0.00181	0.00175	0.00169	0.00164	0.00159	0.00154	0.00149	0.00144	0.00139
3	0.00135	0.00131	0.00126	0.00122	0.00118	0.00114	0.00111	0.00107	0.00104	0.001
3.1	0.00097	0.00094	0.0009	0.00087	0.00084	0.00082	0.00079	0.00076	0.00074	0.00071
3.2	0.00069	0.00066	0.00064	0.00062	0.0006	0.00058	0.00056	0.00054	0.00052	0.0005
3.3	0.00048	0.00047	0.00045	0.00043	0.00042	0.0004	0.00039	0.00038	0.00036	0.00035
3.4	0.00034	0.00032	0.00031	0.0003	0.00029	0.00028	0.00027	0.00026	0.00025	0.00024
3.5	0.00023	0.00022	0.00022	0.00021	0.0002	0.00019	0.00019	0.00018	0.00017	0.00017
3.6	0.00016	0.00015	0.00015	0.00014	0.00014	0.00013	0.00013	0.00012	0.00012	0.00011

z	0	0.01	0.02
0	0.5	0.49601	0.49202
0.1	0.46017	0.4562	0.45224
0.2	0.42074	0.41683	0.41294
0.3	0.38209	0.37828	0.37448
0.4	0.34458	0.3409	0.33724
0.5	0.30854	0.30503	0.30153
0.6	0.27425	0.27093	0.26763
0.7	0.24196	0.23885	0.23576
0.8	0.21186	0.20897	0.20611
0.9	0.18406	0.18141	0.17879
1	0.15866	0.15625	0.15386
1.1	0.13567	0.1335	0.13136
1.2	0.11507	0.11314	0.11123
1.3	0.0968	0.0951	0.09342
1.4	0.08076	0.07927	0.0778
1.5	0.06681	0.06552	0.06426
1.6	0.0548	0.0537	0.05262
1.7	0.04457	0.04363	0.04272
1.8	0.03593	0.03515	0.03438
1.9	0.02872	0.02807	0.02743
2	0.02275	0.02222	0.02169
2.1	0.01786	0.01743	0.017
2.2	0.0139	0.01355	0.01321
2.3	0.01072	0.01044	0.01017
2.4	0.0082	0.00798	0.00776

The *p*-value is less than the preset alpha level. It is safe—relative to the risk measured by the Type I error rate—to conclude that H_A is likely true. There is enough evidence to suggest that more than 65% of the entire student population would have identified stress as a primary sleep impediment, had they all been asked. The university now has enough evidence to commit funds to address stress on their own campus.

Single-Sided with a Less-than Alternative

Suppose instead of a greater-than alternative, you had a less-than alternative. That is, suppose your hypothesis looked like this:

$$H_0 : p \geq p_0$$

$$H_A : p < p_0$$

The form of this hypothesis is said to be single-sided with a less-than alternative. This just means the alternative hypothesis is making a claim about the parameter being less than some number, as opposed to larger than or just not equal to some number. The steps involved in the testing of this form of hypothesis are nearly identical to those just demonstrated. There is one important twist that is necessary only because of how the standard score table has been arranged.

1. Compute the associated standard score;
2. If the standard score is:

   a. Negative, then take the **absolute value** of the standard score to the standard score table and look up the *p*-value from that;

   b. Positive, then take the original standard score to the standard score table and look up the *p*-value; the *p*-value you want is **one minus** the one you just looked up.

3. Compare the *p*-value from 2. to the preset alpha level. If the *p*-value is smaller, then it is safe to conclude H$_A$ is likely true. If the *p*-value is larger, then it is too risky to say that.

Double-Sided with a Not-Equal Alternative

Finally, for completeness, we should record how you would generate a *p*-value for a hypothesis that looks like the following, using a table of *p*-values (commonly) indexed like the one in this book:

$$H_0 : p = p_0$$
$$H_A : p \neq p_0$$

The form of this hypothesis is said to be double-sided. There are many reasons why this kind of test is preferred to either of the one-sided tests, but that discussion is beyond the scope of our discussion here. The steps involved in the testing of this form of hypothesis are nearly identical to those for the greater-than alternative. Here's the twist:

1. Compute the associated standard score;
2. Take the absolute value of the standard score to the standard score table and look up the *p*-value; **multiple that *p*-value by 2.** That is the *p*-value for the double-sided form of the test.
3. Compare the *p*-value to the preset alpha level. If the *p*-value is smaller, then it is safe to conclude H$_A$ is likely true. If the *p*-value is larger, then it is too risky to say that.

Example: Stress and Sleep Revisited

Suppose you are testing:

$$H_0 : p = 0.65$$
$$H_A : p \neq 0.65$$

In the example above we found a standard score of $z = 2.11$. Having taken that to the standard score table, we found a p-value of 0.01743. Therefore, the p-value for testing this hypothesis is $0.01743 \times 2 = 0.03486$. Therefore, it is still possible to reject the null in favor of the alternative.

This is, in a sense, all we need to know about the basic mechanics of hypothesis testing, provided our primary focus is on the intelligent consumption of testing results like those that appear routinely in the media. Granted, the actual mechanics can vary a lot depending, in part, on the parameter that is under scrutiny, the size of the sample, etc. But the logic of the testing process is largely the same. That having been said, tests about a population mean are so common that it is worth our time to be aware of those details.

When the Parameter Is a Mean

Drum Corps International sets field performance guidelines for all their competitive participants. Setup time, which occurs before a field performance, has to be practiced carefully in order to meet the Drum Corps' time constraints. These time limits change from year to year, but the World Class division typically allots 3 minutes to set up, 2 minutes to do a pre-show, and then 12 minutes to perform the main show. Suppose Carolina Crown, the defending 2013 World Class division champions, has practiced the setup 30 times. They have achieved an average setup time of 3 minutes and 4 seconds with a standard deviation of 15.5 seconds.

Suppose you wanted to test the hypothesis that the true mean time for setup is more than 3 minutes. Your hypothesis would then be of the form:

$$H_0 : \mu \leq \mu_0$$

$$H_A : \mu > \mu_0$$

where $\mu_0 = 180$ seconds (3 minutes), and μ can be thought of as the true average setup time that the Crown would experience over a very large number of attempts.

A standard score is still our ticket to a p-value for this kind of hypothesis. It just looks different now than it did when we were focused on a proportion, not a mean.

> The **standard score** associated with a test of hypothesis about **a population mean** is
> $$z = \frac{\bar{x} - \mu_0}{\sqrt{\frac{s^2}{n}}}.$$

where:

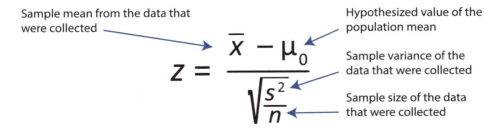

Once again, we have to admit to taking liberties with the actual form of this statistic. No matter, though, since all we need is a simplied version for our purposes.

The three steps needed to compute a *p*-value for the test of hypothesis are *exactly* the same as above. All that has changed is how the standard score is computed:

1. Compute the associated standard score;

2. Take the standard score to the standard score table and look up the *p*-value;

3. Compare the *p*-value to the preset alpha level. If the *p*-value is smaller, then it is safe to conclude H_A is likely true. If the *p*-value is larger, then it is too risky to say that.

Example: Drum Corps Revisited

In the motivating example, we assumed Crown had achieved a sample mean of 3 minutes and 4 seconds (184 seconds), with a sample standard deviation of 15.5 seconds. If we are testing the following hypothesis, the steps involved are very easy.

$$H_0 : \mu \leq 180$$

$$H_A : \mu > 180$$

We simply have to compute $z = \dfrac{184 - 180}{\sqrt{\dfrac{(15.5)^2}{30}}} = 1.41$. A quick check of the standard score table

reveals that a *z* of 1.41 corresponds to a *p*-value of 0.07927. Since 0.07927 is greater than the preset alpha level of 0.05, it is not safe to assume that the true average setup time is more than three minutes. This should not be much solace for Crown since it just affirms that there is not enough evidence to conclude that they will be in violation of the time constraints on average. It does not offer any evidence to support that they will *not* be in violation.

The other forms of this test are handled exactly as are those for the proportion.

Less-Than Alternative

$$H_0 : \mu \geq 180$$

$$H_A : \mu < 180$$

Compute *z* just as above.

- If *z* is negative, take the absolute value of *z* to the standard score table to produce the *p*-value appropriate for this hypothesis.

- If *z* is positive, take *z* to the standard score table, find that *p*-value, and subtract it from 1 to produce the *p*-value appropriate for this hypothesis.

Not-Equal-To Alternative

$$H_0 : \mu = 180$$

$$H_A : \mu \neq 180$$

Compute *z* just as above. Take the absolute value of z to the standard score table to produce a preliminary *p*-value. Multiply that *p*-value by 2 to obtain the *p*-value appropriate for this hypothesis.

It is worth repeating that there are many other elementary hypotheses that you'd see in a traditional first course on statistics. Your instructor may wish to show you more of those. However, this book is focused on a conceptual understanding of hypothesis testing. As such, we don't really gain anything by extending the list of possible hypotheses. You should simply anticipate that as the hypotheses change, the way in which the *p*-value is generated changes. That's fun to learn about. But the overarching logic by and large stays exactly the same.

TAKE-HOME POINTS

- Testing a simple hypothesis involves the key computation of a standard score. This standard score, in turn, is taken to a standard score table to produce a *p*-value that we can compare to our preset Type I error rate.
- The form of the hypothesis matters. In particular, slightly different steps for obtaining the right *p*-value are needed depending on the form of the alternative hypothesis.

Chapter 10 Exercises

Reading Check

Short Answer: Please provide brief, concise answers to the following questions.

1. In the stress example, where did the p_0 of 0.65 come from?

2. In the stress example, where did the 0.68 come from?

3. Compute the standard score for testing a hypothesis about a population proportion, given you have $n = 300$, $\hat{p} = 0.76$, and $p_0 = 0.81$.

4. Find the p-value associated with a z-score of 1.25 if your alternative hypothesis is a $<$.

5. Find the p-value associated with a z-score of 1.25 if your alternative hypothesis is a $\neq$.

Beyond the Numbers

1. One and Done

A sample of $n = 75$ UK students is taken by the *Kentucky Kernel*, and each one is asked, "Does knowing that many of the basketball players Kentucky recruits will go pro after one year affect your sense of attachment to the team?" Suppose 60% in your sample said, "Yes, it does affect my attachment." Is it safe for the *Kernel* to report that a majority of all UK students are likely to feel that way? Decide between H_0: $p \leq 0.50$ and H_A: $p > 0.50$. Assume a Type I error rate of $\alpha = 0.05$.

a. Report the standard score, the p-value, and state what your decision is.

   i. $z = 0.60$; p-value $= 0.420$; reject the null.

   ii. $z = 1.73$; p-value $= 0.042$; reject the null.

   iii. $z = 0.60$; p-value $= 0.420$; fail to reject the null.

   iv. $z = 1.73$; p-value $= 0.042$; fail to reject the null.

b. Regardless of what the p-value is in part a., how would it change if the sample percentage of 60% were based on a sample that is a lot bigger than 75, instead of $n = 75$?

   i.   The p-value would increase.

   ii. The p-value would decrease.

   iii. The p-value would not change.

   iv. It is impossible to know if it would change or not, unless you have a specific n to do the computation with.

c. Suppose the p-value you got in part a. is denoted by the capital letter P. What would be the p-value for testing the hypothesis: $H_0: p \geq 0.50$ and $H_A: p < 0.50$?

    i. abs(P)

    ii. Also P

    iii. $1 - P$

    iv. $P - 1$

2. Party Green

A newspaper took a random sample of 1,200 registered voters and found that 925 would vote for the Green Party candidate for governor. Is this evidence that more than ¾ of the entire voting population would vote for the Green Party candidate? Assume a Type I error rate of $\alpha = 0.05$.

a. What are H_0 and H_A?

b. What is the associated standard score?

c. Test the hypothesis. Report a p-value, state what your decision is, and explain why.

3. Got No Satisfaction?

The CEO of a large electric utility claims that more than 80% of his customers are very satisfied with the service they receive. To test this claim, the local newspaper surveyed 100 customers using simple random sampling. Among the sampled customers, 81% said that they were very satisfied. Do these results provide sufficient evidence to accept or reject the CEO's claim? To answer this question, you will have to test the hypothesis $H_0: p \leq 0.80$ versus $H_A: p > 0.80$. Assume a Type I error rate of $\alpha = 0.05$.

a. Report the standard score, the p-value, and state what your decision is.

    i. $z = 0.81$; p-value $= 0.250$; fail to reject the null.

    ii. $z = 0.80$; p-value $= 0.401$; fail to reject the null.

    iii. $z = 0.25$; p-value $= 0.401$; fail to reject the null.

    iv. $z = 0.40$; p-value $= 0.250$; fail to reject the null.

b. Regardless of what the p-value is in part a., how would it change if the sample percentage based on a sample of 100 customers were larger than 81%?

    i. The p-value would increase.

    ii. The p-value would decrease.

    iii. The p-value would not change.

    iv. It is impossible to know if it would change or not, unless you have a specific sample percentage to do the computation with.

c. Suppose the *p*-value you got in part a. is denoted by the capital letter *P*. What would be the *p*-value for testing the hypothesis: $H_0: p = 0.80$ and $H_A: p \neq 0.80$?

    i.   2 times the absolute value of $(1 - P)$

    ii. Abs$(P/2)$

    iii. $1 - \text{abs}(P)$

    iv. 2 times *P*

4. Cancer Treatment

Patients with advanced cancers of the stomach, bronchus, colon, ovary, and breast were treated with ascorbate. Their survival time post-diagnosis was then monitored. The resulting data set is available at www.statconcepts.com/datasets or in the appendix of this book.

a. How many patients survived more than a year?

    i.   27

    ii. 63

    iii. 43

    iv. 46

b. What is the standard score associated with a test of hypothesis of the form $H_0: p \leq 0.40$ and $H_A: p > 0.40$? *p* represents the proportion of cancer patients treated with ascorbate who will survive more than a year.

    i.   0.27

    ii. 0.63

    iii. 0.43

    iv. 0.46

c. Assume the *p*-value corresponding to the standard score in b. is denoted by the capital letter *P*. If you know that *p*-value is such that you are not able to reject H_0 with a preset Type I error rate of $\alpha = 0.05$, what can you say about what would have happened if you had chosen to start your analysis with a Type I error rate of $\alpha = 0.01$?

    i.   You would have not been able to reject your null.

    ii. You would have been able to reject your null.

    iii. You would have been able to accept your alternative.

    iv. You can't know how this would have turned out unless you know the actual value of *P*.

d. Refer to the cancer data referenced in Question 2. Test the hypothesis that the proportion of breast cancer patients treated with ascorbate who will survive more than a year is larger than 0.75. Assume a Type I error rate of $\alpha = 0.05$. Report a *p*-value, say whether you chose H_A or not, and explain why you made the choice you did. Are you surprised by this outcome? Why or why not?

e. Refer to part d. above. What sample size would have been needed for the results to have produced a *p*-value of 0.025? Show all your work.

5. Vytorin Verified

In a 2008 article, author Alex Berenson found that patients taking the prescription drug Vytorin were statistically significantly more likely (with a preset Type I error rate of 0.01) than those taking a placebo to develop cancer.

It should be noted that after further studies were completed, the FDA concluded that "it is unlikely that Vytorin … increase(s) the risk of cancer or cancer related death." There is currently no mention of cancer on the FDA website's Vytorin entry. But let's look at this initial finding of statistical significance.

Reports differ on exactly how many people participated in the original Vytorin clinical trial, but it appears that 950 were assigned to take Vytorin and 920 were assigned to take a placebo. There were a reported 102 cases of cancer in the Vytorin group and 67 cases of cancer in the placebo group. This means that 7.28% of all placebo group members got cancer.

Let's use *p* to denote the true proportion of all Vytorin users who might develop cancer over time while taking the drug. It would be reasonable to test to see if there is evidence in the data that this unknown proportion is likely bigger than the proportion developing cancer in the placebo group.

$H_0 : p \le 0.0728$

$H_A : p > 0.0728$

a. What proportion of Vytorin users in the study developed cancer?

   i. 102

   ii. 0.0728

   iii. 0.1074

   iv. 0.0107

b. What is the standard score associated with the hypothesis shown?

    i. 1.40

    ii. 0.05

    iii. 4.10

    iv. 2.14

c. Using the standard score table in this book, what is the most you can say about the *p*-value for this hypothesis?

    i. It is smaller than 0.00011.

    ii. It is larger than 0.00011.

    iii. It is larger than 0.01.

    iv. It is between 0.001 and 0.01.

d. Suppose a new study is done with *n* Vytorin patients, where *n* is a lot smaller than 950. If the sample proportion of the Vytorin patients in the new study who developed cancer is the same as the real study with 950 patients, how would the new standard score compare to the one from the real study?

    i. It would be larger.

    ii. It would be the same since the proportion of those developing cancer is the same.

    iii. It would be smaller.

    iv. You can't answer this if you don't know *n*.

6. Mary Jane Brain

Title: Moderation of the Effect of Adolescent-Onset Cannabis Use on Adult Psychosis by a Functional Polymorphism in the Catechol-O-Methyltransferase Gene: Longitudinal Evidence of a Gene X Environment Interaction

Author: A. Caspi, et al.

Source: *Biol Psychiatry* 2005;57:1117–1127.

The authors of this study examined the influence of adolescent marijuana use on adult psychosis as a function of certain genetic variables. In particular, they studied the catechol-O methyltransferase (COMT) gene. This gene is known to govern an enzyme that breaks down dopamine, a brain chemical involved in schizophrenia. The COMT gene is expressed in different ways in different people. The figure below shows the results of the study for individuals with one particular COMT expression. As you can see, there are two groups being studied: one whose members did not use marijuana in their adolescence, and one whose members did. The vertical axis records the percentage of group members who went on to develop schizophrenia.

The difference between the two groups has been declared statistically significant at the $\alpha = 0.05$ level. Use the "No Adolescent Marijuana" group to determine a suitable p_0, and then test the hypothesis:

$H_0: p \leq p_0$

$H_A: p > p_0$

where p is the population proportion (conceptualized) of all adolescent marijuana users who would develop schizophrenia. Show all your work. Are the results statistically significant as claimed? Make sure you report the standard score, a p-value, your conclusion, and a comment on what statistical risk has been taken and controlled for in this testing process.

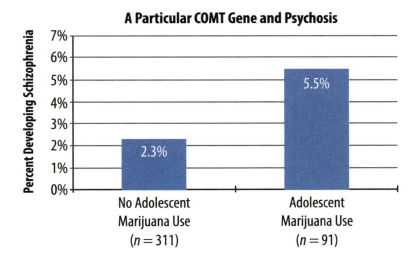

7. Eureka or Not?

Suppose that twenty identical experiments are taking place simultaneously around the world. The researchers are all studying the same drug, which they hope will improve the survival rate of the black-winged peckerwood finch after it has been infected with a particular type of tree mold. The survival rate for those left untreated is unfortunately only 32%. None of the researchers know about the others' work. The table below shows the results from the 20 different studies. In all cases, the Type I error rate was $\alpha = 0.05$ and the hypothesis being tested was:

$H_0: p \leq 0.32$

$H_A: p > 0.32$

a. Use the data from Site 17 to confirm that the null could be rejected. What is the *p*-value associated with the result?

b. Combine all of the studies (*n* = 100 × 20 = 2,000) and test the hypothesis again. Confirm that it cannot be rejected. Report the overall observed survival rate and the *p*-value associated with the overall test.

c. We have a dilemma. Nineteen of the sites don't seek publication because their results are not significant. Site 17 gets published because the results produced there, with an identical experiment, are significant. We know (though the researchers don't) that if we combine the results from all 20 sites, we will not be able to support the alternative. Describe what it means to have a Type I error rate of $\alpha = 0.05$ and explain what has likely happened here in light of that definition.

TABLE 10.1 Finch Survival Rates

Site	Observed Survival Rate with Drug	Number of Finches Studied	Able to Reject $H_0: p \leq 0.32$?
1	0.35	$n = 100$	No
2	0.34	$n = 100$	No
3	0.31	$n = 100$	No
4	0.33	$n = 100$	No
5	0.33	$n = 100$	No
6	0.35	$n = 100$	No
7	0.35	$n = 100$	No
8	0.33	$n = 100$	No
9	0.30	$n = 100$	No
10	0.34	$n = 100$	No
11	0.34	$n = 100$	No
12	0.30	$n = 100$	No
13	0.34	$n = 100$	No
14	0.31	$n = 100$	No
15	0.31	$n = 100$	No
16	0.31	$n = 100$	No
17*	**0.45**	**$n = 100$**	**Yes**
18	0.30	$n = 100$	No
19	0.35	$n = 100$	No
20	0.33	$n = 100$	No

8. Better than Chance?

Suppose that you are developing a new pill designed to help students guess better on yes/no test questions. If students guess totally at random, they have a 50-50 chance of getting it right. You want to show that students perform better after taking your pill so that you can create some interest in crowd-based funding for production and marketing costs. Unfortunately, when testing the pill's effectiveness, you always ended up with 51% of the treatment group getting their yes/no questions correct.

a. Consider the following hypothesis. Complete the entries in the table below for the different sample sizes shown. Remember that $\hat{p} = 0.51$ in **all** cases.

$H_0: p \leq 0.50$
$H_A: p > 0.50$

i. A = 0.20; B = 0.63; C = 2.00; D = 6.32.

ii. A = 0.11; B = 0.35; D = 0.01; D < 0.00011.

iii. A = 0.42; B = 0.26; C = 0.02; D < 0.00011.

iv. A = 0.50; B = 0.51; C = 0.01; D > 1.00.

TABLE 10.2

Sample Size	One-Sided p-Value	Statistically Significant Results? (yes or no)
100	A	No
1,000	B	No
10,000	C	Yes
100,000	D	Yes

b. After recruiting 100,000 people in your study, you were able to report that your results were statistically significant (see Table 10.2). So you decide to begin seeking funding. What are two reasons why you are still likely to have an unconvincing case?

i. Actual difference in treatments is only 10 percent, and the statistical significance is an artifact of increasing sample size.

ii. Actual difference in treatments is only 51 percent, and the statistical significance is an artifact of increasing sample size.

iii. Actual difference in treatments is only 1 percent, and the statistical significance is an artifact of decreasing sample size.

iv. Actual difference in treatments is only 1 percent, and the statistical significance is an artifact of increasing sample size.

9. Crowd Control

Title: Duration of Sleep Contributes to Next-Day Pain Report in the General Population

Author: R. Edwards, et al.

Source: *Pain 137* (2008) 202–207.

The authors of the study referenced above interviewed participants from the general population. These participants reported both the number of hours they slept during the previous sleep period and the frequency of their pain symptoms. Pain was recorded on a five-point scale. A summary of the resulting data is presented in Table 10.3.

a. A comparison of patients in the 0–3 hour category to those in the 11+ hour category is not statistically significant, despite a difference in means of 0.42. However, a comparison of patients in the 5-hour category to those in the 8-hour category *is* statistically significant, even though the difference in means is only 0.19. Why is this the case? Defend your answer.

b. What practical implication does this have for our understanding of testing results?

TABLE 10.3

Sleep (Hours)	Average Pain Rating	Standard Deviation	Sample Size
0–3	1.36	1.51	75
4	1.13	1.36	166
5	0.94	1.29	434
6	0.79	1.11	1,138
7	0.73	1.11	1,568
8	0.75	1.13	1,557
9	0.71	1.09	339
10	1.24	1.4	119
11+	1.78	1.59	66

10. A Tail of Two Hypotheses

Please provide full solutions to the following problems.

a. Patients with advanced cancers of the stomach, bronchus, colon, ovary, and breast were treated with ascorbate. Their survival time post-diagnosis was then monitored. The resulting data set are available at www.statconcepts.com/datasets or in the appendix. Test the hypothesis that the proportion of all cancer patients treated with ascorbate who will survive more than a year is different than 0.40. Assume a Type I error rate of $\alpha = 0.05$. Report a *p*-value, say whether you chose H_A or not, and explain why you made the choice you did.

b. Suppose you want to test a hypothesis about a proportion similar to what you've just done, but you don't know whether to use a two-tailed or a one-tailed test. You do know, however, that you have to have a Type I error rate of 0.05. You absent-mindedly take a look at your data results before forming H_A, and you notice $p > p_0$. So you decide to go with a one-tailed H_A. Why might this be considered cheating? Be very clear when explaining your reasons.

11. Mean from the Corps

Drum Corps International sets field performance guidelines for all their competitive participants. Setup time, which occurs before a field performance, has to be practiced carefully in order to meet Drum Corps' time constraints. These time limits change from year to year, but the World Class division typically allots 3 minutes to set up, 2 minutes to do a pre-show, and then 12 minutes to perform the main show. Suppose Carolina Crown, the defending 2013 World Class division champions, has practiced the setup 30 times. They have achieved an average setup time of 3 minutes and 4 seconds with a standard deviation of 15.5 seconds.

TABLE 10.4 Crown's Setup Times

Attempt	Setup Time
1	200
2	189
3	180
4	168
5	168
6	195
7	167
8	200
9	167
10	210
11	185
12	167
13	190
14	180
15	175
16	175
17	182
18	171
19	202
20	195
21	212
22	203
23	150
24	175
25	200
26	182
27	184
28	165
29	203
30	180

a. Let's suppose Crown's thirty hypothetical setup times are presented in the table below. Use this data and a software package (such as Microsoft Excel or Apple Numbers) to confirm the mean and standard deviation given above.

b. In the chapter we tested the following hypothesis:

$H_0: \mu = 180$ secs

$H_A: \mu \neq 180$ secs

We indicated that the standard score was 1.41, which corresponded (in the two-sided case) to a *p*-value of 0.15854, more or less. Let's confirm that with an online applet that will make some subsequent computations simpler.

▸ Go to http://www.wolframalpha.com/and enter "z-test calculator" in the search field.

▸ Fill out the required fields (make sure you chose two-tailed test).

▸ Hit enter.

▸ Confirm you got the same results as stated above.

c. Let's change the data some. Suppose the *Attempt 23* time was erroneously recorded as 200 and should have been recorded as 150. Re-compute the mean and standard deviation. These are:

i. Mean is 185.67; standard deviation is 14.3343.

ii. Mean is 184; standard deviation is 14.3343.

iii. Mean is 185.67; standard deviation is 15.5.

iv. Mean is 185.67; standard deviation is 205.4722.

d. Re-test the hypothesis:

H_0: $\mu = 180$ secs

H_A: $\mu \neq 180$ secs

Use the same applet as in part a. to compute the new standard score and new p-value. These are:

i. $z = 2.17$; p-value is 0.01.

ii. $z = 2.17$; p-value is 0.03.

iii. $z = 1.27$; p-value is 0.01.

iv. $z = 1.27$; p-value is 0.03.

e. Reflect on your answer to part d. What can you say about how sensitive the tests seem to be to a single typo in the data set?

f. The Cadets Drum and Bugle Corps from Allentown, Pennsylvania, routinely compete with Carolina Crown in the World Class division. They are also subject to the same setup time restrictions. Suppose, based on 30 sample setups, that they have a sample mean of $x = 184$ seconds just like Crown. However, they have a standard deviation of just 5.1 seconds. Test the hypothesis shown.

H_0: $\mu = 180$ secs

H_A: $\mu \neq 180$ secs

What is the two-tailed p-value and the appropriate conclusion for this hypothesis?

i. p-value is > 0.05; reject the null in favor of the alternative.

ii. p-value is > 0.05; fail to reject the null.

iii. p-value is < 0.001; reject the null in favor of the alternative.

iv. p-value is < 0.001; fail to reject the null.

g. Carefully explain the source of any differences in the testing outcomes between the results you found for the Cadets and those for Crown.

12. Double Trouble

Granted this book is not designed to teach you a lot of methods. But if you *were* to need to test a hypothesis for real, then odds are you'd at least need to know how to test one involving two means (or two proportions), instead of just the single-mean (or proportion) test we discussed in the chapter. This problem contains the minimal exposure you need to do that. We are going to use an online applet to do the calculations instead of specifying all the mathematical details. Your instructor may wish to show you the formula(s). They can be somewhat nuanced, so they are usually reserved for more advanced courses.

Suppose we have two populations (or treatments), one with an unknown population mean μ_1 and the other with an unknown population mean μ_2. Testing if these treatments are different is often reduced to a test of whether they are different on average:

$H_0: \mu_1 = \mu_2$

$H_A: \mu_1 \neq \mu_2$

To produce a *p*-value for this kind of hypothesis, here is what we'd need:

Step 1: A value for alpha level, the Type I error rate (typically $\alpha = 0.05$).

Step 2: The following summary values:

- The sample sizes from each of the two treatment groups
- The sample means from each of the two treatment groups
- The sample standard deviations from each of the two treatment groups

Step 3: Access to one of many software applications or online calculators that can compute two-sided *p*-values for tests involving two means. If your instructor does not have a preference, use the GraphPad® online calculator at http://www.graphpad.com/quickcalcs/ttest1/?Format=SD.

Step 4: Compare the two-tailed *p*-value to α (the preset Type I error rate). If the *p*-value is smaller, reject H_0; if it is larger, fail to reject H_0.

The same series of steps and the same applet can be used to test:

$H_0: p_1 = p_2$

$H_A: p_1 \neq p_2$

for two population proportions p_1 and p_2. Just repeat the steps above with the adaptations:

- Sample means are sample proportions and $\hat{p}_1$ and $\hat{p}_2$.
- Sample standard deviations are $\sqrt{\hat{p}_1(1-\hat{p}_1)}$, and $\sqrt{\hat{p}_2(1-\hat{p}_2)}$, respectively.

Please do the following exercises:

Title: Duration of Sleep Contributes to Next-day Pain Report in the General Population

Author: R. Edwards, et al.

Source: *Pain 137* (2008) 202–207.

There is a lot of interest in how disturbed sleep is related to pain perception. Throughout this study, participants from the general population recorded both the number of hours they slept and the frequency of their pain symptoms. The pain symptoms were recorded on a five-point scale: 0 = none of the time, 1 = a little, 2 = some, 3 = most, and 4 = all of the time. A summary of the study results is shown in the table below.

We are interested in testing:

$H_0: \mu_{sc1} = \mu_{sc2}$

$H_A: \mu_{sc1} \neq \mu_{sc2}$

where sc1 and sc2 are just different pairs of sleep categories. In all cases, assume a Type 1 error rate of 0.05.

TABLE 10.5 Pain Relief Results

Sleep (Hours)	Average Pain Rating	Standard Deviation	Sample Size
0–3	1.36	1.51	75
4	1.13	1.36	166
5	0.94	1.29	434
6	0.79	1.11	1,138
7	0.73	1.11	1,568
8	0.75	1.13	1,557
9	0.71	1.09	339
10	1.24	1.4	119
11+	1.78	1.59	66

a. Use the applet above to test the following hypothesis involving the average pain rating for the 0–3 and the 11-hour sleep groups.

$H_0: \mu_{(0-3\ hour)} = \mu_{(11-hour)}$

$H_A: \mu_{(0-3\ hour)} \neq \mu_{(11-hour)}$

The results are not statistically significant. What is the *p*-value?

i.   0.0027

ii. 0.0315

iii. 0.1102

iv. 0.3013

b. Use the applet above (or one of your choice) to test the following hypothesis about the average pain ratings for the 5-hour and the 8-hour sleep groups.

$H_0: \mu_{(5-hour)} = \mu_{(8-hour)}$

$H_A: \mu_{(5-hour)} \neq \mu_{(8-hour)}$

The results are statistically significant. What is the *p*-value?

i. 0.0027

ii. 0.0315

iii. 0.1102

iv. 0.3013

c. Look back at the two problems you just completed. In the first, the two groups had average pain ratings of 1.36 and 1.78 (a difference of 0.42), and this was not statistically significant at the 0.05 level. In the second, the two groups had average pain ratings of 0.94 and 0.75 (a difference of only 0.19), and this *was* statistically significant. Explain how this happened.

Author's note: There can be serious problems with testing lots of pairs of hypotheses from a single study. Doing so can complicate what one means by the Type I error rate. Your instructor may choose to elaborate. A discussion of this kind of issue is well beyond the scope of this book. We have also chosen to ignore discussions of the different considerations of how these kinds of tests might be best done. Such discussions are not consistent with our purpose.

13. New Use for Old Intervals

It is not uncommon to use confidence intervals to make decisions similar to those made in hypothesis testing. To explain this process a bit more formally, we will focus on the following hypothesis.

$$H_0: p = p_0$$

$$H_A: p \neq p_0$$

We will have to surface a little more detail about a confidence interval for a proportion than you saw in Chapter 6. To test the above hypothesis using a confidence interval, you would do the following:

Step 1: Establish a value for the Type I error rate (typically α = 0.05) and **divide that by 2**.

Step 2: Find the standard score from our standard score table that corresponds to table entry of α/2. Call this z^*. So you are reading the table backwards in the sense that you are starting internal to the table and reading the margins. For example, if α = 0.05, then $z^* = 1.96$.

Step 3: Form the confidence interval:

$$\hat{p} \pm z^* \sqrt{\frac{p_0(1 - p_0)}{n}}$$

Step 4: See if p_0 is within this interval.

- **If it is not, you can safely reject H$_0$** with the assumed Type I error rate.
- **If it is, then you have to fail to reject H$_0$.**

Notice, p-values do show up explicitly when using confidence intervals for testing. Also, the confidence interval in Step 3 looks a bit different than what you saw in the chapter on confidence intervals. At that time, there was no context for p_0 since there were no hypotheses to be tested.

Author's Note: Standard score tables are indexed differently in different publications. These instructions apply to how the standard score table is formatted in this book.

a. Patients with advanced cancers of the stomach, bronchus, colon, ovary, and breast were treated with ascorbate. Their survival time post-diagnosis was then monitored. The resulting data are presented in the appendix. Use a confidence interval to test the hypothesis that the proportion of all cancer patients treated with ascorbate who will survive more than a year is different than 0.30. Assume a Type I error rate of $\alpha = 0.05$. Show all your work. Say whether you chose H$_A$ or not, and explain why you made the choice you did.

b. Revisit the question in part a. Use a confidence interval to retest the same hypothesis, but this time use a Type 1 error rate of 0.01. Show all your work. Say whether you chose H$_A$ or not, and explain why you made the choice you did.

Projects

1. Conducting a Formal Survey: Part III

Introduction

This project is a natural progression from Parts I and II, introduced in Chapters 4 and 5.

The scope of what is required depends on whether you are being asked to do this individually or as part of a group. Your instructor will clarify for your class.

From Parts I and II

In Part I you did the following:

Individual Project

- Selected a simple random sample of 50 individuals
- Constructed a single yes/no question that was asked of all 50 of those individuals
- Recorded each sampled individual (by ID, not by name) in a table with a record of whether they answered "Yes" or "No" to your question

Group Project

- ▸ Selected a simple random sample of 150 individuals
- ▸ Constructed a single yes/no question that was asked of all 150 of those individuals
- ▸ Recorded each sampled individual (by ID, not by name) in a table with a record of whether they answered "Yes" or "No" to your question

In Part II you followed up by answering a series of prompts about confidence intervals.

New to Part III

In this installment you have to test one of the following hypothesis. Your choice.

$$H_0: p \leq p_0 \qquad H_0: p = p_0$$
$$\text{or}$$
$$H_A: p > p_0 \qquad H_A: p \neq p_0$$

This installment of your project report should be professionally prepared, typed, and proofread. Your instructor may require that it be added to Parts I and II. Follow her instructions. Your task is to exhibit the following to your readers, complete with full explanation of what you are doing and what it means:

- Explain where you got your p_0 from. You need good reasons. Don't just say you made it up.
- The value of your standard score. Explain to reader how the standard score is computed.
- Your p-value. Explain to the reader how you used the standard score table to produce the p-value.
- Your conclusions. Explain if you are able to accept the alternative or not. Explain the statistical risk involved in whatever decision is made.
- An interpretation of your results in the context of your project.
- A comparison of your inferential results here, with those seen in Part II where you computed a confidence interval. Explain carefully if the results are consistent or not.

All of your explanations and comments MUST be in the context of your specific survey. You are not trying to convince your reader of anything. Rather, you are explaining how to do simple hypothesis testing in the context of your study and how those results compare to the inferential perspective provided by the confidence interval you computed in Part II.

2. Designing Your Own Experiment: Part III

Introduction

This project is a natural progression from Part I, introduced in Chapter 3, and Part II, from Chapter 5.

The scope of what is required depends on whether you are being asked to do this individually or as part of a group. Your instructor will clarify for your class.

From Parts I and II

In Part I you did the following:

Individual Project

▸ Randomized subjects into two groups, formed from a single change in a single variable. You had a minimum of 50 subjects.

▸ Applied the treatments in each group and collected the data

▸ Collected the response data for each subject

Group Project

▸ Randomized subjects into two groups, formed from a single change in a single variable. You had a minimum of 150 subjects.

▸ Applied the treatments in each group and collected the data

▸ Collected the response data for each subject

In Part II, you followed up by constructing confidence intervals for each treatment, which you interpreted; you also offered some comments on whether you thought the treatments really were different based on what you saw in the intervals.

New to Part III

In this installment you will be testing one of the following hypothesis. It will depend on whether you collected data commensurate with proportions or means.

$$H_0: \mu_1 = \mu_2 \qquad H_0: p_1 = p_2$$
$$\text{or}$$
$$H_A: \mu_1 \neq \mu_2 \qquad H_A: p_1 \neq p_2$$

You are going to need the results in Problem 12 (Double Trouble) in this chapter in order to carry out the testing process. If you were not assigned Problem 12, you will want to read it on your own.

This installment of your project report should be professionally prepared, typed, and proofread. Your instructor may require that it be added to Parts I and II. Follow her instructions. Your task is to exhibit the following to your readers, complete with full explanation of what you are doing and what it means:

- The value of your standard score. Explain to reader how the standard score is computed.

- Your *p*-value. Explain to the reader how you produced the *p*-value.

- Your conclusions. Explain if you are able to accept the alternative or not. Explain the statistical risk involved in whatever decision is made.

- An interpretation of your results in the context of your project.

- In Part II you computed a pair of confidence intervals. It turns out that you can't reliably test hypotheses about two means (or proportions) based on whether their individual confidence intervals overlap or not. That is, it is tempting to say that when the individual intervals overlap, the hypothesis of equal parameters cannot be rejected; and when they don't overlap, that hypothesis can be rejected. Actually only ONE of those two statements is true. Google "overlapping confidence intervals and statistical significance," do some research, and tell your reader which one is true and which one isn't.

All of your explanations and comments MUST be in the context of your specific experiment. You are not trying to convince your reader of anything. Rather, you are explaining how to do hypothesis testing in the context of your study and how those results compare to the potentially imperfect inferential perspective provided by the confidence intervals you computed in Part II.

3. Reality Video

Introduction

This chapter discussed procedures for testing simple hypotheses involving a mean or a proportion.

The Assignment

Your instructor may have you do this as individuals or may form you into teams. We'll refer to "teams" below, although you may be a team of one depending on what your instructor wants. Each team is required to do the following:

a. Pick one or more topics from this chapter. Your instructor may want to approve them before you proceed.

b. Construct a three-minute video that teaches the chosen topic to the class. You must carefully and correctly use the language of the chapter in your presentation, and you are required to give an example that is NOT in the chapter.

c. Post the video on YouTube and send the link to your instructor in the manner he stipulates.

11

CHAPTER 11
Hypothesis Testing: Importance of Clinical Significance

Introduction

Much attention is paid to the concept of statistical significance. The goal of this chapter is to call attention to another type of significance—clinical (also called practical) significance—that is just as important to be aware of. Have a look at the *New York Times* article on multivitamins, appearing October 17th, 2012. The article addresses a clinical trial that unfolded over more than a decade. Male doctors were followed during that period of time, and those taking a daily multivitamin were compared to those who took a placebo.

Title: Multivitamin Use Linked to Lowered Cancer Risk

Author: Roni Caryn Rabin

Source: *New York Times*

> After a series of conflicting reports about whether vitamin pills can stave off chronic disease, researchers announced on Wednesday that a large clinical trial of nearly 15,000 older male doctors followed for more than a decade found that those taking a daily multivitamin experienced 8% fewer cancers than the subjects taking dummy pills.
>
> The findings were to be presented Wednesday at an American Association for Cancer Research conference on cancer prevention in Anaheim, Calif., and the paper was published online in the *Journal of the American Medical Association*.
>
> The reduction in total cancers was small but statistically significant, said the study's lead author, Dr. J. Michael Gaziano, a cardiologist at Brigham and Women's Hospital and the VA Boston Healthcare System. While the main reason to take a multivitamin is to prevent nutritional deficiencies, Dr. Gaziano said, "It certainly appears there is a modest reduction in the risk of cancer from a typical multivitamin."

What did the researchers find? Those physicians in the multivitamin ("treatment") group "experienced 8% fewer cancers…." We also see from the article that this difference of 8% was statistically significant. Therefore, we know that a test of hypothesis was done, probably something like this:

$$H_0 : \mu_{\text{Cancer Occurences in Vitamin Group}} \geq \mu_{\text{Cancer Occurences in Placebo Group}}$$

$$H_A : \mu_{\text{Cancer Occurences in Vitamin Group}} < \mu_{\text{Cancer Occurences in Placebo Group}}$$

Further, we know how the formal comparison turned out. The researchers surely adopted a Type I error rate of 0.05, computed a *p*-value from the observed group averages and standard deviations, and found the *p*-value to be smaller than 0.05. That is, we know they were able to conclude that the number of cancers in the vitamin group is most likely less than in the placebo group and make that conclusion with the probabilistic support provided by the α level.

DO YOU CARE?

However, this does not imply that the actual difference is big enough that anyone would care. That is a separate question. That is the question of clinical significance.

> Experimental results are said to be **clinically significant** if the effect of the treatment(s) under study have a genuine effect on daily life.

When it comes to cancer reduction, 8% would likely catch people's attention and may generally be seen as clinically significant. That question is not specifically addressed in the article, though; admittedly, the author does refer to the effect as "small."

The definition of clinical significance is undeniably vague in a subject area that is mostly devoted to things crisp and mathematical. In fact, attempts at precise assessments of clinical significance can quickly become a confusing quagmire of practical advice, theory-based admonishments, and general misunderstanding. A large part of the confusion revolves around whether it is even rational to have post hoc methods of quantifying clinical significance, or if ante hoc methods are the only valid ones. This is too much for us to worry about in this course, and it is still a topic that even those in the statistics profession don't always agree on.

Nonetheless, some understanding of practical significance is important if we are going to be intelligent consumers of statistical information.

Informal Assessment of Practical Significance

In the simplest situation, an assessment of practical or clinical significance is just based on what the experts (or those interested) in the area subjectively consider to be big enough to have a genuine effect on daily life. Weight loss sometimes falls into that category. Consider the following example.

Example: Mental Illness and Obesity

The *New York Times* study shown here addresses weight loss in people with serious mental illness. The article calls into question the difference between statistical significance and clinical significance.

Title: A Battle Plan to Lose Weight

Author: Catherine Saint Louis

Source: *New York Times*

People with serious mental illnesses, like schizophrenia, bipolar disorder, or major depression, are at least 50% more likely to be overweight or obese than the general population. They die earlier, too, with the primary cause being heart disease.

Yet diet and exercise usually take a back seat to the treatment of their illnesses. The drugs used, like antidepressants and antipsychotics, can increase appetite and weight.

"Treatment contributes to the problem of obesity," said Dr. Thomas R. Insel, the director of the National Institute of Mental Health. "Not every drug does, but that has made the problem of obesity greater in the last decade."

It has been a difficult issue for mental health experts. A 2012 Dartmouth review of health promotion programs for those with serious mental illness concluded that of 24 well-designed studies, most achieved statistically significant weight loss, but very few achieved "clinically significant weight loss."

While we don't have an abundance of information from the article, we do see in the last paragraph that 24 well-designed studies of weight loss programs for the mentally ill were scrutinized. We are told that "most achieved statistically significant weight loss, but very few achieved 'clinically significant' weight loss." The authors say elsewhere in the article that in this context, 'clinically significant' typically means more than 5% of one's body weight. To be clear, statistical significance is not being questioned. Hypotheses were formed, data were collected, and presumably a *p*-value was computed and found to be less than a preset Type I error rate for most of the studies. At issue is whether the actual weight loss that was seen in the studies is enough for experts in the area to say it would have a genuine effect on the daily lives of those enrolled in the studies.

Alas, even experts fail to agree some of the time. In the April 2016 issue of *Obesity*, author Robert Ross argues:

> Although it is true that for most health outcomes a 5% weight loss is associated with benefit, it is also true that for adults who adopt and sustain physical activity combined with a healthful diet, significant health benefits are achieved in association with very modest (less than 3%) or even no weight loss. Strong evidence in support of this notion comes from the Diabetes Prevention Study wherein participants who increased physical activity for about 150 min per week but did not achieve "clinically significant weight loss" reduced diabetes incidence by 44%, observations consistent with the findings of the Diabetes Prevention Program. The findings from these and numerous randomized controlled trials consistently report that physical activity is associated with benefit for many health outcomes with or without weight loss.

> It is suggested that the 5% clinically significant weight loss threshold for effective obesity management is "here to stay." While the veracity of this contention will continue to be the source of debate among academics, its interpretation for many practitioners and obese adults may well be that, for effective obesity

management, substantial weight loss is required for benefit. This would be an unfortunate interpretation and one that may lead obese adults who are unable to make the major changes in behavior required to sustain a 5% weight loss in today's environment to become frustrated and consequently discontinue physical activity and/or the consumption of a healthy diet.

Let's look at another example from an on-line reflection that can be found under the recurring moniker of the Skeptical Scalpel. You can search on the title and date and likely find the entire article.

Example: Sleep Apnea

Title: Statistical versus Clinical Significance: They Are Not the Same

Author: Skeptical Scalpel

Source: *MedPage Today*

MedPage Today featured an article about the beneficial effects of daytime wearing of compression stockings on obstructive sleep apnea. The premise was that increased edema in the neck could be caused by fluid coming from the legs when patients were in the supine position at night. Twelve patients who served as their own controls wore compression stockings for a week and then no stockings for a week alternating. The stockings lowered the amount of fluid in the neck by 60%, a statistically significant difference. So far, so good.

This resulted in another highly statistically significant finding, which was a 36% reduction in episodes of apnea [cessation of breathing] and hypopnea [inadequate breathing]. Sounds good, right? The problem is that the average number of episodes of apnea/hypopnea decreased from 48 per hour to 31 per hour. Patients experiencing more than 30 episodes of apnea/hypopnea per hour are classified as having severe obstructive sleep apnea. This means that the treatment only put the patients in the low range of severe obstructive sleep apnea. They still would require maximum therapy. Is a reduction in apnea/hypopnea episodes that do not move the patient out of the severe category really clinically significant? It does not seem so to me.

Bottom line: Although there was a statically significant improvement in the number of apnea/hypopnea episodes when compression stockings were worn, it does not appear that there was a clinically significant improvement in the course of obstructive sleep apnea.

In the article the author tells us about a study focused on the daytime wearing of compression stockings and the effect that activity had on obstructive sleep apnea. You can see from the article that the study concluded that the stockings "lowered the amount of fluid in the neck by 60%, a statistically significant difference."

The author goes on to quote the study, noting that "this resulted in another highly statistically significant finding, which was a 36% reduction in episodes of apnea…." The reason the Skeptical Scalpel is not happy with this study is that even with the 36% reduction in apnea episodes, the average number in an hour was still 31. This average remained in the category of severe obstructive sleep apnea, requiring maximum therapy, which is defined as having more than 30 episodes per hour. The author surmises that he would only be convinced that the intervention (compression stockings) was worthwhile if it was able to move the (average) patient into a lower category of apnea classification. Hence, the author is not at all convinced the results are practically significant even though they were found to be statistically significant.

We are not surprised that people are going to disagree about what actually constitutes a change that is big enough to have a real impact on the daily lives of the subjects participating in the study. While this kind of discussion might be non-mathematical, it can still be very difficult to navigate. Complexity aside, a larger point has already surfaced by way of these two examples:

> **Having statistically significant results does not imply that those same results are clinically significant**.

Likewise, arguments can be made that results might be practically significant even if they aren't statistically significant. The next example illustrates this point.

Example: Coffee and Pregnancy

The following excerpt is from a January 21st, 2008, article in the *San Francisco Chronicle*. Only a small portion of the article is shown here. In this study the miscarriage rate in pregnant women was scrutinized in relation to their caffeine intake. The study interviewed 1,063 women.

Title: Best Evidence yet of Caffeine-Miscarriage Link: Kaiser Researchers Find 1½ Cups of Coffee Doubles Risk

Author: Catherine Saint Louis

Source: *San Francisco Chronicle*

> Women in early pregnancy who drink a cup and a half of coffee every day are at greater risk of miscarriage than those who stay away from caffeine, according to a Kaiser Permanente study out of Oakland.
>
> The study, published today in the *American Journal of Obstetrics and Gynecology*, is one of the largest to look at the connection between caffeine and miscarriage and the first to interview women about their caffeine habits before they've actually suffered a miscarriage.
>
> "I would probably not even recommend a cup a day, based on this. It's not a huge risk, but it's a real effect," said Dr. Aaron Caughey, a perinatologist as UCSF.
>
> Even now, no one knows for sure how much caffeine is too much. The Kaiser study showed a small and statistically insignificant increase in risk for women who had less than 200 mg of caffeine every day. But that includes a wide group of caffeine

drinkers, doctors noted, and it's possible the women who drink, say, a cup of coffee—with roughly 150 mg of caffeine—increase their risk of miscarriage while those who drink one soda—with 30 to 50 mg of caffeine—do not.

Of interest to us is the juxtaposition of Dr. Caughey's statement with the last paragraph. There was a small, but not statistically significant, increase in risk for women who drink less than 200 mg of caffeine per day. As noted in the article, that is more than a cup of coffee. Still yet, Dr. Caughey is careful to say that he would not even recommend a cup a day. While on the one hand this is just rational, conservative reasoning on the part of the physician, it does illustrate one other point of importance to us:

 Having statistically insignificant results may not mean the results are clinically insignificant.

Perhaps Dr. Caughey was simply expressing what many of us might feel. It is just not worth taking the risk of having a miscarriage, even if the results of the study suggested that less than 200 mg of caffeine was, in a manner of speaking, not found to be "statistically unsafe." He was obviously still worried about the practical impact on his patients of even one cup of coffee a day.

Structured Assessment of Practical Significance

Minimal Clinically Important Difference

Of course, not all assessments of practical significance are ad hoc discussions about how much of a difference in treatments is enough of a difference. In particular, it is becoming more common to see clinical experiments employ something called the "minimal clinically important difference" (MCID) as a way of quantifying practical significance. MCID can be defined as follows:

The **minimal clinically important difference (MCID)** is the smallest change in a treatment outcome that would result in a way the patient was managed, given its side effects, costs, and inconveniences.

Estimating MCID values typically requires an extensive amount of research, and they are usually going to be very specific to the type of study being performed, the patient population being addressed, and the types of instruments used to measure outcomes of interest such as pain, range of motion, or perceived patient comfort. So getting this sort of more structured assessment of practical significance comes at a cost. Let's look at some examples of areas where it has been possible to estimate MCID values.

Example: Coronary Disease Rehabilitation

Title: Determining the minimal clinically important difference for the six-minute walk test and the 200-meter fast-walk test during cardiac rehabilitation program in coronary artery disease patients after acute coronary syndrome.

Authors: Gremeaux; Troisgros; Benaim, S; Hannequin; Laurent; Casillas; and Benaim, C.

Source: Arch Phys Med Rehabil, 2011.

Coronary artery disease patients who have suffered acute coronary syndrome (a category of heart issues including heart attack) often end up in a kind of cardio rehabilitation. One of two common tasks to complete is the six-minute walk test (6MWT) designed to see how far the patient can walk in six minutes. To evaluate the practical effectiveness of the rehabilitation, professionals in this area have turned to MCID values. One study was done on 81 patients and used both patients and physiotherapists to provide ratings of perceived change in walking ability. This led to a recommendation of an MCID of 25 meters for the 6MWT.

The result reported in this research study was very specific to the type of disease, the type of rehabilitation, and the way in which the progress was being evaluated. The authors concluded that "this result will help physicians interpret 6MWT change and help researchers estimate sample sizes in further studies using 6MWT as an endpoint. " We'll have more to say about the last part of this sentence at the end of this chapter.

Example: Orthopaedics

Title: Minimal clinically important differences for American Orthopaedic Foot and Ankle Society Score in Hallux Valgus Surgery.

Authors: Chan; Chen; Zainul-Abidin; Ying; Koo; Rikhrai

Source: Foot Ankle Int., 2017.

This study followed 446 patients with a relatively common condition of the foot and toes known as Hallux Valgus. The researcher looked at an outcome scale known as the American Orthopaedic Foot and Ankle Society score, which can be as high as 100, indicating no symptoms or impairments. This is a physician-based score and not a patient-based outcome score. The smallest difference that physicians suggested patients would perceive as beneficial on this scale was reported as the rather wide range from 7.9 to 30.2.

Example: Oral Cancer

Title: Minimal clinically important differences in quality of life scores of oral cavity and oropharynx cancer patients.

Authors: Binenbaum; Amit; Billan; Cohen; Gill

Source: Ann Surg Oncol, 2014.

This study addressed 1,011 patients with the goal of developing an MCID value regarding quality of life. In part, this research looked at three quality of life evaluation tools, including the University of Washington Quality of Life Questionnaire (UW-QOLQ). The UW-QOLQ is scored on a scale of 0 to 100. The smallest difference that patients perceived as beneficial on this quality of life scale was around 13 points. The researchers concluded that this level of change tended to happen after about one year post treatment.

How are these kinds of values used in practice? Look back at the oral cancer example. If researchers were testing a hypothesis regarding a new intervention that was designed to increase quality of life for oral cancer patients, it would not be enough to just report whether that intervention was statistically significantly better than, say, the current and standard intervention. With a large enough number of patients, a difference of only a few points on the UW-QOLQ might well show up as statistically significant. However, the article above suggests that any change less than 13 points on that quality of life scale would not be judged practically meaningful to the patient.

While none of us are likely to be in the position of designing a clinical trial, we do need to be able to assess the import of hypothesis testing results. In some cases, knowing what an MCID value is and how it affects the larger interpretation of those testing results will help with that assessment. At the least, the three examples we've just seen give us some idea of how complex the task of quantifying practical significance can be.

The Idea of Effect Size

Well before there were discussions of MCID values, practitioners were reporting something called effect sizes. While a full discussion of effect size is beyond the scope of this book, a brief discussion will be useful. Suppose we are confronted with a simple two-mean hypothesis such as the one shown here:

$$H_0: \mu_{\text{Treatment 1}} = \mu_{\text{Treatment 2}}$$

$$H_A: \mu_{\text{Treatment 1}} \neq \mu_{\text{Treatment 2}}$$

Regardless of how the hypothesis testing turns out (whether we did it or someone else did), it makes sense to quantify just how far apart, on average, the actual observed treatment values are. One obvious way of doing this is to report the difference in the sample means collected for each treatment. A slightly more sophisticated way would be to then scale that difference by an estimate of the standard deviations from the two treatment groups, which, for simplicity, we will assume to be very similar. This is one version of what is formally known as effect size, and it is commonly known as *Cohen's d*.

Cohen's d (associated with a two-mean hypothesis of the form shown above) is given by:

$$\text{Cohen's d} = \frac{\bar{x}_1 - \bar{x}_2}{s_p}, \text{ where}$$

$$s_p = \sqrt{\frac{(n_1 - 1)s_1^2 + (n_2 - 1)s_2^2}{n_1 + n_2 - 1}}$$

and

- $\bar{x}_1$ is the average of the Treatment 1 values
- $\bar{x}_2$ is the average of the Treatment 2 values
- s_1 is the standard deviation of the Treatment 1 values
- s_1^2 is the variance of the Treatment 1 values
- s_2 is the standard deviation of the Treatment 2 values
- s_2^2 is the variance of the Treatment 2 values
- n_1 is the number of Treatment 1 values
- n_2 is the number of Treatment 2 values

Effect size, defined this way, has an obvious interpretation. Since Cohen's d is the difference in treatment means divided by a pooled estimate of a common standard deviation, it measures how many of those standard deviations apart the treatments means are. How many is "enough" to be practically significant is a subjective assessment to be sure. Still yet, this kind of assessment allows for some structure to be brought to an otherwise unstructured problem.

Example: Sleep and Pain

Title: Duration of Sleep Contributes to Next-day Pain Report in the General Population

Author: R. Edwards, et al.

Source: *Pain 137* (2008) 202–207.

There is a lot of interest in how disturbed sleep is related to pain perception. Throughout the study referenced, participants from the general population recorded both the number of hours they slept and the frequency of their pain symptoms. The pain symptoms were recorded on a five-point scale: 0 = none of the time, 1 = a little, 2 = some, 3 = most, and 4 = all of the time. A test of hypothesis was performed to compare the pain ratings for persons who slept 5 hours with those who slept 8 hours. The sample means and standard deviations are shown in Table 11.1.

$H_0: \mu_{(5\text{-hour})} = \mu_{(8\text{-hour})}$

$H_A: \mu_{(5\text{-hour})} \neq \mu_{(8\text{-hour})}$

TABLE 11.1 Pain Relief Results

Sleep (Hours)	Average Pain Rating	Standard Deviation	Sample Size
5	0.94	1.29	434
8	0.75	1.13	1,557

The results can be shown to be statistically significant with a *p*-value of 0.0027. So it is safe to reject the null hypothesis in favor of the alternative. The effect size in this case is estimated to be the following:

Cohen's d $= \dfrac{0.94 - 0.75}{1.16} = 0.164$, since

$$s_p = \sqrt{\dfrac{(434-1)(1.29)^2 + (1557-1)(1.13)^2}{434 + 1557 - 1}} = 1.16$$

Therefore, the observed differences in the means amounted to only about 0.16 of a standard deviation. That's not very much. What's more, both of the mean values are in the same category on the rating scale between "none of the time" and "a little." This may temper one's excitement about the difference being statistically significant.

From Effect Size to Sample Size

When an experiment is being planned, it is more common, and some would argue more legitimate, to have some estimate of effect size *prior* to any data being collected and then to use that to estimate how many subjects one needs to recruit in order to be able to detect an effect at that level as statistically significant. This gets complicated very quickly, not so much because of the computations, but more because of the complexity of the underlying justifications for such an approach and a genuine debate over the usefulness of what you end up with when all is said and done. We are going to stay well away from those controversies and just briefly outline how Cohen's d is used in this way. Here's what we will need:

- An a priori measure of Cohen's d. Remember no data have been collected.

- An agreement on the Type I error rate that will be used in the subsequent hypothesis test. This is typically taken to be 0.05 as usual.

- An agreement on the Type II error rate that is tolerable. That is, an estimate of how likely it will be in the subsequent hypothesis test to fail to select H_A as likely the truth, when that would have been the right choice (a false negative). This is typically taken to be a rather liberal value of 0.20.

Often, when experiments are being planned, researchers will use the subjective classifications of effect size that Cohen established in the 1980s. See Table 11.2.

TABLE 11.2

Cohen's Thresholds for d			
Small	**Medium**	**Large**	**Very Large**
0.20	0.50	0.80	1.30

A difference in treatment averages that is one-half of a standard deviation would be considered a "medium" effect size to Cohen, whereas a difference that was 1.3 standard deviations would be considered "very large." Obviously, the smaller the desired effect size one wants to have a chance at detecting, the larger the sample sizes will be. Another way to think about this though is that if you are only practically interested in effect sizes that are at least "medium" then you'd like to have a way to set up your hypothesis testing paradigm so as to show up as statistically significant only if it also produces results that you would find practically significant. No reason to waste money on subject recruitment beyond what you'd need for that. To be fair, this is not likely the way that most researchers would think about Cohen's d out there in the real world of experimentation and clinical trials. Statistical significance is still the goal, for good or bad. But there is little to gain with spending money on hundreds of subjects for a study that finds a very small effect as statistically significant, but is too small to make any real difference in patients' lives.

Our final task in this chapter is much simpler. We need to make sure we can use any one of a number of free applets to estimate appropriate sample sizes if we start with Cohen's d value. The following examples illustrate how this can be done.

Example: Cohen's Sample Size Computation

Suppose a scientist at a major research institution is studying a new cannabis-based pain medication. The research plan calls for the recruitment of patients who will ultimately be randomly split into two groups. Group 1 will be given a placebo, and after one hour, group members will rate the effectiveness of the pain relief on a scale of 1 to 100. Similarly, Group 2 will be given the new cannabis-based drug, and after an hour, those group members will rate the pain relief effectiveness. The researcher wants to test the following:

$H_0: \mu_{Treatment} = \mu_{Placebo}$

$H_A: \mu_{Treatment} \neq \mu_{Placebo}$

The first order of business is to decide how many subjects need to be recruited. Let's suppose the researcher has decided on the following:

- The hypothesis will be tested at an alpha level of 0.05.

- The relevant effect size that she hopes to detect is medium. That is, she wants to be able to detect a difference in means that is around one-half the common standard deviation. So she has adopted a Cohen's d of 0.5.

- She is willing to risk making a Type II error (failing to say H_A is true when it really is) up to two times in 10. That is, she is willing to have a Type II error rate as high as 0.20.

The appropriate treatment group sizes can then be determined by any number of online sample size calculators. Search for "sample size calculators for two-sample tests for means." Be prepared to be witty in how you use them. For example, here are two that are readily available:

- If we use https://www.danielsoper.com/ statcalc/calculator.aspx?id=47, then we just have the obvious fields to fill out, then hit calculate. Notice it is estimated that the researcher will need to have about 64 subjects in each of her two groups (treatment and placebo). The Type II error rate is 1 – power level, so enter a 0.80 for power.

Please enter the necessary parameter values, and then click 'Calculate'.

Anticipated effect size (Cohen's *d*): 0.5
Desired statistical power level: 0.8
Probability level: 0.05

Calculate!

Minimum total sample size (one-tailed hypothesis): **102**
Minimum sample size per group (one-tailed hypothesis): **51**
Minimum total sample size (two-tailed hypothesis): **128**
Minimum sample size per group (two-tailed hypothesis): **64**

- If we use https://www.stat.ubc.ca/~rollin/stats/ssize/n2.html, then we just need to enter *any* "mu1," "mu2," and "sigma" values that will produce a Cohen's d ratio of 0.50. So we can enter 2, 1, and 2 for those values (respectively) and hit calculate. We will get the following. Again, notice it is estimated that the researcher will need to have about 63 subjects in each of her two groups (treatment and placebo).

- ● Calculate Sample Size (for specified Power)
- ○ Calculate Power (for specified Sample Size)

Enter a value for mu1: 2
Enter a value for mu2: 1
Enter a value for sigma: 2

- ○ 1 Sided Test
- ● 2 Sided Test

Enter a value for α (default is .05): .05
Enter a value for desired power (default is .80): .80
The sample size (for each sample separately) is: 63

Calculate

It is also becoming common for research studies that are lucky enough to have access to MCID values for their area of experimentation to start with those and work backwards to sample size computations in a way that is similar to what we have just illustrated for Cohen's d. We will not be concerned with that in this book, however.

TAKE-HOME POINTS

- Statistical significance is a mathematical way of assessing whether treatment differences are big enough to be considered unlikely to have happened by chance; whereas practical significance addresses whether the observed difference is big enough that it is practically worth caring about.

- Minimally clinically important differences and Cohen's d are two ways that practical significance are assessed.

- Cohen's d can be used when designing an experiment, prior to any data being collected, to estimate appropriate numbers of subjects needed.

Chapter 11 Exercises

Reading Check

Short Answer: Please provide brief, concise answers to the following questions.

1. Distinguish practical or clinical significance from statistical significance.

2. Refer back to the multivitamins and cancer study. Suppose, for convenience, that 7,500 of the older male doctors followed were taking vitamins, and the other 7,500 were taking dummy pills. If there was an 8% reduction in total cancers between the vitamin group and the dummy pill group, what was the reduction in the average number of cancers between the two groups? Your answer won't be a number but will be in terms of a variable.

3. Recall the mental illness and obesity study. Suppose the average weight of subjects in a test group was 180 pounds before the group participated in a health promotion program. If the results were to be considered practically significant (on the average), what would the average weight of the test group subjects have to be when the health promotion program completed?

4. Regarding the article on coffee and pregnancy, in what sense can we say that the Kaiser study found 200 mg or less of coffee to be "statistically safe" to consume during pregnancy? What is the problem with this kind of conclusion?

5. If you want to use Cohen's d as an ad hoc quantification of effect size for the results of a two-mean test of hypothesis, what do you need to have at your disposal?

Beyond the Numbers

1. Effect Size Matters

From Time Health

While researching a drug treatment for depression, German drugmaker Boehringer Ingelheim (BI) GmbH, claimed to have stumbled upon a libido-enhancing treatment for women. The drug, flibanserin, apparently did not lift women's moods, but during the study for depression, participants taking flibanserin reported higher sexual appetites on measures of well-being. This prompted the company to conduct clinical trials specifically investigating flibanserin's usefulness in treating hypoactive sexual desire disorder (that is, low libido). The flibanserin findings are based on the study of 1,378 women who had been in a monogamous relationship for 10 years on average. The women were randomly assigned to take 100 mg of flibanserin or a placebo daily to record daily whether they had sex and whether it was satisfying.

Women in the flibanserin group self-reported an average of 2.8 sexually satisfying events in the four-week baseline period. In the final four weeks of the 24-week study period, those women reported an average of 4.5 sexually satisfying events, a more than 50% increase. Women in the placebo group reported an average increase from 2.7 events to 3.7.

The difference between flibanserin and the placebo— reported as about 0.8 sexually satisfying events—was found to be statistically significant, the drug company said. The side effects from the drug, which included dizziness and fatigue, among others, were mild to moderate and transient.

Read more: http://content.time.com/time/health/article/0,8599,1939884,00.html

a. Where does the statement "about 0.8 sexually satisfying events" come from?

   i. It's the difference between the 4.5 and the 3.7.

   ii. It's the difference between the 3.5 and the 2.7.

   iii. It's the result of computing Cohen's d for the full data set (which we don't have).

   iv. It's the difference between the 4.5 and the 2.8 and averaged over the two groups.

b. We don't know the variance of the measurements in the flibanserin group, but let's assume it is 2.3; similarly, we will assume that the variance in the placebo group was 1.5. Likewise, we don't know exactly how the 1,378 women were divided, but let's assume 700 were in the flibanserin group and 678 were in the placebo group. If 0.8 is the difference in the average number of sexually satisfying events between the flibanserin and the placebo group, what is the value for Cohen's d in this study?

   i. 0.80

   ii. 1.58

   iii. 0.58

   iv. 0.38

c. What was the difference in the change from baseline for "number of sexually satisfying events" between flibanserin and the placebo? Support a claim that this would have been more appropriate for the article to have reported on.

2. The Economics of No Significance

Title: US Supreme Court: Statistical Significance Not Needed in Drug Lawsuits

Author: Heidi Ledford

Source: Nature.com, March 23, 2011. http://blogs.nature.com/news/2011/03/us_supreme_court_statistical_s.html

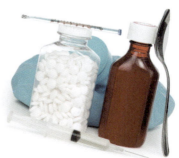

In a unanimous decision, the US Supreme Court ruled yesterday that a pharmaceutical company may be required to notify investors of safety reports regarding its products, even if those reports do not rise to the level of statistical significance.

Investors sued the company, based in Scottsdale, Arizona, arguing that it should have notified them earlier about reports that some of its popular zinc gluconate cold medications may have robbed some users of their sense of smell.

When news of the possible link finally became public, Matrixx stock plummeted. Matrixx tried to shoot down the lawsuit by arguing that the adverse event reports it received about Zicam were not statistically significant. But Judge Sonia Sotomayor, writing for the court, said that test would be too stringent. "Both medical experts and the Food and Drug Administration (FDA) rely on evidence other than statistically significant data to establish an inference of causation," she wrote. "It thus stands to reason that reasonable investors would act on such evidence."

a. This is clearly a case wherein results that were not statistically significant were judged by the Supreme Court to be practically significant. What practical significance was of primary interest *to the court* in this article?

   i. At issue is the practical importance of users who risked losing their sense of smell after having used Zicam, even if those numbers were too small to be statistically significant.

   ii. At issue is the practical importance of investors who stood to lose large amounts of money over unreported safety issues, even if those numbers were too small to be statistically significant.

   iii. At issue is the practical importance of users who risked losing their sense of smell after having used Zicam, even though those numbers were big enough to be statistically significant.

   iv. At issue is the practical importance of investors who stood to lose large amounts of money over unreported safety issues, even though those numbers were big enough to be statistically significant.

b. Access the article by searching on the title. In the article, Matrixx's attorneys argued that companies receive "many such reports" of the existence of adverse events and that they "vary widely in quality." Hence, Matrixx argued that making all of them pubic would unnecessarily frighten consumers. What was Judge Sotomayor's response to this argument? What did her response have to do with statistical significance and practical significance?

3. The *N* Crowd

Suppose a scientist at a major research institution is studying a new cannabis-based pain medication. The research plan calls for the recruitment of patients who will ultimately be randomly split into two groups. Group 1 will be given a placebo, and after one hour, group members will rate the effectiveness of the pain relief  on a scale of 1 to 100. Similarly, Group 2 will be given the new cannabis-based drug, and after an hour, those group members will rate the pain relief effectiveness. The researcher wants to test the following:

$$H_0: \mu_{\text{Treatment}} = \mu_{\text{Placebo}}$$

$$H_A: \mu_{\text{Treatment}} \neq \mu_{\text{Placebo}}$$

Initially, only 25 volunteers were available for each group. By the end of the month, however, there were 25,000 volunteers in each group, making this the largest (completely made up) clinical trial in history. Group treatment means and standard deviations are shown in the table to the right, for each week, as the number of participants increased. Both Cohen's d for a *p*-value are computed for the data from the first week. You might want to make sure you can repeat those.

TABLE 11.3

	1st Week	2nd Week	3rd Week	4th Week
Sample size in each group (*N*)	25	250	2,500	25,000
Standard deviation in each group	20	20	20	20
Treatment 1 mean (placebo)	500	500	500	500
Treatment 2 mean (new drug)	501	501	501	501
Cohen's d	0.05			
p-value	0.8604			

a. Compute Cohen's d for each of the final three weeks. Which of the following graphs is correct for a plot of Cohen's d over all four weeks?

i.
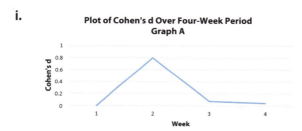

iii.
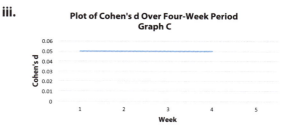

ii.
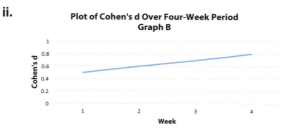

iv.
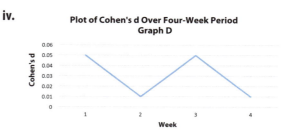

b. Compute the *p*-value for testing the two-sided hypothesis shown for each of the final three weeks. Unless your instructor tells you otherwise, you may use the convenient GRAPHPAD® applet to complete your calculations. This applet is available at http://www.graphpad.com/quickcalcs/ttest1/?Format=SD. Which of the following graphs is correct for a plot of the *p*-values over all four weeks?

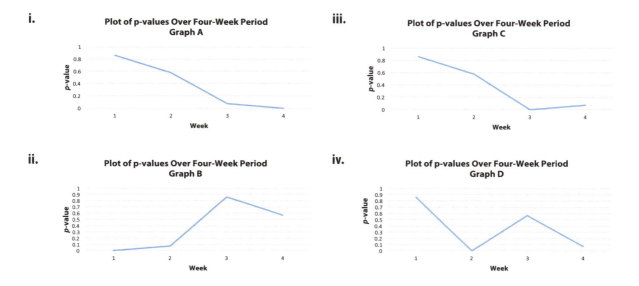

i. Plot of p-values Over Four-Week Period — Graph A

iii. Plot of p-values Over Four-Week Period — Graph C

ii. Plot of p-values Over Four-Week Period — Graph B

iv. Plot of p-values Over Four-Week Period — Graph D

c. What does this exercise have to say about the relationship between sample size and statistical significance? Be very specific in your answer.

d. What are the possible challenges to correct interpretation of the results from a study that had enormous sample sizes? Explain your reasoning.

4. Multivitamins and Cancer

The following *New York Times* article on multivitamins appeared on October 17th, 2012. The article addresses a clinical trial that unfolded over more than a decade. Male doctors were followed during that period of time, and those taking a daily multivitamin were compared to those who took a placebo.

Title: Multivitamin Use Linked to Lowered Cancer Risk

Author: Roni Caryn Rabin

Source: *New York Times*

After a series of conflicting reports about whether vitamin pills can stave off chronic disease, researchers announced on Wednesday that a large clinical trial of nearly 15,000 older male doctors followed for more than a decade found that those taking a daily multivitamin experienced 8% fewer cancers than the subjects taking dummy pills.

> The findings were to be presented Wednesday at an American Association for Cancer Research conference on cancer prevention in Anaheim, Calif., and the paper was published online in the *Journal of the American Medical Association*.
>
> The reduction in total cancers was small but statistically significant, said the study's lead author, Dr. J. Michael Gaziano, a cardiologist at Brigham and Women's Hospital and the VA Boston Healthcare System. While the main reason to take a multivitamin is to prevent nutritional deficiencies, Dr. Gaziano said, "It certainly appears there is a modest reduction in the risk of cancer from a typical multivitamin."

We aren't told how many doctors were in each group, but let's assume for sake of this exercise that there were 7,500 doctors in the vitamin group and 7,500 in the placebo group.

a. The results say that there were 8% fewer TOTAL cancers in the vitamin group than in the placebo group. If there were 1000 total cancers reported in the placebo group, what was the difference in the *average* number of cancers reported between the two groups?

   i. About 0.50

   ii. About 80

   iii. About 1500

   iv. About 0.01

b. Suppose you could plan this study all over again. Given the seriousness of cancer, you decide you want to be able to detect a small effect size (Cohen's d), with a Type I error rate of 0.05 and a Type II error rate of 0.20. What percentage of the 7500 that were studied in each group would actually have been needed? Assume you were conducting a two-sided test.

   i. About 5%

   ii. About 50%

   iii. About 0.5%

   iv. About 500%

5. A Measure of Effect for Proportions

We only discussed Cohen's d in the context of a test of hypothesis about two means. What if, instead, you had tested a hypothesis about two proportions (below)? This exercise will walk you through one (of many) ad hoc measures of "effect" that is used in that specific context.

$H_0: p_1 = p_2$

$H_A: p_1 \neq p_2$

Let's revisit the Vytorin study that appeared earlier in the book.

In a 2008 article, author Alex Berenson found that of 950 patients assigned to take the prescription drug Vytorin, 102 developed cancer. Another 920 were assigned to take a placebo and 67 in that group developed cancer. So the sample proportion developing cancer in the Vytorin group was $\hat{p}_{\text{Vytorin}} = 0.11$, and the similar sample proportion in the placebo group was $\hat{p}_{\text{Placebo}} = 0.07$.

a. What is a reasonable interpretation of the ratio $\dfrac{\hat{p}_{\text{Vytorin}}}{1 - \hat{p}_{\text{Vytorin}}}$?

   i.   It's the odds of developing cancer in the Vytorin group.

   ii. It's the odds of not developing cancer in the Vytorin group.

   iii. It's the proportion of subjects who developed cancer in the Vytorin group.

   iv. It's the proportion of subjects who did not develop cancer in the Vytorin group.

b. What is the value of the ratio $\dfrac{\frac{\hat{p}_{\text{Vytorin}}}{1 - \hat{p}_{\text{Vytorin}}}}{\frac{\hat{p}_{\text{Placebo}}}{1 - \hat{p}_{\text{Placebo}}}}$?

   i.   0.12

   ii. 1.64

   iii. 0.65

   iv. 0.08

c. The ratio $\dfrac{\left[\frac{\hat{p}_1}{1 - \hat{p}_1}\right]}{\left[\frac{\hat{p}_2}{1 - \hat{p}_2}\right]}$ will sometimes be used as a kind of "effect size" measure for a hypothesis of the form $H_0: p_1 = p_2$ versus $H_A: p_1 \neq p_2$. Offer a careful explanation of what this ratio measures and in what sense it can be said to measure "effect."

Projects

1. Effect Size and Error Rates

Introduction

This mini-project is appropriate as an in-class activity or a homework assignment. It can be configured either as an individual assignment or a group assignment. Follow your instructor's instructions. The goal of the activity is to better understand how sample size calculations based around Cohen's d value vary with Type I and Type II error rates.

We are going to focus only on the two-mean hypothesis shown below:

$H_0: \mu_1 = \mu_2$

$H_A: \mu_1 \neq \mu_2$

Collecting Your Data

Find the sample sizes recommended to detect effect sizes (Cohen's d) of 0.20 and 0.50, with the varying Type I and Type II error rates shown. We are assuming that the sample sizes for groups 1 and 2 are the same, so you only need to report that common size. One entry is completed in each table for you. Make sure you can complete those entries before proceeding further.

TABLE 11.4 Sample Sizes for $d = 0.50$

Type II Error Rate	Type I Error Rate			
	0.05	0.10	0.20	0.30
0.05				
0.10				
0.20	64			
0.30				

TABLE 11.5 Sample Sizes for $d = 0.20$

Type II Error Rate	Type I Error Rate			
	0.05	0.10	0.20	0.30
0.05				
0.10				
0.20	394			
0.30				

Displaying Your Data

You should construct the following four plots on the SAME set of axes. Make sure you use software to create the plots and that all axes and plots are clearly labeled and distinguished.

- For a Cohen's d of 0.50, plot sample size (y-axis) versus Type I error rate (x-axis).
- For a Cohen's d of 0.50, plot sample size (y-axis) versus Type II error rate (x-axis).
- For a Cohen's d of 0.20, plot sample size (y-axis) versus Type I error rate (x-axis).
- For a Cohen's d of 0.20, plot sample size (y-axis) versus Type II error rate (x-axis).

Interpret What You Have Found

Look at what your plots show you. Give detailed explanations of the following:

a. How does sample size vary with Type I error rate? Is this reasonable or not? Explain.

b. How does sample size vary with Type II error rate? Is this reasonable or not? Explain.

c. How does sample size vary with the value of Cohen's d? Is this reasonable or not? Explain.

2. Designing Your Own Experiment: Part IV

Introduction

This project is a natural progression from Part I, introduced in Chapter 3, Part II, from Chapter 5, and Part III from Chapter 10.

The scope of what is required depends on whether you are being asked to do this individually or as part of a group. Your instructor will clarify for your class.

From Parts I, II, and III

In Part I you did the following:

Individual Project

▸ Randomized subjects into two groups formed from a single change in a single variable. You had a minimum of 50 subjects.

▸ Applied the treatments in each group and collected the data

▸ Collected the response data for each subject

Group Project

▸ Randomized subjects into two groups formed from a single change in a single variable. You had a minimum of 150 subjects.

▸ Applied the treatments in each group and collected the data

▸ Collected the response data for each subject

In Part II, you followed up by constructing confidence intervals for each treatment, which you interpreted; you also offered some comments on whether you thought the treatments really were different based on what you saw in the intervals.

In Part III you tested one of the following hypothesis. It depended on whether you collected data commensurate with proportions or means.

$$H_0: \mu_1 = \mu_2 \qquad H_0: p_1 = p_2$$
$$\text{or}$$
$$H_A: \mu_1 \neq \mu_2 \qquad H_A: p_1 \neq p_2$$

New to Part IV

In this installment you are required to summarize the effect size you found in Part III. How you do this will depend on what you tested.

■ If you tested the hypothesis about the means (on the left), then you will have direct access to the necessary information for computing Cohen's d as a measure of effect size.

■ If you tested the hypothesis about proportions (on the right), then you will need to make sure you understand Exercise 5, above, and then compute the ratio $\dfrac{\left[\dfrac{\hat{p}_1}{1-\hat{p}_1}\right]}{\left[\dfrac{\hat{p}_2}{1-\hat{p}_2}\right]}$ as a measure of

effect size.

This final installment of your project report should be professionally prepared, typed, and proofread. Your instructor may require that it be added to Parts I, II, and III. Follow her instructions. Your task is to exhibit the following to your readers:

- The value you have computed for your effect size.
- A complete interpretation of what that effect size value means in the context of your specific experiment.

3. Reality Video

Introduction

This module explained the difference between practical significance and statistical significance and discussed MCID values and Cohen's d as methods that have been used to quantify practical, or clinical significance. Now, we want to give you a chance to both show you know what you are doing and to help teach others.

The Assignment

Your instructor may have you do this as individuals or may form you into teams. We'll refer to "teams" below, although you may be a team of one depending on what your instructor wants. Each team is required to do the following:

a. Pick one or more topics from this chapter. Your instructor may want to approve them before you proceed.

b. Construct a three-minute video that teaches the chosen topic to the class. You must carefully and correctly use the language of the chapter in your presentation, and you are required to give an example that is NOT in the chapter.

c. Post the video on YouTube and send the link to your instructor in the manner he stipulates.

12

CHAPTER 12
More than One Variable: Association and Correlation

Introduction

"Correlation" is a word you are probably already familiar with. We always use it in the context of measurements on at least two variables. For example, we might all say we know that height and weight are correlated, gender and pay scale are correlated, and how much you exercise is correlated with your overall level of health. These are all sensible uses of the word in the vernacular. This chapter will help us learn why we need to be more careful how we use the word. The goal is more than pedantic. Our human inferences can be all too easily led astray in the absence of a broad understanding of how to assess the broader idea of *association*.

Acquiring some simple language at this point is useful.

> Two variables are **associated** if certain values of one of the variables tends to be paired with certain values of the other variable.

Measurement Levels

The way we think about association, how we know when we need to call it that and not correlation instead, and how we quantify the degree of association are all more easily clarified if we understand that data can basically be measured on three different levels.

> **Nominal** data are data measurements that can't be located on a number line and ordered in an unambiguous way.

For example, gender is most naturally measured as male, female, or non-identifying. These are not the kinds of measurements that can be located on a number line and ordered the way numbers can be ordered. Similarly, if a survey collects information on whether a respondent ate breakfast or not (Yes or No), those responses are not numbers with an unambiguous location on a number line.

> **Ordinal** data are data measurements that can be located on a number line and ordered in an unambiguous way, but there is no fixed unit of measurement associated with the numbers.

Suppose a survey asks you to rate your satisfaction on a scale of 1 to 10 with the most recent iPhone app you downloaded. If you rate the app with a 3, you are not communicating any obvious units of satisfaction. While it is clear that this is a better rating than a 1, it would not be sensible to say it was a rating that was three times as good. At least there is no unambiguous way of making that kind of statement.

> **Ratio** data are data measurements that have a fixed unit of measurement where one can sensibly interpret the ratio of two different measurements.

Ratio data are just regular numerical measurements such as height, weight, GPA, yearly income, number of tumors present, number score on your first statistics exam, etc. If you weigh 100 lbs. and your little brother weighs 50 lbs., then not only is it clear you weigh more than your little brother, it is quite unambiguous to say that you weigh twice as much.

There is a fourth measurement level that legitimately falls between ordinal and ratio, but it adds very little to a practical discussion of association, so we have not mentioned it.

We are going to start our more formal discussion of association with ratio data since it will take us to places that are likely the most familiar. We are going to limit our entire discussion to variable pairs that are either each ratio level or each nominal level. We aren't going to discuss ordinal data or what to do when levels are mixed within variable pairs.

Association for Ratio Data

Graphical Representation of Association

Any association between two variables that are ratio level—like height and weight—can often easily be seen in a *scatterplot*. Likely, you saw scatterplots (may have called them "*x-y* plots") a long time ago, probably in middle school. A scatterplot is easy to construct and we will remind you how in the next example. However, if you find that you don't remember anything about them at all, it would be a good idea to review. Constructing the plot is not a college-level task, but understanding what the plot tells us about association is dependent on being able to construct such a plot.

Example: DTP and Autism

There is an on-going debate regarding possible links between vaccines containing thimerosal and the onset of autism. The data set shown here lists the percentages of California children who received 4 doses of DTP by their second birthday and the number of autism cases logged in California's Department of Developmental Services' regional service center system.

Just below we have constructed a scatterplot with DPT Coverage on the horizontal ("x") axis and Number of Autism Cases on the vertical ("y") axis. Each point ("dot") on the plot corresponds to a pair of these values. For instance, to plot the first point in the table, we have to locate 50.9 along the horizontal axis and move up 176 along the vertical axis. Chances are, this is all very obvious to you. As long as both numbers in the data pairs are ratio data, it is possible to unambiguously locate them on the horizontal and vertical axes, and, hence, unambiguously locate the point on the plot.

TABLE 12.1

Year	DTP Coverage (%)	Number of Autism Cases
1980	50.9	176
1981	55.4	201
1982	52.1	212
1983	47.7	229
1984	48.9	246
1985	54.3	293
1986	54.1	357
1987	55.3	347
1988	60.9	436
1989	62.2	522
1990	65.9	663
1991	67.3	823
1992	69.8	1042
1993	73.6	1090
1994	75.7	1182

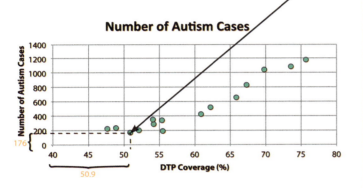

We are now ready for more language.

A scatterplot exhibits a **positive association** between the two variables being plotted if the points in that plot have an upward trend as we look from left to right.

A scatterplot exhibits a **negative association** between the two variables being plotted if the points have a downward trend as we look from left to right.

If there is no trend at all in the scatterplot, then it is said to exhibit **no association.**

If a scatterplot exhibits a positive or negative trend, the **strength of the association** is visually assessed by how tightly clustered the points are about the trend.

Examples

- **DTP and Autism**—The scatterplot shown above shows a positive association that appears to be quite strong.
- **Highway MPG and Carbon Footprint**—Sixty-nine 2015 cars were compared on two measurements: their highway miles-per-gallon and their carbon footprint. The data are in the table below, following the scatterplot. The scatterplot is downwards to the right, suggesting that cars with higher highway mpg, have a lower carbon footprint. That's not surprising. The association is negative and appears to be relatively strong.

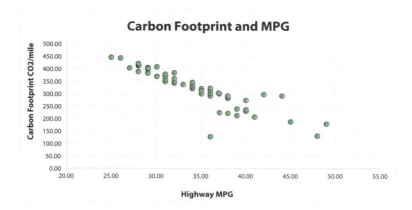

TABLE 12.2

HWY MPG	CO₂/Mile	HWY MPG	CO₂/Mile	HWY MPG	CO₂/Mile	HWY MPG	CO₂/Mile
48.00	133.00	36.00	298.00	32.00	348.00	30.00	408.00
49.00	178.00	35.00	302.00	34.00	346.00	29.00	403.00
45.00	188.00	37.00	303.00	31.00	355.00	29.00	403.00
41.00	209.00	37.00	308.00	31.00	354.00	28.00	413.00
39.00	215.00	37.00	304.00	31.00	349.00	28.00	417.00
37.00	225.00	36.00	310.00	32.00	362.00	28.00	422.00
38.00	224.00	36.00	310.00	32.00	361.00	28.00	417.00
36.00	129.00	36.00	307.00	31.00	362.00	26.00	443.00
36.00	129.00	35.00	317.00	31.00	358.00	25.00	446.00
40.00	237.00	36.00	316.00	30.00	371.00		
40.00	237.00	36.00	311.00	30.00	371.00		
40.00	239.00	35.00	323.00	31.00	369.00		
39.00	242.00	35.00	314.00	31.00	369.00		
44.00	293.00	36.00	322.00	29.00	385.00		
42.00	298.00	36.00	322.00	28.00	390.00		
40.00	275.00	34.00	324.00	32.00	385.00		
38.00	282.00	34.00	330.00	32.00	385.00		
38.00	284.00	32.00	344.00	31.00	379.00		
36.00	292.00	34.00	338.00	29.00	399.00		
38.00	293.00	33.00	338.00	27.00	405.00		

■ **Sodium and Fat in Burgers**—Seven burgers had both their fat and sodium content measured. The data are shown in the accompanying table and graphed in a scatterplot below.

Notice that some burgers with lower fat content (relatively speaking!) have higher sodium content, but there are also some burgers with lower fat content that also have lower sodium content. So there is no tendency to see any particular kind of pairing between the two variables. Evidently, fat content and sodium content are not very strongly associated.

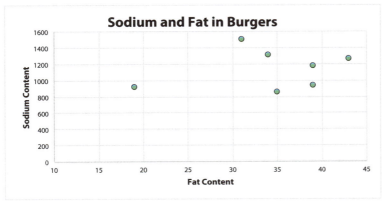

TABLE 12.3

Burger Brand	Fat	Sodium
A	19	920
B	31	1500
C	34	1310
D	35	860
E	39	1180
F	39	940
G	43	1260

■ **Consumer Goods and Life Expectancy**—The number of TVs per person and the average life expectancy were recorded for thirty-eight different countries. The raw data are in Table 12.4, just after a scatterplot of those data. The plot does not follow a straight-line trend. Rather, it is curvilinear. Still yet, the association between Number of TVs/Person and Life Expectancy is positive and appears to be relatively strong.

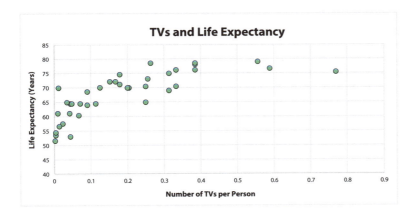

TABLE 12.4

Country	Life Expectancy	TV/Person	Country	Life Expectancy	TV/Person
Argentina	70.5	0.250	Morocco	64.5	0.048
Bangladesh	53.5	0.003	Myanmar(Burma)	54.5	0.002
Brazil	65	0.250	Pakistan	56.5	0.014
Canada	76.5	0.588	Peru	64.5	0.071
China	70	0.125	Philippines	64.5	0.114
Colombia	71	0.179	Poland	73	0.256
Egypt	60.5	0.067	Romania	72	0.167
Ethiopia	51.5	0.002	Russia	69	0.312
France	78	0.385	South Africa	64	0.091
Germany	76	0.385	Spain	78.5	0.385
India	57.5	0.023	Sudan	53	0.043
Indonesia	61	0.042	Taiwan	75	0.312
Iran	64.5	0.043	Thailand	68.5	0.091
Italy	78.5	0.263	Turkey	70	0.200
Japan	79	0.556	Ukraine	70.5	0.333
Kenya	61	0.010	United Kingdom	76	0.333
Korea-North	70	0.011	United States	75.5	0.769
Korea-South	70	0.204	Venezuela	74.5	0.179
Mexico	72	0.152	Vietnam	65	0.034

Numerical Assessment of Association

The scatterplot of DTP Coverage and Number of Autism cases clearly shows a relationship that is positive and fairly strong. But how strong is it? If we want to use a single number to answer that question, the most common such summary is provided by the correlation coefficient, which is almost always denoted by the letter "r."

The complex expression you see below is one way that the correlation coefficient can be calculated. You will probably recall from an earlier mathematics course that the symbol "Σ" means "sum of."

$$\text{Correlation coefficient } r = \frac{\left[n \sum_{i=1}^{n}(x_i y_i) \right] - \left[\sum_{i=1}^{n}(x_i) \right]\left[\sum_{i=1}^{n}(y_i) \right]}{\sqrt{\left[n \sum_{i=1}^{n}(x_i)^2 - \left(\sum_{i=1}^{n}(x_i) \right)^2 \right]\left[n \sum_{i=1}^{n}(y_i)^2 - \left(\sum_{i=1}^{n}(y_i) \right)^2 \right]}}$$

The correlation coefficient is always between –1 and 1. Provided r is computed in an appropriate context, it will always admit the following interpretations:

- r will be negative if the association in the scatterplot is negative.

- r will be positive if the association in the scatterplot is positive.

- The closer r is to 1 or –1 the more tightly packed the points are about the linear trend.

- If r is close to 0 (positive or negative), then the association between the two variables in the scatterplot is likely very weak.

As a measure of association, r is *only* appropriate for ratio data whose scatterplot is roughly linear with no data points that are wildly different from the bulk of the data. It is not immediately appropriate for ratio data with curved scatterplots, scatterplots that have outliers, or non-ratio data.

We will illustrate on a small data set in the next example, but it is not so important anymore to know how to compute these kinds of things by hand.

Example: DTP and Autism Revisited

TABLE 12.5

DTP Coverage "x"	Number of Autism Cases "y"	"xy"	"x²"	"y²"
50.9	176	8958.4	2590.81	30976
55.4	201	11135.4	3069.16	40401
52.1	212	11045.2	2714.41	44944
47.7	229	10923.3	2275.29	52441
48.9	246	12029.4	2391.21	60516
54.3	293	15909.9	2948.49	85849
54.1	357	19313.7	2926.81	127449
55.3	347	19189.1	3058.09	120409
60.9	436	26552.4	3708.81	190096
62.2	522	32468.4	3868.84	272484
65.9	663	43691.7	4342.81	439569
67.3	823	55387.9	4529.29	677329
69.8	1042	72731.6	4872.04	1085764
73.6	1090	80224	5416.96	1188100
75.7	1182	89477.4	5730.49	1397124
$\Sigma x = 894.1$	$\Sigma y = 7819$	$\Sigma xy = 509037.8$	$\Sigma x^2 = 54443.51$	$\Sigma y^2 = 5813451$

$$r = \frac{[15 \times 509037.8] - [894.1][7819]}{\sqrt{[15 \times 54443.51 - (894.1)^2][15 \times 5813451 - (7819)^2]}} = 0.961655$$

Recall, just from looking at the scatterplot of these data, it was clear that the relationship was positive and quite strong. The correlation coefficient quantifies the direction and strength with a value that is positive and very close to 1.

The following tip shows you the command needed to find the correlation coefficient in Excel. The operative function is *correl*. See below. Typically, you'd only calculate *r* with software or an applet of some type.

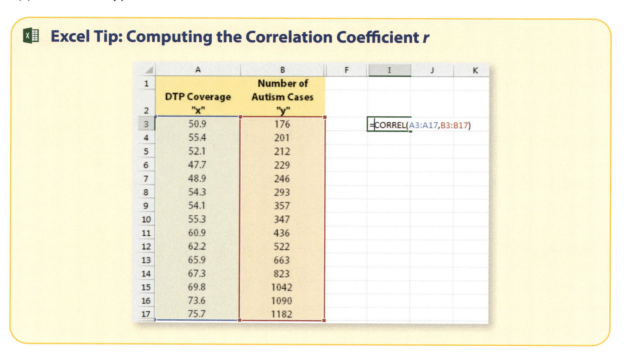

Examples Revisited

Let's look back at the correlation coefficients for the examples we used to start this chapter.

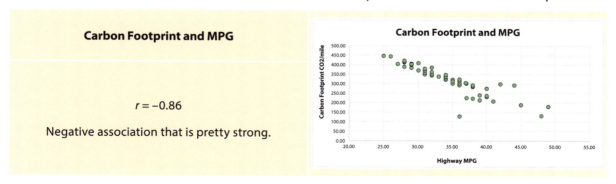

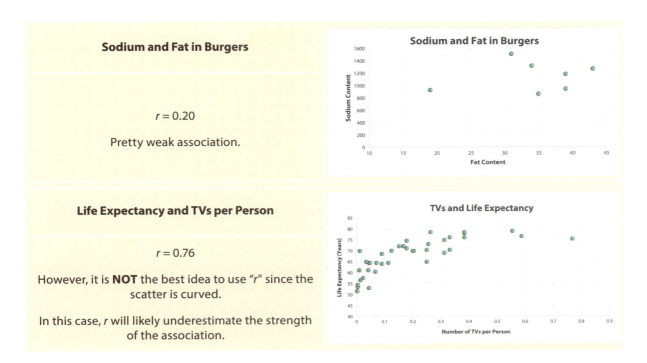

Sodium and Fat in Burgers	Sodium and Fat in Burgers
$r = 0.20$ Pretty weak association.	

Life Expectancy and TVs per Person	TVs and Life Expectancy
$r = 0.76$ However, it is **NOT** the best idea to use "r" since the scatter is curved. In this case, r will likely underestimate the strength of the association.	

Transforming a Curvilinear Scatter

What are your options for quantifying correlation if your scatterplot is curvilinear like the example above? There are actually several. There is an entire area within statistical science known as "regression" that can deal with that situation quite nicely—at least most of the time. However, one can still often get away with just reporting the correlation coefficient providing the data have been transformed first. In fact, there have been entire books written on how to best "straighten out" (linearize) a scatterplot so that the correlation coefficient makes sense. You will have the opportunity to deal with this in more detail in some of the homework problems at the end of this chapter. However, a cursory accounting of what is often done can be provided here.

- **To linearize a curvilinear scatterplot,** it is often effective to take the square root or logarithm of one or both of the variables, provided that is possible.

Example—TVs and Life Expectancy

If we plot Life Expectancy versus the Square Root of the Number of TVs per Person, notice how the slight bend that was in the original scatterplot above—toward the left, in that plot—has been notably diminished. In fact, the following plot exhibits enough of a straight-line trend that summarizing the association with the correlation coefficient probably makes sense. When r is recomputed for the transformed data, we see that r has changed from the original value of 0.76 to 0.85. That's not a huge increase, but it is consistent with our intuition that r was probably underrepresenting the association in the original data.

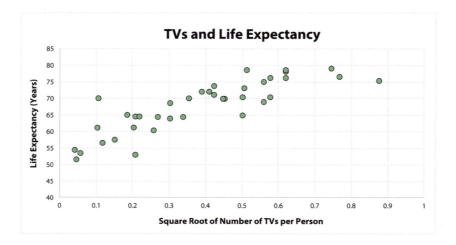

The following tip shows you the command needed to find the square root in Excel. The operative function is *sqrt*. See below. As with any spreadsheet, you only need to enter this value in one cell; then you can drag that entry down to do all the other computations automatically.

Excel Tip: Transforming with a Square Root

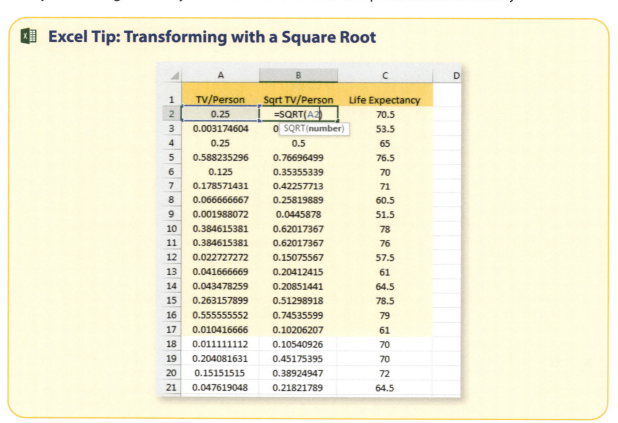

Association for Nominal Data

Graphical Representation of Association

Not all pairs of variables can be plotted in a scatterplot, as we know. For example, you can't use a scatterplot to plot presence or absence of a tattoo versus whether or not the respondent had hepatitis. The following table cross-classified 626 individuals from the University of Texas Southwestern Medical Center according to their tattoo status and their hepatitis status.

While you couldn't use a scatterplot on these data, you could use a nested bar chart. But what would you plot? One way you could approach the question of association would be to ask "are you more likely to have hepatitis if you have a tattoo than if you don't?" So it would make sense to calculate percentages for each row of Table 12.6.

TABLE 12.6

	Hepatitis	No Hepatitis
Tattoo	25	88
No Tattoo	22	491

TABLE 12.7

	Hepatitis	No Hepatitis
Tattoo	22.12%	77.88%
No Tattoo	4.29%	95.71%

We can then plot these percentages in a nested bar chart. See below.

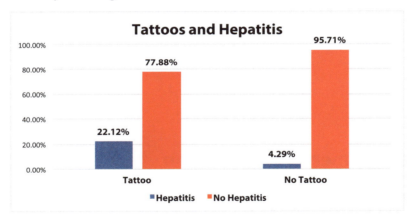

What can you see from this plot? Of those people who had tattoos, about 22% had developed hepatitis. Of those who did not have tattoos, only about 4.25% developed hepatitis. So on the surface, therefore, there seems to be an association between having a tattoo and having hepatitis.

The following tip shows you how to construct this kind of plot in Excel.

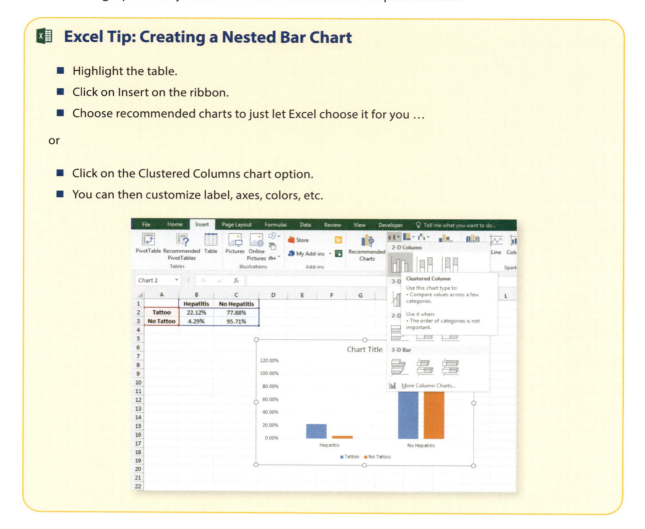

Excel Tip: Creating a Nested Bar Chart

- Highlight the table.
- Click on Insert on the ribbon.
- Choose recommended charts to just let Excel choose it for you …

or

- Click on the Clustered Columns chart option.
- You can then customize label, axes, colors, etc.

Numerical Assessment of Association

If we have nominal level data that are cross-classified into row and column categories like the tattoo data, then Cramer's V Coefficient (V) is often used to quantify the level of the association. V depends on something called the chi-squared value. The expression you see below is one way that the chi-squared value can be calculated. It has to be computed on the counts themselves, not on the percentages.

Chi-squared Value $\chi^2 = \sum \left[\dfrac{(\text{observed} - \text{expected})^2}{\text{expected}} \right]$

This sum is over all entries in the table. To compute it, we need the actual counts (observed) and an estimate of what the count would be if the two variables were completely unassociated (expected). We will practice computing it below. Cramer's V is a simple adaptation of the chi-squared value.

Cramer's V Coefficient $V = \sqrt{\dfrac{\chi^2}{n(m-1)}}$

where *n* is the total number of subjects or observations, and m is the minimum of the number of rows and columns in the table.

Cramer's *V* Coefficient will always admit the following interpretations:

- It will always be between 0 and 1.
- The closer to 1 it is, the stronger the association.

It is common to use the following characterization of the strength of Cramer's *V*, though this varies in practice.

- Above 0.40 means very strong association
- 0.30 to 0.40 means strong association
- 0.20 to 0.30 means moderate association
- 0.10 to 0.20 means weak association
- Below 0.10 means little or no association

Example: Tattoos and Hepatitis Revisited

The chi-square value has to be calculated on the original counts, not on any percentages derived from those counts. The expected values in each cell of the table are computed by taking the row totals times the column totals, and dividing by the grand total.

TABLE 12.8

	Hepatitis	No Hepatitis	Totals
Tattoo	25 (8.48)	88 (104.52)	113
No Tattoo	22 (38.52)	491 (474.48)	513
Totals	47	579	626

For instance, the expected value that will be paired with the 25 (upper left cell in table) is computed as

$$(113 \times 47)/626 = 8.48$$

All the other expected values are computed similarly and are shown in parentheses by the actual counts. Therefore:

$$X^2 = \frac{(25 - 8.48)^2}{8.48} + \frac{(88 - 104.52)^2}{104.52} + \frac{(22 - 38.52)^2}{38.52} + \frac{(491 - 474.48)^2}{474.48}$$

$$= 42.45$$

In this example, *n* = 626 and *m* = min (2,2) = 2. Therefore, Cramer's *V* is:

$$V = \sqrt{\frac{42.45}{626\,(1)}} = 0.26$$

If we adhere to the scale above, then this value suggests a moderately strong relationship between having a tattoo and having hepatitis.

Excel Tip: Computing Cramer's V

There is no one-function way to computer Cramer's *V* in Excel from a table. You can do the following:

- Enter the 2×2 table of counts; let's say they are in the four-cell range M19:N20.
- Create the corresponding 2×2 table of expectations; let's say in the four-cell range M24:N25.
- Enter the following code in an empty cell: = CHIINV(CHISQ.TEST(M19:N20,M24:N25),1)
- Hit return.
- This will give you the chi-square value. It is then easy to compute Cramer's *V*.

If your table only has 2 columns and 2 rows of counts, as in our example above, then there is a trick you can use.

- Create two new variable columns in Excel that will have *n* data rows. So in the tattoo example there will be 626 rows of data.
- For the tattoo example, 25 of those rows will have a 1 in the first new column and a 1 in the second new column (tattoo and hepatitis).
- 88 rows will have a 1 in the first new column and a 0 in the second (tattoo, no hepatitis).
- 22 rows will have 0 in the first column, and a 1 in the second.
- 491 will have 0's in both columns.
- Now if you just compute = *correl* on this data range (the two new columns), the value will be Cramer's *V*! This is **not** an efficient way to compute Cramer's *V*, but it does make an important connection with the correlation coefficient.

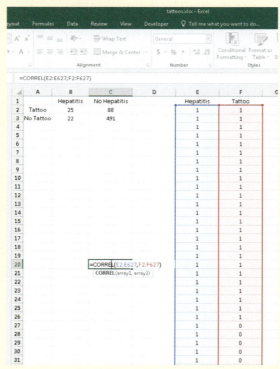

While Excel is really cumbersome for this particular kind of computation, there are many online applets (e.g. http://vassarstats.net/newcs.html) that will compute Cramer's *V* directly. These kinds of online calculators come and go, but you are sure to always be able to find one. Your challenge is just to always be curious and look when the need arises.

Example: Cramer's *V* for Tattoos and Hepatitis Using an Online Calculator

- Access the calculator at http://vassarstats.net/newcs.html.
- Specify the number of rows (2) and columns (2).
- Enter the data in the four cells.
- Hit Calculate.
- A lot of things are calculated but one of them is Cramer's *V*.

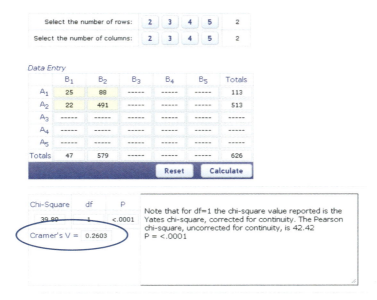

TAKE-HOME POINTS

- Simple scatterplots are a useful visual means of addressing association between two ratio-level variables.

- The correlation coefficient is the most common numerical measure of the strength of a straight-line relationship between two ratio-level variables.

- Computing the correlation coefficient between two ratio-level variables with a curvilinear scatterplot is not appropriate and can have a direct effect on your human inference about the strength of the actual association.

- Bar charts are a useful visual means of addressing association between two nominal-level variables.

- Cramer's *V* is a common numerical measure of the strength of the association between two nominal-level variables.

Chapter 12 Exercises

Reading Check

Short Answer: Please provide brief, concise answers to the following questions.

1. Give an example not in the assigned reading of two variables that would exhibit a strong negative association.

2. Give an example not in the assigned reading of two variables that would exhibit a weak positive association.

3. Describe a graphical way of assessing association for variables that cannot be plotted in a scatterplot.

4. Give an example of two variables that can't be exhibited in a scatterplot but would nonetheless exhibit a strong association.

5. What does a correlation coefficient of −0.95 suggest about the strength of the relationship between the two variables in question?

6. What does such a correlation coefficient of 0.10 suggest about the scatterplot of the two variables in question?

7. What does a Cramer's *V* Coefficient of 0.35 suggest about the strength of the relationship between the two nominal variables in question?

8. Refer to the Sodium and Fat in Burgers example. What would surely happen to the correlation coefficient if Burger Brand A were removed, and it was re-computed on just the remaining data?

9. What is the maximum possible value of Cramer's *V* Coefficient?

10. What is the range of possible values for the correlation coefficient?

Beyond the Numbers

1. Data Leveling

Identify the correct measurement level for each of the variables described below.

Please read the following excerpt:

Source: *Journal of Biometrics & Biostatistics*

Author: Shaffi Ahamed Shaikh

Date: December 2011

> The British Regional Heart Study was a cohort of 7735 men aged 40–59 years randomly selected from general practices in 24 British towns, with the aim of identifying risk factors for ischemic heart disease. Of the 7718 men who provided

information on smoking status, 5899(76.4%) had smoked at some point in their lives. Over the subsequent 10 years, 650 of these 7718 men (8.4%) had a myocardial infarction (MI).

a. What is the measurement level of the two variables "smoking status" and "myocardial infarction?"

i. Both are nominal.

ii. Both are ratio.

iii. Myocardial Infarction is ratio, and smoking status is nominal.

iv. Myocardial Infarction is nominal, and smoking status is ratio.

Please read the following excerpt:

Source: *BBC News Magazine* https://www.bbc.com/news/magazine-20356613

Author: Charlotte Prichard

Date: November 2012

Franz Messerli of Columbia University started wondering about the power of chocolate after reading that cocoa was good for you. One paper suggested regular cocoa intake led to improved mental function in elderly patients with mild cognitive impairment, a condition which is often a precursor to dementia, he recalls.

"There is data in rats showing that they live longer and have better cognitive function when they eat chocolate, and even in snails you can show that the snail memory is actually improved," he says.

So Messerli took the number of Nobel Prize winners in a country as an indicator of general national intelligence and compared that with the nation's chocolate consumption.

b. What is the measurement level of the two variables "number of Nobel Prize winners" and "nation's chocolate consumption?"

i. Both are nominal.

ii. Both are ratio.

iii. Number of Nobel Prize winners is ratio, and chocolate consumption is nominal.

iv. Number of Nobel Prize winners is nominal, and chocolate consumption is ratio.

2. Anscombe's Activity

These data were created by F.J. Anscombe in 1973 to remind us of the importance of plotting our data. You will see these data in other exercises. We are using the notation for the variables that Anscombe used.

TABLE 12.9

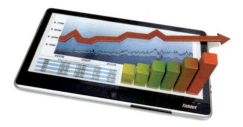

Obs	$x1$	$y1$	$x4$	$y4$
1	10	8.04	8	6.58
2	8	6.95	8	5.76
3	13	7.58	8	7.71
4	9	8.81	8	8.84
5	11	8.33	8	8.47
7	6	7.24	8	5.25
6	14	9.96	8	7.04
8	4	4.26	19	12.5
9	12	10.84	8	5.56
10	7	4.82	8	7.91
11	5	5.68	8	6.89

a. Which of the following is the correct scatterplot of variable $y1$ on the vertical axis versus variable $x1$ on the horizontal axis?

i.
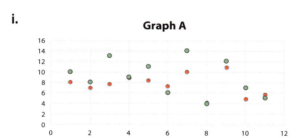

ii.
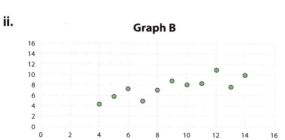

iii.

iv.
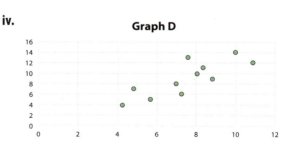

b. Compute the correlation coefficient for the variables $x1$ and $y1$. What is it?

    i. 0.82

    ii. −0.82

    iii. 0.67

    iv. −0.67

c. Use software (e.g. Excel, online applet) to construct a scatterplot of $y4$ (vertical axis) versus $x4$ (horizontal axis). Describe the shape.

d. Compute the correlation coefficient for the variables $x4$ and $y4$. What is it?

    i. 0.82

    ii. −0.82

    iii. 0.67

    iv. −0.67

e. Reflect on your answers to parts a–d. What are the implications for the information contained in the correlation coefficient as a simple statistical summary of association?

3. Mortality and Global Warming

For this exercise, you will need to access the "Child Mortality" and "CO$_2$ Emissions" data set. The 2006 version of these data were collected for 191 countries, and archived by Dr. Hans Rosling. They are in the back of this book and are also available at www.statconcepts.com/datasets, where they can be downloaded as a text or Excel file.

a. Construct a scatterplot of Child Mortality (vertical axis) versus CO$_2$ Emissions (horizontal axis). Make sure all axes are labeled, and there is a title on the plot. You have to use software (e.g. Excel or an online applet) to construct the plot.

b. What kind of association do you see in the scatterplot?

    i. Negative, extremely weak

    ii. Positive, moderately strong

    iii. Positive, extremely weak

    iv. Negative, moderately strong

c. Use software to compute the correlation coefficient for these two variables. What is it?

    i. 0.44

    ii. −0.44

    iii. 0.80

    iv. −0.80

d. Is it wise to use the correlation coefficient "*r*" to summarize the association? Why or why not?

   i. Yes, because the scatterplot is essentially a straight line

   ii. Yes, because the data are ratio level

   iii. No, because the data are ratio level

   iv. No, because the scatterplot is curvilinear and not a straight line

4. Global Warming Transformed

For this exercise, you will need to access the "Child Mortality" and "CO_2 Emissions" data set. The 2006 version of these data were collected for 191 countries, and archived by Dr. Hans Rosling. They are in the back of this book, and also available at www.statconcepts. com/datasets, where they can be downloaded as a text or Excel file. The scatterplot for the original pair of variables "Child Mortality" and "CO_2 Emissions" is shown below.

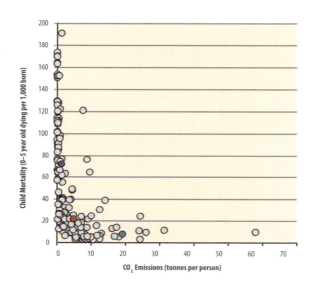

a. Construct a scatterplot of $\log_{10}$ (Child Mortality) versus $\log_{10}$ (CO_2 Emissions). How does this plot compare to the one for the original variables shown here? You need to be detailed with your answer.

b. What kind of association do you see in your newly constructed scatterplot?

   i. Negative, extremely weak

   ii. Positive, moderately strong

   iii. Positive, extremely weak

   iv. Negative, moderately strong

c. Use software to compute the correlation coefficient for these two variables. What is it?

i. 0.44

ii. −0.44

iii. 0.80

iv. −0.80

d. Is it wise to use the correlation coefficient "*r*" to summarize the association between this new pair of variables? Why or why not?

i. Yes, because the scatterplot is essentially a straight line

ii. Yes, because the data are ratio level

iii. No, because the data are ratio level

iv. No, because the scatterplot is curvilinear and not a straight line

5. Heptathletes

Finish data for two 1992 Olympic Heptathlon events are shown here.

TABLE 12.10

Name	Hurdles (seconds)	Javelin (meters)	Name	Hurdles (seconds)	Javelin (meters)
Joyner-Kersee	12.85	44.98	Lesage	13.75	41.28
Nastase	12.86	41.3	Nazaroviene	13.75	44.42
Dimitrova	13.23	44.48	Aro	13.87	45.42
Belova	13.25	41.9	Marxer	13.94	41.08
Braun	13.25	51.12	Rattya	13.96	49.02
Beer	13.48	48.1	Carter	13.97	37.58
Court	13.48	52.12	Atroshchenko	14.03	45.18
Kamrowska	13.48	44.12	Vaidianu	14.04	49
Wlodarczyk	13.57	43.46	Teppe	14.06	52.58
Greiner	13.59	40.78	Clarius	14.1	45.14
Kaljurand	13.64	47.42	Bond-Mills	14.31	43.3
Zhu	13.64	45.12	Barber	14.79	0
Skjaeveland	13.73	35.42	Chouaa	16.62	44.4

a. Which of the following is the correct scatterplot of Javelin Distance on the vertical axis versus Hurdles Time on the horizontal axis?

i.

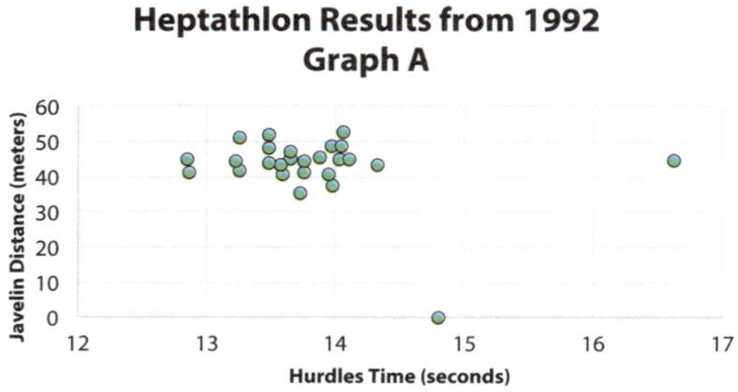
Heptathlon Results from 1992
Graph A
Javelin Distance (meters)
Hurdles Time (seconds)

iii.

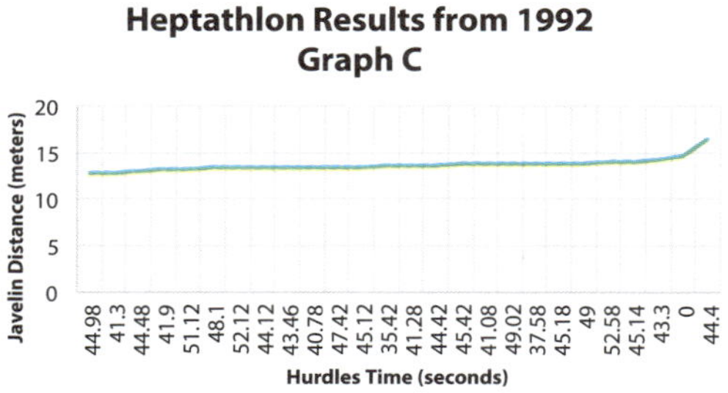
Heptathlon Results from 1992
Graph C
Javelin Distance (meters)
Hurdles Time (seconds)

ii.

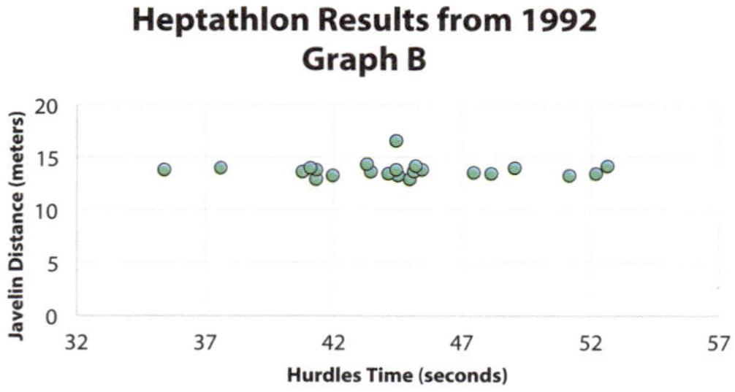
Heptathlon Results from 1992
Graph B
Javelin Distance (meters)
Hurdles Time (seconds)

iv.

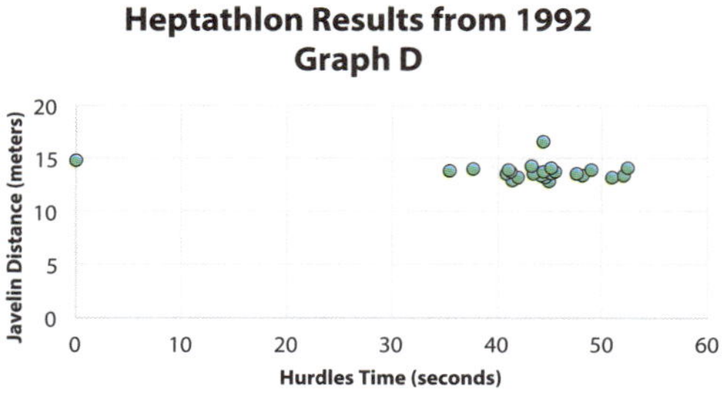
Heptathlon Results from 1992
Graph D
Javelin Distance (meters)
Hurdles Time (seconds)

b. Compute the correlation coefficient for the variables Javelin Distance and Hurdles Time. What is it?

i. −0.461

ii. −0.252

iii. < 0.001

iv. −0.025

c. Remove Barber from the data set and compute the correlation coefficient for the variables Javelin Distance and Hurdles Time. What is it?

i. −0.461

ii. −0.252

iii. < 0.001

iv. −0.025

d. Put Barber back in the data set and remove Chouaa. Re-compute the correlation coefficient for the variables Javelin Distance and Hurdles Time. What is it?

i. −0.461

ii. −0.252

iii. < 0.001

iv. −0.025

e. Reflect on your answers to parts b–d. What are the implications for the information contained in the correlation coefficient as a simple statistical summary of association?

6. Cage the Pools

To the right are data on the number of people who drowned by falling into a swimming pool and the number of films Nicolas Cage appeared in (1999 through 2009).

a. Which of the following is the correct scatterplot of Drowning Deaths on the vertical axis versus Number of Cage Films on the horizontal axis?

TABLE 12.11

Year	Number of People Who Drowned by Falling into a Swimming Pool	Number of Films Nicolas Cage Appeared In
1999	109	2
2000	102	2
2001	102	2
2002	98	3
2003	85	1
2004	95	1
2005	96	2
2006	98	3
2007	123	4
2008	94	1
2009	102	4

i.

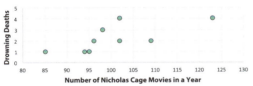

Cage Movies and Drowning Deaths Graph A

ii.

Cage Movies and Drowning Deaths Graph B

iii.

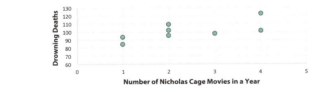

Cage Movies and Drowning Deaths Graph C

iv.

Cage Movies and Drowning Deaths Graph D

b. Compute the correlation coefficient for the variables Drowning Deaths and Number of Cage Films. What is it?

i. −0.33

ii. 0.67

iii. 0.81

iv. 0.33

7. Smoking and Heart Health

Please read the following excerpt:

Source: *Journal of Biometrics & Biostatistics*

Author: Shaffi Ahamed Shaikh

Date: December 2011

The British Regional Heart Study was a cohort of 7735 men aged 40–59 years randomly selected from general practices in 24 British towns, with the aim of identifying risk factors for ischemic heart disease. Of the 7718 men who provided information on smoking status, 5899(76.4%) had smoked at some point in their lives. Over the subsequent 10 years, 650 of these 7718 men (8.4%) had a myocardial infarction (MI). The following 2×2 table shows the number and percentage of smokers and non-smokers who developed and did not develop an MI over the 10 year period.

TABLE 12.12

	Myocardial Infarction in 10 years	No Myocardial Infarction in 10 years
Smoked	563	5336
Never Smoked	87	1732

a. Which of the following plots is most appropriate for assessing the association between Smoking Status and Myocardial Infarction in 10 Years?

i.
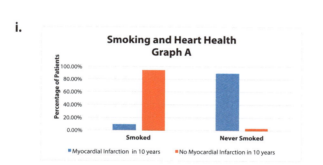

iii.

ii.
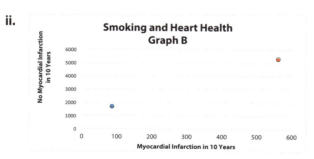

iv.
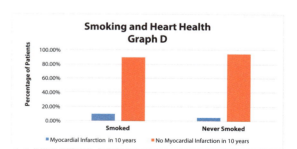

b. What is Cramer's *V* for these data?

i. 40.25

ii. 0.073

iii. 59.59

iv. 0.110

8. Hip Fractures

Please read the following excerpt:

Source: *Journal of Biometrics & Biostatistics*

Author: Shaffi Ahamed Shaikh

Date: December 2011

A total of 1327 women aged 50–81 years with hip fractures, who lived in a largely urban area in Sweden, were investigated in an unmatched case-control study. They were compared with 3262 controls with the same age range randomly selected from the national registry. The objective was to determine whether women currently taking postmenopausal hormone replacement therapy (HRT) were less likely to have hip fractures than those not taking it. The results were given in a 2×2 table format and show the number of women who were current users of HRT and those who had never used or formerly used HRT.

TABLE 12.13

	Current HRT	Never HRT
Hip Fracture	40	1287
No Hip Fracture	239	3023

a. Construct a plot that allows an effective visual assessment of the association between Hip Fractures and HRT. Comment on the degree of association.

b. Compute Cramer's *V* for these data. Interpret as a measure of association in the context of these data.

Projects

1. Ball Park or Bull's Eye?

Introduction

This mini-project is appropriate as an in-class activity or a homework assignment. It can be configured either as an individual assignment or a group assignment. Follow your instructor's instructions. The goal of the activity is to better understand how sample size affects Cramer's *V*, but may not affect another common measure of association. You will be working with the counts in the following table that cross-classifies 626 individuals from the University of Texas Southwestern Medical Center according to their tattoo status and their hepatitis status.

TABLE 12.14

	Hepatitis	No Hepatitis
Tattoo	25	88
No Tattoo	22	491

The Odds Ratio

Another popular measure of association for nominal level data is called the "odds ratio." In the above table, if you have a tattoo, the odds of you developing hepatitis are 25-to-88 or 25/88 = 0.284. Similarly, if you do not have a tattoo, the odds of you developing hepatitis are 22 to 491 = 0.045. So the "odds ratio" is 0.284/0.045 or 6.31. Hence, the odds of developing hepatitis if you have a tattoo are about 6.3 times greater than if you don't have a tattoo. Of course, the actual odds of developing hepatitis as seen in these numbers are (25 + 22)/(88 + 491) = 0.08, chances in 100.

Of course, we can change the actual numbers in the chart and produce the same odds ratio of 7. Let's just reduce the No Hepatitis group.

TABLE 12.15

	Hepatitis	No Hepatitis
Tattoo	25	A
No Tattoo	22	B

That is, we want to know what A and B can be so that:

$$6.31 = \frac{25/A}{22/B} = \frac{25B}{22A} \Rightarrow \frac{B}{A} = \frac{22 \times (6.31)}{25} = 5.55$$

So as long as B = 5.55(A), the odds ratio from the table above will be the same. But Cramer's V will not stay the same, as B and A change.

Generating and Plotting Values

Compute all the missing entries in the following table.

TABLE 12.16

A	B	Odds Ratio	Cramer's V
88	491	6.31	
	400	6.31	
	350	6.31	
	300	6.31	
	250	6.31	
	200	6.31	
	150	6.31	
	100	6.31	

Use software to produce the following overlaid plots on one set of axes:

- Cramer's V (vertical axis) versus total sample size;

- Horizontal lines for *each* of the categories used to interpret Cramer's V as seen in the beginning of this chapter. For example, a horizontal line at 0.30 indicates the break point from weak to moderate association for Cramer's V.

Interpret What You Have Found

Look at what your plots show you. Give detailed explanations of the following:

a. How does Cramer's V vary as total sample size decreases? Explain.

b. Why is what you have found relevant if someone has to choose between using Cramer's *V* and the Odds Ratio for measuring association? What advice would you offer for interpreting these kinds of measures? Explain.

What You Turn In

A statement of the problem, the completed table, your fully labeled plot(s), and your detailed explanations all need to be organized into a typed document that reads like a report, not a homework assignment, and turned in to your supervisor. Your instructor will tell you how she wants that document submitted.

2. Drug Deaths and Income

Introduction

This mini-project is appropriate as an in-class activity or a homework assignment. It can be configured either as an individual assignment or a group assignment. Follow your instructor's instructions. The goal of the activity is to collect and analyze the association between the per capita income and the drug overdose mortality rate for each of the fifty states. Make sure your income data and mortality data are for the same year. Any year in the last ten that you can find is fine.

Find And Plot Values

Do the research that is needed to fill out the entire table on the next page. Finding the appropriate data is part of the challenge.

Use software to produce the following scatterplot of Overdose Death Rate (vertical axis) versus Per Capita Income (horizontal axis). Make sure the plot is fully labeled.

Summarize the Association

Use the correlation coefficient to summarize the association between the two variables. If the scatterplot is not sufficiently linear, make sure you transform the data so that the correlation coefficient is appropriate. If there are data points that are potentially overly influencing the correlation coefficient, remove them and re-compute the value of *r*. In all cases, and at all stages, explain what the correlation tells us about the direction and strength of the association.

What You Turn In

A statement of the problem, the completed table, your fully labeled plot, and your detailed explanations about what you did and why all need to be organized into a typed document that reads like a report, not a homework assignment, and turned in to your supervisor. Make sure you ultimately address the question of whether per capita income and drug overdose death rates seem to be associated. Your instructor will tell you how she wants the final document submitted.

TABLE 12.17

				Year				
State	Per Capita Income	Overdose Death Rate	State	Per Capita Income	Overdose Death Rate	State	Per Capita Income	Overdose Death Rate
Alabama			Louisiana			Ohio		
Alaska			Maine			Oklahoma		
Arizona			Maryland			Oregon		
Arkansas			Massachusetts			Pennsylvania		
California			Michigan			Rhode Island		
Colorado			Minnesota			South Carolina		
Connecticut			Mississippi			South Dakota		
Delaware			Missouri			Tennessee		
Florida			Montana			Texas		
Georgia			Nebraska			Utah		
Hawaii			Nevada			Vermont		
Idaho			New Hampshire			Virginia		
Illinois			New Jersey			Washington		
Indiana			New Mexico			West Virginia		
Iowa			New York			Wisconsin		
Kansas			North Carolina			Wyoming		
Kentucky			North Dakota					

3. Reality Video

Introduction

This chapter discussed association and how it should be visualized and quantified, depending on what kind of data one has. Now, we want to give you a chance to both show you know what you are doing and to help teach others.

The Assignment

Your instructor may have you do this as individuals or may form you into teams. We'll refer to "teams" below, although you may be a team of one depending on what your instructor wants. Each team is required to do the following:

a. Pick one or more topics from this chapter. Your instructor may want to approve them before you proceed.

b. Construct a three-minute video that teaches the chosen topic to the class. You must carefully and correctly use the language of the chapter in your presentation, and you are required to give an example that is NOT in the chapter.

c. Post the video on YouTube and send the link to your instructor in the manner she stipulates.

13 CHAPTER 13
Association and Causation: More than Meets the Eye

Introduction

The point of this chapter is deceptively simple: two variables could be highly associated, but no causal link exists between them. You'd think that any chapter whose point can be reduced to one sentence should perhaps be a one-sentence chapter! However, the confusion over association and causation runs deep and persists; as such, it deserves more than a few words of explanation and clarification. Let's start by establishing a little street credibility. Just how widespread is the confusion between association and causation?

> Journalists are constantly being reminded that "correlation doesn't imply causation;" yet, conflating the two remains one of the most common errors in news reporting on scientific and health-related studies.
> —Rebecca Goldin, Aug. 19, 2015 SENSE about SCIENCE USA

> A source of confusion about causation is that news reports about research findings often suggest causation when they should not. A causal claim may be easier to understand—compare "seat belts save lives" with "the use of seat belts is associated with lower mortality"—because it presents a cause (seat belts) acting directly (saving lives). It seems to tell a more compelling story than a correlational claim, which can come across as clumsy and indirect.
> —Nick Barrowman, Summer 2014 The New ATLANTIS

> Just a quickie. You've heard what follows many times before from us. It is one of the most common flaws in health care journalism. Studies that show a statistical association between two things do not necessarily prove that one thing causes another to occur. We saw that principle violated several times in news coverage of several different studies today—regarding coffee and sleep.
> —Gary Schwitzer, July 2017 HEALTH NEWS REVIEW.ORG

> "Correlation is not causation" is a statistics mantra. …. Despite embodying an important truth, the phrase has not caught on in the wider world. It's easy to see why. Our preconceptions and suspicions about the way things work tempt us to make the leap from correlation to causation without any hard evidence.
> —Nathan Green, January 2012 The Guardian

Our purpose here is to reinforce the importance of not conflating correlation, more generally association, with causation. We will do that primarily through the use of multiple examples. Let's start by understanding what spurious correlation is.

Spurious Curious

Example: Beer and Sex

In April 2000 the Centers for Disease Control did a study at the state level, recording for each state the Beer Tax (horizontal axis; scale suppressed) and the Gonorrhea Rate for that state (vertical axis, scale suppressed). These two variables are plotted in the scatterplot seen here.

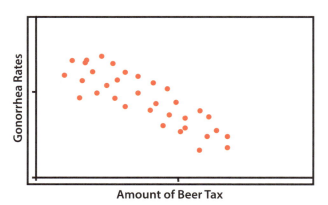

The two variables are clearly negatively associated, with states that have higher beer taxes also having lower gonorrhea rates. Since there are no scales on the axes, we can't easily assess the strength of the association, but the existence of an association is quite clear.

There's nothing tricky about that part. The tricky part is what we do with that information. Let's see what the CDC did.

Take another look at the scatterplot. Again, since we don't have access to the original data, we can't put the appropriate scales on the horizontal and vertical axes, but it is clear that for every unit moved to the right on the Beer Tax axis, there is a reduction in the Gonorrhea Rates. Using elementary methods of a statistical procedure known as regression, the CDC was able to ascertain that for every twenty-cent move to the right on the Beer Tax axis, there would be a drop of 8.9% on the Gonorrhea Rates axis.

This is a mathematical fact that legitimately follows from the scatterplot. There is no reason to believe that the CDC, staffed with really smart people, did this calculation incorrectly. The issue was not the calculation, but the inference the CDC attached to it. The CDC released a statement claiming that if states just increased their beer tax by 20 cents on a six-pack, then they could reduce their overall gonorrhea rates by 8.9%. This implicitly assumes a *cause and effect* relationship between Beer Tax and Gonorrhea Rates. It assumes a kind of "dial and respond" relationship whereby dialing *up* the Beer Tax will lead to a guaranteed dialing *down* response of Gonorrhea Rates.

What is the CDC's argument? Basically, that "young persons who drink alcohol may be more likely than persons who abstain to participate in high-risk sexual activity, such as unprotected sexual intercourse or multiple sexual partners." That's not an unreasonable argument.

What is one potential problem with this argument though? You will recall the discussion of confounding in Chapter 3. That was when the presence of a third variable might complicate the cause-and-effect connection one wanted to make between an explanatory variable and a response. There is a very similar idea here.

> **Spurious correlation** is when two variables have a definite association, but no underlying cause and effect. This often happens as the result of each of those being the effect of a third variable that is really the cause of both.

That is, to accept the CDC inference on face value, we are being asked to believe the following:

The problem is that it may be that a third variable, say some genetic propensity for risky behavior, is causing both of these behaviors. As such, there may be no causal relationship that directly links alcohol consumption and risky sexual behavior.

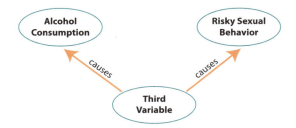

This is exactly the argument that Canadian researchers Anindya Sen and May Luong made regarding the CDC report. Indeed, they argued for the possibility of a genetic predisposition:

> "… both alcohol consumption and participation in risky sexual practices might be driven by an underlying proclivity for high-risk behavior."

This did not change the CDC's argument, however. Sophisticated statistical methods for mitigating the effects of a third variable, such as high-risk behavior, were performed, and there was still evidence that a rise in beer prices would likely lead to a decrease in gonorrhea rates. Still yet, others have challenged the argument because twenty cents is such a small part of the disposable income of even a young drinker that it would simply be lost in the jingle of extra coins in the pocket and not lead to enough excessive drinking to create social consequences.

Our point is not to take sides. Rather, our job is to understand just how complicated it can be to infer cause and effect from association. We all may have, as suggested above, almost a knee-jerk response to immediately use cause and effect to explain association, but we have to be very careful about doing this too hastily and creating a completely incorrect idea of what is actually causing what.

Example: Cage the Pools

While silly examples are, well, silly, they can serve us well in making subtle points a little clearer. An often-discussed case of correlation involves data on the number of people who drowned by falling into a swimming pool, and the number of films Nicolas Cage appeared in (1999 through 2009). Those data are in Table 13.1.

If we compute the correlation coefficient between drowning deaths and Cage films, we find that $r = 0.67$. That's a pretty substantial correlation. But no one in full command of their senses would say either one of these variables causes the other one. Even our reflexive need to explain using the language of causation would likely stay in full check in this example.

To be fair, some have argued, perhaps for fun, perhaps not, that the real explanation goes like this. Every time a Cage film came out, legions of Cage fans would get together for a screening party around a pool, and with a little risky behavior thrown in, more than usual fall into the pool, and tragedy strikes. You get the idea. It is really hard to accept that as an explanation that makes sense. In fact, it is not at all clear that there is any underlying third variable that could explain the correlation.

TABLE 13.1

Year	Number People Who Drowned by Falling into a Swimming Pool	Number of Films Nicolas Cage Appeared In
1999	109	2
2000	102	2
2001	102	2
2002	98	3
2003	85	1
2004	95	1
2005	96	2
2006	98	3
2007	123	4
2008	94	1
2009	102	4

In most cases, it won't be nearly as clear as in the Cage and drowning example that an observed correlation is not direct evidence of causation. Better to just be very careful and appropriately skeptical when we make inferences based on correlational results. Let's revisit the real examples from Chapter 12, most of which are a little harder to assess than the one we just saw.

Examples: Correlation Examples Revisited

TABLE 13.2

Autism and DTP Cases

$r = 0.96$

Causation is subject of emotional ongoing debate. Most scientists reject direct causation in this case, but not all do.

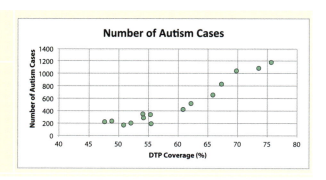

TABLE 13.2 Continued

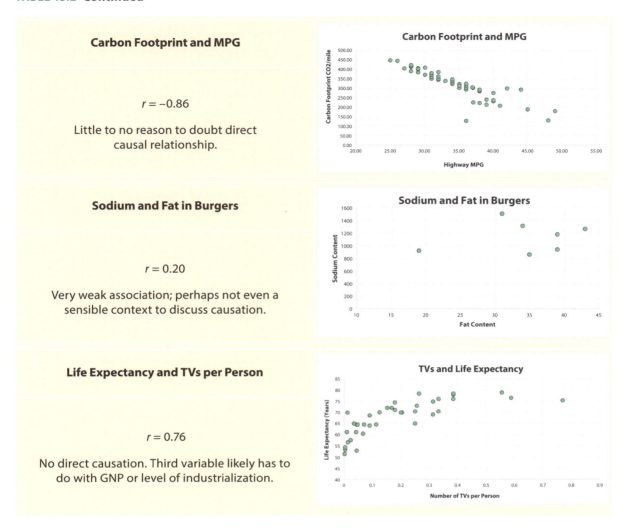

Carbon Footprint and MPG

$r = -0.86$

Little to no reason to doubt direct causal relationship.

Sodium and Fat in Burgers

$r = 0.20$

Very weak association; perhaps not even a sensible context to discuss causation.

Life Expectancy and TVs per Person

$r = 0.76$

No direct causation. Third variable likely has to do with GNP or level of industrialization.

We don't need a scatterplot to discuss causation, of course. If we have variables that can't be plotted in that manner, then we can't use the correlation coefficient to summarize the extent of the association, but the fidelity of the link between association and causation is no less pressing. For example, consider the plot shown below that compares the percent of time studying for boys and girls. It is clear that there is an association. Girls tend to study more according to the chart. But does "being a girl" cause more studying and "being a boy" cause less?

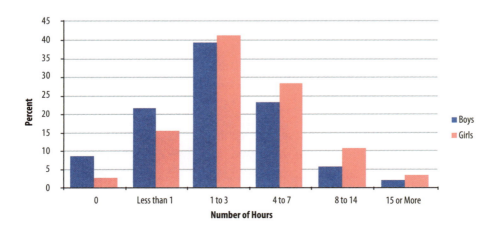

Some have argued that this is indeed the case, in part because girls bring genetic predispositions toward paying attention and interest in learning. Others argue that the difference is largely socially induced, perhaps early on by elementary school teachers who reward girls with higher grades in part because of better handwriting or neater presentations. The theory is that this creates an environment where girls "expect" to do well and boys don't, which would invariably affect motivation to study. While these arguments are compelling, they are not the final word.

Correlation as Evidence of Causation

The point so far has been simple. Correlation does not imply causation, and we need to always be skeptical about any claims of causation from data that only exhibit association. Not unlike confounding, there can sometimes be a third variable that is responsible for the association being observed, and one that could, in fact, be the real key to any meaningful causal relationships.

However, strong evidence of association may suggest causation, and that may be well worth considering seriously even if causation has not been established beyond a doubt. In fact, we want to resist the temptation to too quickly dismiss any correlational evidence of causation as inadequate. This may be a mistake that is just as serious as over-interpreting correlation. Let's look at some examples.

Examples: Diet Soda and Weight Gain

We all understand that diet soda contains chemicals with health effects that are not fully understood. But at least you won't gain weight drinking them, right? Well, not so fast, say researchers at the Utah Health Sciences Center. After a very careful study that adjusted for age, sex, and ethnicity, they found a strong association between incidence of becoming overweight and consumption of diet sodas. A brief online commentary was provided by the so-called Digital Bits Skeptic.

Authors: Digital Bits Skeptic

A study at the Utah Health Sciences Center found that the more diet soda in a person's diet, the more likely that person was to become overweight or obese. "After adjusting for age, sex, and ethnicity, Williams found that regular soft drinks were no longer significantly linked to the incidence of becoming overweight or obese, but diet soft drinks were."

The media will pick up a press release about this study and report that diet soda can make you fat.

…

www.cartoonstock.com

Of the people that consume a lot of fast food, those that are starting to experience weight gain are more likely to choose diet soda than those that haven't experienced weight gain. So, diet soda consumption is probably highly correlated with people who have a slowing metabolism and a habit of eating fast food. Perhaps, it is not the diet soda that it the problem, but the type of restaurant where large portions of diet soda often accompany the food.

Just as the Digital Bits Skeptic predicted, the media *did* run with headlines claiming that diet soda makes you fat. Certainly one can offer some well-reasoned objections to a claim of direct causation between the amount of diet sodas consumed and body mass index. We might take the approach of the Digital Bits Skeptic and argue, quite reasonably, that people with a poor diet—say those who eat a lot of fast food and are already overweight and getting heavier—are more likely to turn to the consumption of diet soda.

So perhaps it is not the diet soda that is the problem but the types of food and the portion sizes that accompany the consumption of lots of diet soda. That is a really good argument and in the absence of other evidence would be really convincing.

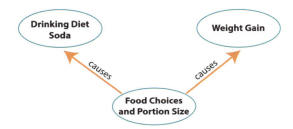

It turns out there *was* other evidence that surfaced, however. Researchers at the San Diego Health Science Center did a controlled experiment on mice. The mice were randomly divided into two groups with one group getting chow with the popular sweetener aspartame added.

Title: New Study Is Wake-Up Call for Diet Soda Drinkers

Authors: Ryan Jaslow

Source: CBS News

Researchers divided mice into two groups, one of which ate food laced with the popular sweetener aspartame. After three months, the mice eating aspartame-chow had higher blood sugar levels than the mice eating normal food. The authors said in a written statement their findings could "contribute to the associations observed between diet soda consumption and the risk of diabetes in humans."

"Artificial sweeteners could have the effect of triggering appetite but unlike regular sugars they don't deliver something that will squelch the appetite," Sharon Fowler, obesity researcher at UT Health Science Center at San Diego and a co-author on both of these studies, told the Daily Mail. She also said sweeteners could inhibit brain cells that make you feel full.

Researchers have since speculated that artificial sweetener might have directly had the effect of triggering appetite or even inhibiting brain cells that tell you when you are full. In this case, the evidence provided by a strong correlation between drinking diet sodas and weight gain ended up being a legitimate preview of an actual causal link.

Perhaps the most famous example of mounting correlational evidence eventually triggering a causal claim is smoking and lung cancer. For decades, tobacco companies avoided legal troubles and warning labels by using the "correlation does not imply causation" argument. Even the father of modern statistical inference, Sir Ronald Fisher—himself a smoker—defended this argument. Patterns of cancer and smoking emerged and strengthened over time, however. As noted by Lucas and Harris in an article on the nature of evidence (*Int. J. Environ Res Public Health*, 2018, August):

> The first indication that smoking may increase the risk of lung cancer came from ecological studies, showing that mortality rates for lung cancer in the UK began to increase from approximately 1925, around 15 years after the increase in consumption of tobacco and cigarettes. A similar pattern has been repeated over time and location—rising prevalence of smoking is mirrored by rising prevalence (and mortality) of lung cancer.

Eventually this persistent correlational evidence rose to a sufficient level that the real possibility of a cause and effect link between smoking and cancer could no longer be responsibly ignored. While experiments had long been completed on mice that showed a causal link, such experiments on humans would never be ethically possible. So the traditional cause and effect evidence provided by experimentation would never have been possible. It took the mounting correlational evidence to tip the scales and change the social and medical perception of smoking safety.

Lucas and Harris (above) present a complex and interesting discussion of the nature of evidence and what is generally accepted now as "enough" to suggest causality. You will have the opportunity to read more about this in a project that is outlined at the end of this chapter.

Simpson's Paradox

Sometimes the presence of a third variable driving the association between two others can have profound consequences on the interpretation of association in 2 x 2 tables. Let's start with an example.

Example: Breakfast and Biology

In a study done by Gregory W. Phillips from Blinn College, and reported in *Bioscience* (2005), the performance of 1259 students on a biology exam was recorded, along with whether the students had eaten breakfast the morning of the exam. In the 2×2 table to the right, we have grouped the students further by whether they made a C or better on the exam.

TABLE 13.3

	Breakfast	No Breakfast	Totals
C or above on Exam	600	220	820
Below C on Exam	225	214	439
Totals	825	434	1259

Of the 825 students who ate breakfast, 600 got a C or above on the exam. That's 72.73% of those students. In contrast, of the 434 students who did not eat breakfast, 220 got a C or above. That is only 50.69% of those students. Hence, it is evident that eating breakfast is associated with getting a C or above on this biology exam. Is this any evidence of causation? That question has to be answered much more carefully.

Let's suppose we have data on whether those same students stayed up all night watching *Game of Thrones* episodes, or went to bed before 10 p.m. To be clear, this part of the data is completely made up. The original data table shown above might subdivide as follows:

TABLE 13.4

	Breakfast	No Breakfast	Totals
C or above on Exam	600	220	820
Below C on Exam	225	214	439
Totals	825	434	1259

Watched *Game of Thrones* all night

	Breakfast	No Breakfast	Totals
C or above on Exam	90	160	250
Below C on Exam	180	160	340
Totals	270	320	590

Went to bed by 10 p.m.

	Breakfast	No Breakfast	Totals
C or above on Exam	460	110	570
Below C on Exam	95	4	99
Totals	555	114	669

For those who watched *Game of Thrones* all night, we see the following:

- Of those who ate breakfast, 33.33% got a C or better on the exam.
- Of those who didn't eat breakfast, 50.00% got a C or better on the exam.

For those who went to bed before 10 p.m.:

- Of those who ate breakfast, 82.88% got a C or better on the exam.
- Of those who didn't eat breakfast, 96.49% got a C or better on the exam.

So in BOTH of the sub-tables, those who ate breakfast did worse as a group than those who did not eat breakfast. So the association seen in the original table reversed direction when broken down into sub-tables by whether the students watched *Game of Thrones* all night or went to bed by 10 p.m! In fact, if we just categorize the original data by that variable alone, we have Table 13.5.

TABLE 13.5

	Up all Night	In Bed by 10 p.m.	Totals
C or above on Exam	250	570	820
Below C on Exam	340	99	439
Totals	590	669	1259

We can see from the table that:

- Of those who were up all night, 42.37% got a C or better on the exam.
- Of those who were in bed by 10, 85.20% got a C or better on the exam.

So it seems that whether one was in bed early the night before the exam might be the variable that is really having the effect on the exam performance. This is an example of what is known as Simpson's Paradox.

> **Simpson's paradox** refers to the situation when associations seen in a 2 × 2 table change, or even reverse direction, when that table is broken down into sub-tables.

Extreme cases of Simpson's paradox are not that common, but there are a few famous ones, and you'll see one of those in the exercises. Even when the cross-classifying into sub-tables has a less dramatic effect than it did in this manufactured example, it is not that hard to be mistaken about what is actually causing the effect when you are looking at data in a 2 × 2 table.

TAKE-HOME POINTS

- Association doesn't imply causation because of other variables that may be influencing the relationships.
- It is always a good idea to look for these other variables and other explanations.
- Be aware that strong association may well be evidence of causation.
- The association seen between variables in a 2 × 2 table may change or even reverse direction when that table is broken into sub-tables by a third variable.

Chapter 13 Exercises

Reading Check

Short Answer: Please provide brief, concise answers to the following questions.

1. The CDC has claimed that gonorrhea rates and beer taxes are associated. What sort of association did they suggest? What did Sen and Luong have to say about the possibility of cause and effect in this example?

2. Explain the position of the Digital Bits Skeptic with respect to weight gain and diet soda.

3. Explain how the work by the researchers at the UT Health Science Center at San Diego counters the argument of the Digital Bits Skeptic.

4. Give an example of two variables that are clearly associated but likely have no causal relationship at all.

5. Give an example of two variables that are clearly associated and likely have a causal relationship.

Beyond the Numbers

1. Biology for Breakfast

Please read the following excerpt and look at the plot that follows:

Title: Does Eating Breakfast Affect the Performance of College Students on Biology Exams?

Author: Gregory W. Phillips, Blinn College

Source: *Bioscience* 30, no. 4 (2005): 15–19.

Abstract   This study examined the breakfast eating habits of 1,259 college students over an eleven- year period to determine if eating breakfast had an impact upon their grade on a General Biology exam.

…

Results and Discussion   This study showed that students who ate breakfast had a higher success rate on General Biology exams than those students who did not eat breakfast. This finding supports earlier research, which indicated that eating breakfast affects student performance.

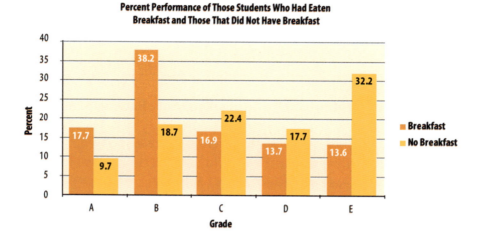

Percent Performance of Those Students Who Had Eaten Breakfast and Those That Did Not Have Breakfast

a. The data revealed that 65.53 percent self-identified that they had eaten some type of breakfast. How many had eaten breakfast?

   i.   1259

   ii. 17.7

   iii. 825

   iv. 648

b. Of those who didn't eat breakfast, about how many got a "C"?

   i.   97

   ii. 73

   iii. 434

   iv. 22

c. Explain why the graph shows clear evidence of association.

d. Suggest and defend two possibilities for a third variable that may be creating the illusion of cause and effect between "Eating Breakfast" and "Performance on Biology Exam."

2. Mortality and Global Warming

On November 14, 2007, the Kentucky Legislature held hearings on global warming with speakers Christopher Walker Monckton and James Taylor of the Heartland Institute. Have a look at the graph shown here. The graph records the CO_2 emissions and the child mortality rates for each of 192 countries. It is allegedly typical of the exhibits that Monckton and Taylor showed.

For example, India is the purple circle centered at 1.3 tonnes of CO_2 per person (*x*-value) and 72 child/infant deaths per 1,000 births (*y*-value). The red circle represents China, and the green circle represents the United States.

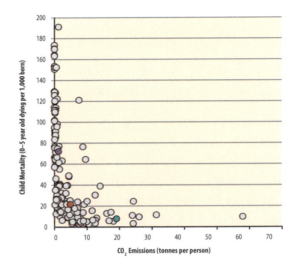

a. Reportedly, the Legislature was told that this graph shows that global warming is good because it *causes* a decrease in infant mortality. Is this an accurate assessment? Explain.

3. Some Bran Flakes on My Doughnut, Please

Please read the following excerpt:

Title: Whole Grains Do a Heart Good

Author: Ed Edelson

Source: HealthDay News, Monday October 22, 2007. http://www.washingtonpost.com/wp-dyn/content/article/2007/10/22/AR2007102201449.html

> Diets rich in whole grains, fruits, vegetables, and even a little alcohol may help ward off heart woes, new studies show. In one study, regular consumption of whole-grain breakfast cereal cut the risk of heart failure for male American physicians.
>
> Compared to those who ate no whole-grain cereal, men who consumed two to six servings per week saw their risk of heart failure fall by 21%, while those who ate seven or more servings per week reaped a 29% reduction in risk, the researchers reported in the October 22 issue of the *Archives of Internal Medicine*.

a. Does this study suggest that eating whole-grain cereal is associated with heart failure risk? If so, what is the nature of the suggested association? Is there evidence of causation? Explain.

4. College Sex at Berkeley

Please read the following excerpt:

Title: Sex Bias in Graduate Admissions: Data from Berkeley

Author: P.J. Bickel, E.A. Hammel, and J. W. O'Connell

Source: *Science* 187, no. 4175 (1975): 398–404.

Some years back, the University of Berkeley found evidence that there was sex discrimination in admission to graduate school. The tables below represent data summarized from the original study. In the following, we are going to think of Berkeley as divided into two colleges. College A represents the Engineering College at UC Berkeley, and College B represents the English College at UC Berkeley. Granted, we have simplified the actual situation, but it will make our point effectively.

The admission numbers for both colleges combined are in Table 13.6.

a. What is the admission rate for men?

   i.   0.30
   ii. 0.25
   iii. 0.45
   iv. 0.62

TABLE 13.6 Both Colleges Combined

Gender	Not Admitted	Admitted	Totals
Men	1493	1198	2,691
Women	1278	557	1,835

b. What is the admission rate for women?

   i.   0.30
   ii. 0.25
   iii. 0.26
   iv. 0.45

c. Look at the answers you gave to parts a.–b. What association do you see between admission success to the university and gender? Explain.

College A's admission numbers are in Table 13.7.

d. What is the admission rate for men to College A?

   i.   0.80
   ii. 0.25
   iii. 0.26
   iv. 0.62

TABLE 13.7 College A (College of Engineering)

Gender	Not Admitted	Admitted	Totals
Men	520	865	1,385
Women	27	106	133

e. What is the admission rate for women to College A?

    i. 0.80

    ii. 0.25

    iii. 0.26

    iv. 0.62

f. Look at the answers you gave to parts d.–e. What association do you see between admission success and gender in College A? Explain.

College B's admission numbers are in Table 13.8.

TABLE 13.8 College B (College of English)

Gender	Not Admitted	Admitted	Totals
Men	973	333	1,306
Women	1251	451	1,702

g. What is the admission rate for men to College B?

    i. 0.80

    ii. 0.25

    iii. 0.26

    iv. 0.62

h. What is the admission rate for women to College B?

    i. 0.80

    ii. 0.25

    iii. 0.26

    iv. 0.62

i. Look at the answers you gave to parts g.–h. What association do you see between admission success and gender in College B? Explain.

j. Is this exercise an example of Simpson's Paradox or not? Defend your answer.

5. Race and the Death Penalty

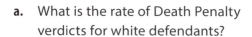

Please read the following excerpt and look at the plot that follows:

Title: Racial Discrimination in Criminal Sentencing: A Critical Evaluation of the Evidence with Additional Evidence on the Death Penalty.

Author: Gary Kleck, Florida State University

Source: *American Sociological Review* 46, no. 6 (1981): 783-805.

In what follows we only present summary data from this study. Professor Fleck was interested, in part, in studying the cases of 326 victims of "nonprimary" homicides. A nonprimary homicide is one in which the defendant and the victim did not know each other, except in a vague way. That is, they're not lovers, relatives, or friends. He found that these 326 cases cross-classified by race of defendant as in Table 13.9.

a. What is the rate of Death Penalty verdicts for white defendants?

    i. 17.5 out of 100

    ii. 5.8 out of 100

    iii. 10.2 out of 100

    iv. 11.9 out of 100

TABLE 13.9 Imposition of the Death Penalty by Race

	White Defendant	Black Defendant
Death Penalty	19	17
No Death Penalty	141	149
Total Cases	160	166

b. What is the rate of Death Penalty verdicts for black defendants?

    i. 17.5 out of 100

    ii. 5.8 out of 100

    iii. 10.2 out of 100

    iv. 11.9 out of 100

c. Look at the answers you gave to parts a.–b. What association do you see between a verdict of death and race? Explain.

The table above was then broken into two tables, depending on whether the victim of the crime was black or white. Table 13.10 contains the results for when the victim was white.

TABLE 13.10 Imposition of the Death Penalty by Race of Defendant: White Victim

	White Defendant	Black Defendant
Death Penalty	19	11
No Death Penalty	132	52
Total Cases	151	63

d. What is the rate of Death Penalty verdicts for white defendants when the victim was white?

    i.   17.5 out of 100

    ii. 12.6 out of 100

    iii. 0 out of 100

    iv. 5.8 out of 100

e. What is the rate of Death Penalty verdicts for black defendants when the victim was white?

    i.   17.5 out of 100

    ii. 12.6 out of 100

    iii. 0 out of 100

    iv. 5.8 out of 100

f. Look at the answers you gave to parts d.–e. What association do you see between a death verdict and race, when the victim is white? Explain.

Finally, the overall data were subset for the cases when the victim was black.

g. What is the rate of Death Penalty verdicts for white defendants when the victim was black?

    i.   17.5 out of 100

    ii. 12.6 out of 100

    iii. 0 out of 100

    iv. 5.8 out of 100

TABLE 13.11 Imposition of the Death Penalty by Race of Defendant: Black Victim

	White Defendant	Black Defendant
Death Penalty	0	6
No Death Penalty	9	97
Total Cases	9	103

h. What is the rate of Death Penalty verdicts for black defendants when the victim was black?

    i.   17.5 out of 100

    ii. 12.6 out of 100

    iii. 0 out of 100

    iv. 5.8 out of 100

i. Look at the answers you gave to parts g.–h. What association do you see between a death verdict and race when the victim is black? Explain.

j. Is this exercise an example of Simpson's Paradox or not? Defend your answer.

Projects

1. Experts' Advice

This mini-project is appropriate for a homework assignment. Follow your instructor's instructions. The goal of the activity is to become better acquainted with the advice and guidance offered by experts who are concerned with causation and association.

Choose and Find

Choose one of the two articles and find it online or in your library. Often, just searching on the title will locate the article.

- Distinguishing Association from Causation: A Backgrounder for Journalists. Kathleen Meister, October 29, 2007. (Any article written FOR journalists on this topic can be substituted with your instructor's permission).

- On the Nature of Evidence and 'Proving' Causality: Smoking and Lung Cancer vs. Sun Exposure, Vitamin D and Multiple Sclerosis. Lucas and Harris, August 12, 2018. (Any article written on the topic of evidence and causality can be substituted with your instructor's permission).

Read and Summarize

Read your chosen article in depth. Create a powerpoint that summarizes what you have read. The powerpoint has to have a minimum of 15 slides. You should NOT just grab text from the article and create the slides. The slides should show you have assimilated the material you read and are now capable of explaining it to others.

What You Turn In

Turn in your powerpoint. Follow your instructor's lead. She may want you to convert the powerpoint to a pdf prior to submission.

2. I Got Your Simpson Right Here

Introduction

This mini-project is appropriate for a homework assignment or, with help from your instructor getting started, an in-class assignment. Follow your instructor's instructions. The goal of the activity is to become better acquainted with Simpson's Paradox.

Write and Erase and Write

Simpson's Paradox is not really a paradox. Still yet, that doesn't make it any less important to understand. With a marginal comfort level with fractions, and a lot of concentration, you should be able to generate your own examples of Simpson's Paradox.

Your instructor may have you do this as individuals or may form you into teams. We'll refer to "teams" below, although you may be a team of one depending on what your instructor determines. Each team is required to do the following:

- **Main Table**

  - ▸ Create a 2×2 table that cross-classifies two variables. These data should be completely made up by you.

  - ▸ Exhibit the proper fractions to describe the association between the two variables in the main table.

- **Sub-Tables**

  - ▸ Make up a third variable and subset your Main Table into two 2×2 sub-tables based on binary values of this third variable.

  - ▸ Exhibit the proper fractions to describe the association between the two variables in each of the sub-tables.

  - ▸ Create the sub-tables in such a way that the association seen in the sub-tables is inconsistent when what was seen in the Main Table.

What You Turn In

Turn in your Main Table, both sub-tables, and a well-formed summary of what the associations were in all cases, and in what sense they were inconsistent. Tables should be prepared with software, and the final submission should be typed. DO NOT use examples that you have found elsewhere.

3. Reality Video

Introduction

This chapter discussed association and causation, and the complexity of how they differ. Now, we want to give you a chance to both show you know what you are doing and to help teach others.

The Assignment

Your instructor may have you do this as individuals or may form you into teams. We'll refer to "teams" below, although you may be a team of one depending on what your instructor wants. Each team is required to do the following:

a. Pick one or more topics from this chapter. Your instructor may want to approve them before you proceed.

b. Construct a three-minute video that teaches the chosen topic to the class. You must carefully and correctly use the language of the chapter in your presentation, and you are required to give an example that is NOT in the chapter.

c. Post the video on YouTube and send the link to your instructor in the manner he stipulates.

APPENDIX

Introduction

Quick Reference

Term	Chapter	Page
Alternative Hypothesis	9	194
Associated	12	273
Baye's Rule	9	166
Chi-square value	12	284
Clinical significance	11	252
Coefficient of Variation	2	22
Cohen's d	11	258
Confidence coefficient	5	101
Confidence interval	5	99
Confidence interval for population mean	5	102
Confidence interval for population proportion	5	102
Confidence interval interpretation	5	103
Confounding	3	46
Correlation coefficient	12	278
Cramer's V	12	284
Empirical Rule	6	123
Experiment	3	45
Explanatory variable	3	46
False negative rate	8	157
False positive rate	8	157
Histogram	2	23
Hypothesis testing	9	192
Lurking variable	3	47
Margin of error	5	101
Mean	2	18
Median	1	18
Minimal clinical important difference	11	256
Negative association	12	275
Negative predictive value	9	165
Nominal data	12	273
Non-sampling error	7	141
Null hypothesis	9	194

Tables

VALIDATION OF THE STANDARDIZED FIELD SOBRIETY TEST BATTERY AT BACS BELOW 0.10%: DATA SET

Case	HGN	OLS	WAT	Total FST	Actual BAC	Case	HGN	OLS	WAT	Total FST	Actual BAC
229	0	0	1	1	0	130	2	1	1	4	0.07
254	0	0	1	1	0.02	297	4			4	0.08
66		1		1	0.067	271	2	0	2	4	0.1
142	2	0	0	2	0.005	119		2	2	4	0.121
217	2	0	0	2	0.03	296	2	1	2	5	0.01
191	2			2	0.034	88	2	1	2	5	0.02
182	2	0	0	2	0.038	176	4	1	0	5	0.02
109	2			2	0.04	295	2	1	2	5	0.02
259	2	0	0	2	0.04	137		2	3	5	0.03
199	0	1	1	2	0.048	288	2	1	2	5	0.03
113	2	0	0	2	0.05	13	0	1	4	5	0.043
22	2			2	0.06	54	4	1	0	5	0.046
67	2		1	3	0.022	20	2		3	5	0.049
145	0	3	0	3	0.03	164	2	1	2	5	0.07
53	2	0	1	3	0.032	99	2	2	1	5	0.12
15	2	0	1	3	0.04	223			5	5	0.18
287	2	0	1	3	0.04	91	2	2	2	6	0.017
89	2	1	0	3	0.05	111	2	3	1	6	0.027
123	0	1	2	3	0.05	31	2	3	1	6	0.039
258	2	1	0	3	0.053	29	4	1	1	6	0.05
35	2	0	2	4	0	76	4	1	1	6	0.05
11	2	1	1	4	0.01	260	6			6	0.07
247	4			4	0.016	86	6			6	0.088
6	2	0	2	4	0.02	8	6			6	0.09
294	2	0	2	4	0.02	212	6			6	0.11
231	2	0	2	4	0.03	114	4	1	1	6	0.116
74	2	1	1	4	0.04	79	6			6	0.12
214	2	1	1	4	0.04	198	6			6	0.2
58	4			4	0.05	59	6			6	0.22
14	2	1	1	4	0.058	107	6			6	0.23
34	4	0	0	4	0.058	221	6			6	0.23
12	2	0	2	4	0.06	24	6			6	0.26
211	2	1	1	4	0.06	157	6			6	0.33
232	4	0	0	4	0.06	126	4	1	2	7	0.04
293	2	2	0	4	0.06	289	2	1	4	7	0.04

Case	HGN	OLS	WAT	Total FST	Actual BAC	Case	HGN	OLS	WAT	Total FST	Actual BAC
75	2	4	1	7	0.048	282	6	4		10	0.08
52	4	2	1	7	0.05	112	6	1	3	10	0.09
2	2	2	3	7	0.06	37	6	1	3	10	0.1
263	2	3	2	7	0.07	116	6	1	3	10	0.1
149	6	1		7	0.13	234	4	2	4	10	0.11
117		3	4	7	0.14	124	6	3	1	10	0.142
165	2	3	3	8	0.02	115	6	4		10	0.21
27	4	2	2	8	0.05	227	4	4	3	11	0.06
219	2	3	3	8	0.07	185	6	3	2	11	0.07
213	4	2	2	8	0.08	224	4	4	3	11	0.07
291	2	4	2	8	0.08	135	3	3	5	11	0.078
131	4	2	2	8	0.082	28	6	2	3	11	0.08
72	4	2	2	8	0.09	50	4	4	3	11	0.08
166	6	1	1	8	0.09	298	6	3	2	11	0.085
292	4	2	2	8	0.09	152	6	2	3	11	0.104
125	4	0	4	8	0.091	205	6	3	2	11	0.11
280	6	1	1	8	0.1	96	6	3	2	11	0.12
134	6	1	1	8	0.102	144	6	3	2	11	0.12
62	6	1	1	8	0.15	200	6	1	4	11	0.14
56	6	1	1	8	0.16	257	6	3	2	11	0.14
189	6	1	1	8	0.173	286	6	4	1	11	0.15
71	2	3	3	8	0.19	159	6	1	4	11	0.16
84	2	5	2	9	0	133	6	3	2	11	0.176
104	3	3	3	9	0.037	80	6	3	2	11	0.3
183	6	1	2	9	0.04	30	4	5	3	12	0.05
187	2	6	1	9	0.05	16	6	1	5	12	0.069
226	6	1	2	9	0.058	32	6	4	2	12	0.076
121	6	2	1	9	0.06	1	4	2	6	12	0.089
186	4	4	1	9	0.063	235	6	3	3	12	0.09
106	6	2	1	9	0.08	204	6	3	3	12	0.11
140	4	4	1	9	0.08	220	6	3	3	12	0.11
163	6	3	0	9	0.08	238	6	4	2	12	0.11
193	2	3	4	9	0.1	23	6	4	2	12	0.12
10	6	3		9	0.12	97	6	3	3	12	0.12
77	6	3		9	0.156	101	6	3	3	12	0.12
93	6		3	9	0.2	281	6	4	2	12	0.12
47	6		3	9	0.25	188	6	5	1	12	0.128
209	6	2	1	9	0.32	70	6	4	2	12	0.131
122	4	2	4	10	0.017	38	6	3	3	12	0.14

Appendix

Case	HGN	OLS	WAT	Total FST	Actual BAC	Case	HGN	OLS	WAT	Total FST	Actual BAC
127	6	2	2	10	0.028	128	6	4	4	14	0.1
178	6	0	4	10	0.07	208	4	7	3	14	0.1
57	6	4	2	12	0.14	262	6	3	5	14	0.1
85	6	4	2	12	0.14	4	6	4	4	14	0.12
206	6	4	2	12	0.14	103	6	3	5	14	0.12
240	6	3	3	12	0.17	242	6	4	4	14	0.12
250		4	8	12	0.18	5	6	5	3	14	0.13
170	6	2	4	12	0.19	64	6	4	4	14	0.13
94	6	6		12	0.22	261	6	4	4	14	0.14
132	6	4	2	12	0.27	268	6	5	3	14	0.14
148	6	7	0	13	0.05	78	6	4	4	14	0.15
129	4	7	2	13	0.07	181	6	4	4	14	0.15
252	6	4	3	13	0.08	190	6	3	5	14	0.15
120	6	4	3	13	0.09	194	6	5	3	14	0.15
241	6	4	3	13	0.09	82	6	1	7	14	0.16
278	6	4	3	13	0.1	255	6	3	5	14	0.16
65	6	3	4	13	0.11	83	6	4	4	14	0.2
154	4	6	3	13	0.11	108	6	5	3	14	0.2
201	4		9	13	0.11	158	6	5	3	14	0.22
87	6	7		13	0.12	207	6	3	5	14	0.28
167	6	4	3	13	0.12	55	6	6	3	15	0.1
19	6	4	3	13	0.13	173	6	6	3	15	0.1
98	6	5	2	13	0.132	230	6	4	5	15	0.1
17	6	4	3	13	0.14	249	6	6	3	15	0.1
46	6	3	4	13	0.14	279	6	4	5	15	0.11
228	6	5	2	13	0.14	45	6	3	6	15	0.12
233	6	4	3	13	0.14	184	6	5	4	15	0.12
237	6	4	3	13	0.16	51	6	3	6	15	0.13
102	6	3	4	13	0.17	44	6	5	4	15	0.14
68	4	6	3	13	0.19	143	6	4	5	15	0.15
277	6	4	3	13	0.19	265	6	4	5	15	0.15
284	6	3	4	13	0.19	100	6	5	4	15	0.17
81	6	3	4	13	0.2	285	6	5	4	15	0.17
162	6	3	4	13	0.23	275	6	4	5	15	0.2
175	4	6	4	14	0.07	273	6	4	5	15	0.22
196	6	2	6	14	0.074	171	6	4	5	15	0.23
153	6	3	5	14	0.098	40	6	3	6	15	0.24
105	6	4	4	14	0.1	33	6	9	1	16	0.11

Case	HGN	OLS	WAT	Total FST	Actual BAC	Case	HGN	OLS	WAT	Total FST	Actual BAC
156	6	7	3	16	0.11	172	6	7	5	18	0.14
266	6	7	3	16	0.11	49	6	4	8	18	0.15
18	6	6	4	16	0.12	216	6	7	5	18	0.15
169	6	7	3	16	0.12	270	6	8	4	18	0.15
174	6	7	3	16	0.13	195	6	7	5	18	0.16
251	6	4	6	16	0.14	256	6	6	6	18	0.16
264	6	4	6	16	0.14	239	6	4	8	18	0.17
7	6	7	3	16	0.15	276	6	4	8	18	0.19
25	6	6	4	16	0.16	210	6	4	8	18	0.2
218	6	4	6	16	0.19	110	6	4	8	18	0.21
244	6	5	5	16	0.21	95	6	4	8	18	0.23
236	6	5	5	16	0.22	168	6	4	8	18	0.24
141	6	7	4	17	0.1	272	6	4	8	18	0.27
9	6	6	5	17	0.11	36	6	5	8	19	0.14
26	6	5	6	17	0.12	179	6	6	7	19	0.14
222	6	6	5	17	0.12	243	6	9	4	19	0.15
147	6	5	6	17	0.13	63	6	5	8	19	0.16
225	6	6	5	17	0.13	139	6	8	5	19	0.16
269	6	7	4	17	0.17	203	6	7	6	19	0.17
290	6	4	7	17	0.17	160	6	7	6	19	0.19
42	6	5	6	17	0.18	215	6	7	6	19	0.19
146	6	6	5	17	0.18	202	6	6	7	19	0.22
69	6	5	6	17	0.2	3	6	4	9	19	0.23
177	6	3	8	17	0.2	118	6	9	4	19	0.23
136	6	6	5	17	0.219	90	6	12	2	20	0.12
274	6	4	7	17	0.24	245	6	6	8	20	0.12
92	6	4	7	17	0.26	43	4	8	8	20	0.13
161	6	3	8	17	0.29	155	6	9	5	20	0.165
61	6	6	5	17	0.32	138	6	7	7	20	0.19
246	6	9	3	18	0.07	253	6	7	8	21	0.17
39	6	7	5	18	0.1	41	6	8	7	21	0.22
21	6	7	5	18	0.11	283	6	7	8	21	0.24
48	6	8	4	18	0.12	192	6	9	7	22	0.14
180	6	7	5	18	0.13	151	6	8	8	22	0.18
267	6	5	7	18	0.13	248	6	9	8	23	0.23
73	6	9	3	18	0.14	197	6	12	8	26	0.17

Cancer Survival Data Table

This table is referenced in Chapter 10 of this workbook. Please reference as necessary to complete the related Beyond the Numbers. Please note that there are more data sets located at www.statconcepts.com/datasets.

Survival Time (number of days) Post-Diagnosis

Stomach Cancer	Bronchus Cancer	Colon Cancer	Ovarian Cancer	Breast Cancer
124	81	248	1,234	1,235
42	461	377	89	24
25	20	189	201	1,581
45	450	1,843	356	1,166
412	246	180	2,970	40
51	166	537	456	727
1,112	63	519		791
46	64	455		1,804
103	155	406		3,460
876	859	365		719
146	151	942		
340	166	776		
396	37	372		
	223	163		
	138	101		
	72	20		
	245	283		

Child Mortality Rate and CO$_2$ Emissions

This table is referenced in Chapter 12 of this workbook. Please reference as necessary to complete the related Beyond the Numbers. Please note that there are more data sets located at www.statconcepts.com/datasets.

Country	Child Mortality (0–5 Years Old Dying per 1,000 Born)	CO$_2$ Emissions (Tonnes per person)
Afghanistan	114.5	0.0447674078
Albania	21.2	1.342792013
Algeria	25.1	3.213169256
Andorra	4	6.839939571
Angola	191.1	1.308848969
Antigua and Barbuda	12.3	5.006748909
Argentina	16.5	4.38612968
Armenia	22.3	1.427325515
Australia	5.5	18.19043099
Austria	4.6	8.659639646
Azerbaijan	48.1	4.500681609
Bahamas	18.3	7.18475043
Bahrain	11	24.02689968
Bangladesh	63.4	0.3383501821
Barbados	20.2	5.059113166
Belarus	8.7	6.323492475
Belgium	4.8	10.19806569
Belize	20.9	1.417945547
Benin	113.9	0.4923540303
Bhutan	57.7	0.5815093849
Bolivia	55.1	1.616906602
Bosnia and Herzegovina	8.8	7.267229183
Botswana	63.1	2.424634582
Brazil	21.5	1.849542336
Brunei	8.6	13.02037624
Bulgaria	15	6.36437645
Burkina Faso	150.7	0.09303204353
Burundi	129.4	0.02501885445
Cambodia	57.9	0.3013984632
Cameroon	119.8	0.2132781232
Canada	6	16.86231777
Cape Verde	27.5	0.6439917452
Central African Republic	153.7	0.05561625501
Chad	173.6	0.04036061966

Country	Child Mortality (0–5 Years Old Dying per 1,000 Born)	CO_2 Emissions (Tonnes per person)
Chile	9.1	3.808340726
China	21.8	4.879028386
Colombia	21.1	1.44026659
Comoros	91.8	0.183178187
Congo, Dem. Rep.	170.2	0.04033503578
Congo, Rep.	111.7	0.4021796335
Cook Is	13.1	3.357240958
Costa Rica	10.4	1.753913518
Cote d'Ivoire	128.1	0.3895562687
Croatia	6.5	5.226528118
Cuba	6.5	2.432374468
Cyprus	4.6	7.429369477
Czech Republic	5	11.96774619
Denmark	4.8	10.03060277
Djibouti	96.3	0.592056967
Dominica	13.9	1.601164483
Dominican Republic	32.6	2.234733962
Ecuador	28	2.187681241
Egypt	29.1	2.481030897
El Salvador	21.6	1.126953876
Equatorial Guinea	121.1	7.593759438
Eritrea	66.6	0.1207532679
Estonia	6.5	12.60653841
Ethiopia	101.9	0.07922617529
Fiji	22.3	1.647251481
Finland	3.6	12.55148816
France	4.5	6.227735158
Gabon	77.1	1.414503757
Gambia	91	0.2180199523
Georgia	24.6	1.384029306
Germany	4.6	9.835780459
Ghana	85.7	0.4189189783
Greece	5.2	8.67047558
Grenada	14.7	2.241913099
Guatemala	39.7	0.9741281309
Guinea	128.9	0.1283062635
Guinea-Bissau	152.3	0.1970631147
Guyana	41	1.72475003
Haiti	88.1	0.2235787453
Honduras	29.2	0.9833134529

Country	Child Mortality (0–5 Years Old Dying per 1,000 Born)	CO_2 Emissions (Tonnes per person)
Hungary	7.9	5.686384664
Iceland	3	7.56453274
India	72.1	1.300067332
Indonesia	39.9	1.500915597
Iran	24.2	7.157892726
Iraq	39.8	4.062506807
Ireland	4.9	10.24411158
Israel	5.3	9.922316474
Italy	4.3	7.957277529
Jamaica	20.1	4.457657445
Japan	3.6	9.735405481
Jordan	23	3.772682789
Kazakhstan	30.6	12.54522526
Kenya	93.7	0.2619983112
Kiribati	65.7	0.3136114496
Kuwait	11.5	31.36899175
Kyrgyz Republic	38.1	1.001824647
Lao	94.2	0.2705322014
Latvia	11.7	3.306760631
Lebanon	12.9	3.666257713
Liberia	108.8	0.229047855
Libya	21.7	9.124147244
Lithuania	9.5	4.206329704
Luxembourg	3.3	24.27979649
Macedonia, FYR	13.1	5.353489722
Madagascar	76.7	0.09133401386
Malawi	111.4	0.07224778808
Malaysia	8.1	6.418045513
Maldives	21	2.90228743
Mali	164.2	0.04181136341
Malta	6.8	6.261494001
Marshall Islands	39.8	1.755730065
Mauritania	99.3	0.5353637754
Mauritius	16	2.981538148
Mexico	18.9	4.096585267
Micronesia, Fed. Sts.	46.1	0.5009609342
Moldova	22	1.344294059
Mongolia	39.4	3.653693572
Montenegro	9.8	6.718505312
Morocco	39.1	1.530221857

Country	Child Mortality (0–5 Years Old Dying per 1,000 Born)	CO_2 Emissions (Tonnes per person)
Mozambique	125.7	0.09299725066
Myanmar	64.4	0.2795320021
Namibia	61.5	1.42952464
Nauru	39	14.11230633
Nepal	57	0.09563957891
Netherlands	5.2	10.20794675
New Zealand	6.5	8.099271756
Nicaragua	30.3	0.7862586883
Niger	163.4	0.06074727218
Nigeria	152.5	0.6510210507
Niue	27.8	2.231689998
North Korea	32.3	3.560199223
Norway	3.9	9.478448647
Oman	13	16.16612196
Pakistan	98.4	0.9029698794
Palau	24.7	10.07779055
Panama	22.2	1.998836676
Papua New Guinea	73.4	0.7356917744
Paraguay	26.9	0.6633273403
Peru	26.3	1.268302202
Philippines	35.4	0.7117307682
Poland	7.3	8.367266757
Portugal	4.4	5.587546549
Qatar	9.8	57.98689476
Romania	19.4	4.698971559
Russia	15.6	11.63299408
Rwanda	94.8	0.07301172445
Samoa	18.9	0.8722141703
Sao Tome and Principe	66.5	0.8281172698
Saudi Arabia	14.1	17.44796401
Senegal	91.3	0.4287251586
Serbia	8.4	5.465046998
Seychelles	13.9	8.830519668
Sierra Leone	211.7	0.3049037635
Singapore	2.9	10.63593619
Slovak Republic	9.5	7.178985152
Slovenia	4.1	8.08825007
Solomon Islands	35.3	0.3719806764
Somalia	169.9	0.06734915107
South Africa	76.5	8.789515822

Country	Child Mortality (0–5 Years Old Dying per 1,000 Born)	CO_2 Emissions (Tonnes per person)
South Korea	5.3	9.95950733
Spain	5.5	7.951434227
Sri Lanka	12.6	0.5852171122
St. Kitts and Nevis	11.8	4.711141449
St. Lucia	18.7	2.195582488
St. Vincent and the Grenadines	23.2	1.851545812
Sudan	87.7	0.3882809165
Suriname	25.4	4.851000733
Swaziland	122.3	0.9087893986
Sweden	3.4	5.452381206
Switzerland	5	5.606771687
Syria	18	2.832061721
Taiwan	6.7	11.80150996
Tajikistan	71.6	0.4141739342
Tanzania	83.5	0.1492433546
Thailand	16.8	4.220817479
Timor-Leste	75.4	0.1729939443
Togo	109.9	0.2207993334
Tonga	15.4	1.696461385
Trinidad and Tobago	24.5	24.3478454
Tunisia	21.5	2.308450239
Turkey	21.8	3.787048964
Turkmenistan	64.4	9.619770615
Uganda	101.3	0.09038624379
Ukraine	14	7.034864957
United Arab Emirates	9.6	26.07472134
United Kingdom	5.9	8.952961552
United States	7.9	19.1514501
Uruguay	14.6	1.997825563
Uzbekistan	48.9	4.442455833
Vanuatu	21.4	0.2199052716
Venezuela	17.8	6.262415884
Vietnam	25.2	0.9843563771
West Bank and Gaza	26.2	0.6229229602
Yemen	74.6	0.9766033276
Zambia	120.4	0.1944096102
Zimbabwe	95.8	0.8255348348

Standard Score Table

Example: If a standard score, z, is computed to be 1.73, then one would locate 1.73 in the table (see highlighting) and corresponding p-value would be 0.04182 ≈ 0.04. This would then have to be compared to the preset Type I error rate, often taken to be 0.05. This process only applies for simple hypotheses with a positive z score and a ">" in the alternative. Your instructor will show you how to use this table when z is negative and/or when the alternative is a "≠."

z	0	0.01	0.02	0.03	0.04	0.05	0.06	0.07	0.08	0.09
0	0.5	0.49601	0.49202	0.48803	0.48405	0.48006	0.47608	0.4721	0.46812	0.46414
0.1	0.46017	0.4562	0.45224	0.44828	0.44433	0.44038	0.43644	0.43251	0.42858	0.42465
0.2	0.42074	0.41683	0.41294	0.40905	0.40517	0.40129	0.39743	0.39358	0.38974	0.38591
0.3	0.38209	0.37828	0.37448	0.3707	0.36693	0.36317	0.35942	0.35569	0.35197	0.34827
0.4	0.34458	0.3409	0.33724	0.3336	0.32997	0.32636	0.32276	0.31918	0.31561	0.31207
0.5	0.30854	0.30503	0.30153	0.29806	0.2946	0.29116	0.28774	0.28434	0.28096	0.2776
0.6	0.27425	0.27093	0.26763	0.26435	0.26109	0.25785	0.25463	0.25143	0.24825	0.2451
0.7	0.24196	0.23885	0.23576	0.2327	0.22965	0.22663	0.22363	0.22065	0.2177	0.21476
0.8	0.21186	0.20897	0.20611	0.20327	0.20045	0.19766	0.19489	0.19215	0.18943	0.18673
0.9	0.18406	0.18141	0.17879	0.17619	0.17361	0.17106	0.16853	0.16602	0.16354	0.16109
1	0.15866	0.15625	0.15386	0.15151	0.14917	0.14686	0.14457	0.14231	0.14007	0.13786
1.1	0.13567	0.1335	0.13136	0.12924	0.12714	0.12507	0.12302	0.121	0.119	0.11702
1.2	0.11507	0.11314	0.11123	0.10935	0.10749	0.10565	0.10383	0.10204	0.10027	0.09853
1.3	0.0968	0.0951	0.09342	0.09176	0.09012	0.08851	0.08692	0.08534	0.08379	0.08226
1.4	0.08076	0.07927	0.0778	0.07636	0.07493	0.07353	0.07215	0.07078	0.06944	0.06811
1.5	0.06681	0.06552	0.06426	0.06301	0.06178	0.06057	0.05938	0.05821	0.05705	0.05592
1.6	0.0548	0.0537	0.05262	0.05155	0.0505	0.04947	0.04846	0.04746	0.04648	0.04551
1.7	0.04457	0.04363	0.04272	0.04182	0.04093	0.04006	0.0392	0.03836	0.03754	0.03673
1.8	0.03593	0.03515	0.03438	0.03362	0.03288	0.03216	0.03144	0.03074	0.03005	0.02938
1.9	0.02872	0.02807	0.02743	0.0268	0.02619	0.02559	0.025	0.02442	0.02385	0.0233
2	0.02275	0.02222	0.02169	0.02118	0.02068	0.02018	0.0197	0.01923	0.01876	0.01831
2.1	0.01786	0.01743	0.017	0.01659	0.01618	0.01578	0.01539	0.015	0.01463	0.01426
2.2	0.0139	0.01355	0.01321	0.01287	0.01255	0.01222	0.01191	0.0116	0.0113	0.01101
2.3	0.01072	0.01044	0.01017	0.0099	0.00964	0.00939	0.00914	0.00889	0.00866	0.00842
2.4	0.0082	0.00798	0.00776	0.00755	0.00734	0.00714	0.00695	0.00676	0.00657	0.00639
2.5	0.00621	0.00604	0.00587	0.0057	0.00554	0.00539	0.00523	0.00508	0.00494	0.0048
2.6	0.00466	0.00453	0.0044	0.00427	0.00415	0.00402	0.00391	0.00379	0.00368	0.00357
2.7	0.00347	0.00336	0.00326	0.00317	0.00307	0.00298	0.00289	0.0028	0.00272	0.00264
2.8	0.00256	0.00248	0.0024	0.00233	0.00226	0.00219	0.00212	0.00205	0.00199	0.00193
2.9	0.00187	0.00181	0.00175	0.00169	0.00164	0.00159	0.00154	0.00149	0.00144	0.00139
3	0.00135	0.00131	0.00126	0.00122	0.00118	0.00114	0.00111	0.00107	0.00104	0.001
3.1	0.00097	0.00094	0.0009	0.00087	0.00084	0.00082	0.00079	0.00076	0.00074	0.00071
3.2	0.00069	0.00066	0.00064	0.00062	0.0006	0.00058	0.00056	0.00054	0.00052	0.0005
3.3	0.00048	0.00047	0.00045	0.00043	0.00042	0.0004	0.00039	0.00038	0.00036	0.00035
3.4	0.00034	0.00032	0.00031	0.0003	0.00029	0.00028	0.00027	0.00026	0.00025	0.00024
3.5	0.00023	0.00022	0.00022	0.00021	0.0002	0.00019	0.00019	0.00018	0.00017	0.00017
3.6	0.00016	0.00015	0.00015	0.00014	0.00014	0.00013	0.00013	0.00012	0.00012	0.00011

Three Sample Rubrics for Assessing Group Work

Peer Evaluation Form for Group Work

Name: _____

Section Number: _____

Write the name of each of your group members in a separate column. For each person, indicate the extent to which you agree with the statement on the left, using a scale of 1–4 (1 = strongly disagree; 2 = disagree; 3 = agree; 4 = strongly agree). Total the numbers in each row.

Evaluation Criteria	Yourself	Group Member:	Group Member:	Group Member:	Group Member:	Group Member:
Attends Group Meetings Regularly and Arrives on Time						
Contributes Meaningfully to Group Discussions						
Completes Group Assignments on Time						
Prepares Work in a Quality Manner						
Demonstrates a Cooperative and Supportive Attitude						
Contributes Significantly to the Success of the Project						
Totals						

FEEDBACK ON TEAM DYNAMICS

1. How effectively did your group work?

2. Were the behaviors of any of your team members particularly valuable or detrimental to the team? Explain.

3. What did you learn about working in a group from this project that you will carry into your next group experience?

Adapted from a peer evaluation form development at Johns Hopkins University (October, 2006).

RUBRIC FOR ASSESSING GROUP MEMBERS' ABILITY TO PARTICIPATE EFFECTIVELY AS PART OF A TEAM

Rater: _____

Date: _____

Group Topic: _____

Circle the appropriate score for each criterion for each member of your group.

Member Rated (Be sure to rate yourself, too!)	Listening Skills	Openness to Others' Ideas	Preparation	Contribution	Leadership
	0 1 2 3 4 5	0 1 2 3 4 5	0 1 2 3 4 5	0 1 2 3 4 5	0 1 2 3 4 5
	0 1 2 3 4 5	0 1 2 3 4 5	0 1 2 3 4 5	0 1 2 3 4 5	0 1 2 3 4 5
	0 1 2 3 4 5	0 1 2 3 4 5	0 1 2 3 4 5	0 1 2 3 4 5	0 1 2 3 4 5
	0 1 2 3 4 5	0 1 2 3 4 5	0 1 2 3 4 5	0 1 2 3 4 5	0 1 2 3 4 5
	0 1 2 3 4 5	0 1 2 3 4 5	0 1 2 3 4 5	0 1 2 3 4 5	0 1 2 3 4 5
	0 1 2 3 4 5	0 1 2 3 4 5	0 1 2 3 4 5	0 1 2 3 4 5	0 1 2 3 4 5

Criterion	Excellent (5)	Good (4)	Fair (3)	Needs to Improve (2)	Unacceptable (1)	Missing (0)
Listening Skills	Routinely restates what others say before responding; rarely interrupts; frequently solicits others' contributions; sustains eye contact	Often restates what others say before responding; usually does not interrupt; often solicits others' contributions; makes eye contact	Sometimes restates what others say before responding; sometimes interrupts; sometimes asks for others' contributions; sometimes makes eye contact	Rarely restates what others say before responding; often interrupts; rarely solicits others' contributions; does not make eye contact; sometimes converses with others when another team is speaking	Doesn't restate what others say before responding; often interrupts; doesn't ask for contributions from others; is readily distracted; often talks with others when another team member speaks	Never shows up and never contributes
Openness to Others' Ideas	Listens to others' ideas without interrupting; responds positively to ideas even if rejecting; asks questions about the ideas	Listens to others' ideas without interrupting; responds positively to the ideas even when rejecting	Sometimes listens to others' ideas without interrupting; generally responds to the ideas	Interrupts others' articulation of their ideas; does not comment on the ideas	Interrupts others' articulation of their ideas; makes deprecatory comments and/or gestures	Never shows up and never contributes
Preparation	Always completes assignments; always comes to team sessions with necessary documents and materials; does additional research, reading, writing, designing, implementing	Typically completes assignments; typically comes to team sessions with necessary documents and materials	Sometimes completes assignments; sometimes comes to team sessions with necessary documents and materials	Sometimes completes assignments; sometimes comes to team sessions with necessary documents and materials	Typically does not complete assignments; typically comes to team sessions without necessary documents and materials	Never shows up and never contributes
Contribution	Always contributes; quality of contributions is exceptional	Usually contributes; quality of contributions is solid	Sometimes contributes; quality of contributions is fair	Sometimes contributes; quality of contribution is inconsistent	Rarely contributes; contributions are often peripheral or irrelevant; frequently misses team sessions	Never shows up and never contributes
Leadership	Seeks opportunities to lead; in leading is attentive to each member of the team, articulates outcomes for each session and each project, keeps team on schedule, foregrounds collaboration and integration of individual efforts	Is willing to lead; in leading is attentive to each member of the team, articulates general direction for each session and each project, attempts to keep team on schedule	Will take lead if group insists; not good at being attentive to each member of the team, sometimes articulates direction for sessions, has some trouble keeping team on schedule	Resists taking on leadership role; in leading allows uneven contributions from team members, is unclear about outcomes or direction, does not make plans for sessions or projects	May volunteer to lead but does not follow through; misses team sessions, does not address outcomes or direction for sessions or projects, team members become anarchial	Never shows up and never contributes

GROUP PROCESS QUESTIONS

1. Describe any communication problems within your group, or describe how well members of your group were able to communicate with each other.

2. Did you meet outside of class to establish goals and stay in tune with each other?

3. What worries you most when working in groups?

4. Did you think you did your fair share?

5. Did others do their fair share?

PEER AND SELF EVALUATION RUBRIC

Project/Activity: _____

Please rate your contribution to the group and evaluate each individual group member using a scale of 1–5, with 5 being the highest. Explain your reasons for any assigned ranking less than 4. Make sure you evaluate yourself on the first line.

Ranking	Group Member
_____	_____
_____	_____
_____	_____
_____	_____
_____	_____
_____	_____

Explanations, as needed:

Obtaining Information from *Quickbooks*, Including Reports

4

Practice — Maintenance Activities

5

Practice — Purchases and Cash Disbursements Cycle Activities

6

(continued on the following page)

Practice — Purchases and Cash Disbursements Cycle Activities *(continued)*

6

Practice — Sales and Cash Receipts Cycle Activities

7

(continued on the following page)

Practice — Sales and Cash Receipts Cycle Activities *(continued)*   **7**

Practice — Payroll Cycle and Other Activities   **8**

Recording Transactions, Performing Month-end Procedures, Recording Year-end Adjusting Entries, and Printing Reports

9

New Company Setup

10

Acknowledgments

Greatly prized are the exceptional efforts of Regina Rexrode for word processing in preparation of the manuscript, Patricia Naretta for extraordinary proofreading and other assistance, and Erica Borsum for her valuable comments in testing the entire project.

Finally, the encouragement and continuing support of family, friends, and associates have contributed in large measure to the completion of this book.

This page is intentionally blank.

INTRODUCTION

Steps for Successful Completion of the Project

For the entire project, the following symbol is used to indicate that you are to perform a step using your computer:

Whenever you see this symbol in the left margin, you should complete the related step, which is shown in italics. You should not begin doing an activity with your computer until the symbol is shown.

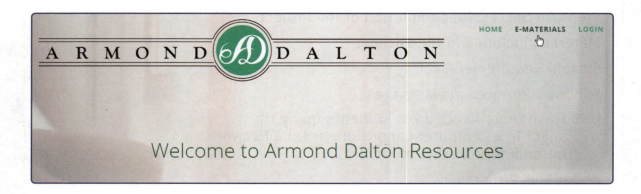

Welcome to Armond Dalton Resources

IN ORDER TO SUCCESSFULLY ACCESS THE SOFTWARE AND DATA REQUIRED FOR THIS PROJECT AND TO PROPERLY BACK UP AND RESTORE FILES, YOU WILL NEED TO CAREFULLY READ AND APPLY THE STEPS DETAILED IN THE E-MATERIALS PROVIDED ON THE ARMOND DALTON RESOURCES WEBSITE.

> *Go to www.armonddaltonresources.com and click on E-Materials (see image above).*
>
> *Use the E-materials drop-down menu and select QuickBooks Pro 2018 5th edition.*
>
> *Select the E-materials that best suit your situation, Personal Computer or Lab Computer.*
>
> *Follow the instructions in the E-materials to register on the Armond Dalton Resources website, download the data sets, follow the link to download the QuickBooks Pro software, and load the Waren Sports Supply, Jackson Supply Company, and Super Office Furniture Plus data sets into QuickBooks.*

Welcome to learning *QuickBooks Pro® 2018*. You will be using these materials to learn how to use an accounting package by following a carefully designed approach.

Materials Included in the Package

The materials that you have purchased are made up of four items:

1. **License Code to Provide Access to the *QuickBooks* software and to the company data files you will need for this project.** The sticker in the lower-left corner of the Instructions book cover includes the one-time-use license code you will need to access the *QuickBooks* software and the company datasets used in this project. The code is hidden by a scratch-off coating in the middle of the sticker. *Gently* remove the scratch-off coating using a coin or something similar. Do not apply too much pressure when removing the coating or you might also damage the license code.

2. **Instructions Book**—This book is the starting point for all assignments. It will guide you through *QuickBooks*. Unless your course instructor informs you otherwise, **you should start with Chapter 1 and proceed through the materials without skipping any parts.**

The software is an education version of the *QuickBooks Pro 2018* program. The education version of the software is similar to the software used commercially by thousands of companies, except that the software is a 160-day trial version.

> The software that you will download contains several educational versions of the *QuickBooks* software: *QuickBooks Pro* (the software that is the subject of this book), *QuickBooks Accountant*, and several different industry editions of *QuickBooks Premier*. The way the software is designed, the user can toggle back and forth between editions after installation. However, the default edition that opens each time the software is opened is *QuickBooks Accountant*, which is virtually the same as *QuickBooks Pro*. Rather than have you toggle to *QuickBooks Pro* each time you open the software, which becomes tedious after several times, you should just leave the screen at *QuickBooks Accountant*. For simplicity, we will refer to the software as *QuickBooks* throughout the remainder of the project.

3. **Reference Book**—The Reference book provides instructions for using each of the windows discussed in these materials. You will be referring to the Reference book frequently in later chapters. You will be instructed when to use the Reference book as you go through the Instructions book.

4. **Student Problems & Cases Book**—The Student Problems & Cases book includes questions, problems, and reporting exercises covering the various concepts and types of transactions in several chapters of the Instructions book. Your instructor will inform you which questions and problems to complete for each chapter and whether you should complete the homework requirements online or using the tear-out pages supplied (consult your instructor).

IMPORTANT WARNINGS . . .

1. **IF THE LICENSE CODE IS ALREADY REVEALED ON THE BOOK YOU PURCHASED, RETURN THE BOOK IMMEDIATELY TO THE BOOKSTORE AND EXCHANGE IT FOR A NEW COPY.** *You can only access the Armond Dalton data sets and Quickbooks software with a new book containing a license code that has not been previously used by another person.*

2. **THE *QUICKBOOKS* SOFTWARE AND THE INITIAL COMPANY BACKUP FILES CAN ONLY BE LOADED ON ONE COMPUTER**, so be careful to choose the computer that you want to use to complete the entire project.

System Requirements

For a complete listing of Systems Requirements visit https://community.intuit.com/articles/1436499-quickbooks-desktop-system-requirements.

The minimum hardware and software requirements for running the *QuickBooks* software include:

Recommended System Configuration

Operating systems supported (for stand-alone Installation)

- **Windows 10,** all editions including 64-bit, natively installed
- **Windows® 8.1 (Update 1)**, all editions including 64-bit, natively installed
- **Windows® 7 SP1**, all editions including 64-bit, natively installed
- **Windows Server 2016**
- **Windows Server® 2012 R2**
- **Windows Server 2011 SP1**
- **Windows Server® 2008 R2 SP1**

Database Servers (for Network/Lab Installation):

- **Windows:** Windows Server 2016, Windows Server 2012 R2, Windows Server 2011 SP1, Windows Server 2008 R2 SP1, Windows 10, Windows 8.1 (Update 1), or Windows 7 SP1 (Enterprise and Professional editions only), natively installed

  Linux: Only supported with *QuickBooks* **Database Server Manager** - OpenSuse 42.2, Fedora 25, Red Hat 7 (Update 3).

Hardware and operating system requirements (client and server)

> MAC computers are not supported. See the technical support section of Armond Dalton's website for suggestions for MAC users: www.armonddalton.com/support-updates/mac-users/

- 2.4 GHz processor minimum
- 4 GB RAM minimum, 8 GB recommended
- Server RAM Requirements
    - 1-5 Users: 8GB RAM
    - 10 Users: 12GB RAM
    - 15 Users: 16GB RAM
    - 20 Users: 20 Users: 20+GB RAM

- Display optimized for 1280 x 1024 screen resolution or higher with up to two extended monitors

- Best optimized for Default DPI setting for a given computer.

- Windows®:
    - U.S. version only
    - Regional Settings are supported when set to English (United States) with keyboard setting to U.S. only
    - Administrator Rights required for the server computer when hosting Multi User Access
    - Natively installed - i.e., installed on a specific system or environment for which it was designed, not installed to run in a virtual environment or emulation software.

- Disk space requirements:
    - 2.5 GB of disk space (additional space required for data files)
    - Additional software: 60 MB for Microsoft® .NET 4.6 Runtime, provided on the *QuickBooks* CD
    - Additional requirements for Intuit Data Protect in *QuickBooks* Connected Services offerings (applies to US only)
        - Require minimum 4.0 GB RAM
        - Twice the size of the largest File set to back up + 100MB or twice the size to restore. The space is only required from the work folder LocalApplicationData + "Intuit\Intuit Data Protect" **Note:** Storing a large .QBW data file on an SSD will greatly speed up the performance and is encouraged.

Software Integration and Compatibility

QuickBooks is capable of integrating with hundreds of third-party applications. The following integrations are provided with *QuickBooks*; additional RAM will enhance the use of these features. See Intuit Marketplace for the most up-to-date list.

- Microsoft® Office:
    - Office 2016 (including Outlook 2016) both on 32 and 64-bit.
    - Office 2010 and Office 2013/365 (including Outlook® 2010 and Outlook® 2013) both on 32 and 64 bit. (**Note:** Office 365 is only supported when it is locally installed, not the web version.)
    - Email Estimates, Invoices, and other forms with Microsoft Outlook 2010-2016, Microsoft Outlook with Office 365, GMail, Yahoo! Mail and Outlook.com, other SMTP-supporting e-mail clients.
    - Preparing letters requires Microsoft® Word 2016, 2013, 2010, or Office 365 (includes 64-bit).
    - Exporting reports requires Microsoft Excel® 2016, 2013, 2010, or Office 365 (includes 64-bit).
    - Contact Synchronization with Microsoft Outlook® requires Outlook® 2010 (32-bit).
    - Synchronization with Outlook® requires *QuickBooks* Contact Sync for Outlook® (the download is available at no charge). Contact Sync does not work with the Microsoft® Business Contact Manager Outlook® plug-in. If a sync is performed, duplicate records could result.

- QuickBooks Point of Sale V12.0, V11.0, V10.0 (applies to US only).

- TurboTax 2016 and 2015 (Personal and Business).

- Lacerte 2016 and 2015 (applies to US only).

- Pro-Series tax years 2016 and 2015 (applies to US only).

- Quicken 2017, 2016, 2015.

- Adobe Acrobat Reader: Business Planner, Payroll and viewing forms require Adobe Acrobat Reader 9.0 or later.

- Payroll and other online features and services require Internet access with at least a 56 Kbps connection speed, 1Mbps recommended. (DSL or cable modem recommended).

- Gmail, Yahoo Email, Windows Mail, Hotmail, and AOL (supported as a plain text version via the Mozilla Thunderbird email application).

- Internet Explorer 11.

Firewall and Antivirus Software Compatibility

QuickBooks 2018 has been tested with the following firewall and antivirus products.

- Windows Server Firewall (all editions)
- Windows 8.1 Firewall (all editions)
- Windows 7 Firewall (all editions)
- Microsoft Security Essentials
- Avast
- AVG
- Symantec
- ESET
- Avira
- Kaspersky
- McAfee
- Bitdefender
- Malwarebytes

> *Note:* In some cases, it may be necessary to adjust settings in these products to ensure the best possible performance with *QuickBooks*. Also note that *QuickBooks* will work with systems running RAID (Redundant Array of Inexpensive Disks), but this is not recommended because performance issues may cause *QuickBooks* to operate slowly.

Windows® Permission Messages

Recent versions of Windows® have more security features than older versions and the permissions messages are one example of this increased security.

Permissions messages may appear when you are working with the software. Whenever you receive a permissions message while attempting to complete a task with *QuickBooks Pro 2018,* give Windows® permission to complete the task.

USB Flash Drive Needed for Students Doing the Project in a Computer Laboratory or On a Network *(Optional for all other students)*

If you are doing the project in a computer laboratory, where the *QuickBooks* software is already installed on the hard drive of each machine, you will need to use a USB flash drive (256 MB is large enough) for saving the data that you process. If you are doing the project in a computer laboratory where the *QuickBooks* software is installed on a network, consult your instructor about whether you should save your data on a USB flash drive or on a designated space on the network. Even if you have a designated space on the network for saving your data, you may still want to back up externally to a USB flash drive as an extra precaution.

If you are using your own computer for the project, you can do the project without a USB flash drive. However, if you are concerned with data loss on your hard drive and you want to back up your data externally, you will also need a USB flash drive.

About This Project

By doing this project you will learn about *QuickBooks*, accounting software which is commercially available and widely used by many small and medium-sized companies throughout the world. More importantly, you will learn what this type of system offers companies to more effectively operate their businesses. Some of the things you will learn include the following:

- How to process economic events with *QuickBooks*.
- The effects of processing transactions on a wide variety of data files in the system that help companies operate their businesses more effectively.
- How a company enters information such as its general ledger, vendor, and customer accounts into the system.
- How to make inquiries of the data files as a part of managing a business.
- Ways to analyze data to improve business decisions.
- The nature of the internal controls included in the system to prevent and detect errors.
- Ways the system helps a company process data and prepare reports efficiently and quickly.

An important distinction between a paper system and computerized system such as *QuickBooks* is the lack of visibility in much of what takes place in a computerized system because it is electronic. The challenge with this lack of visibility is the difficulty understanding what is happening in the system. This lack of understanding in turn causes many users to either use the system improperly or fail to take advantage of all of its capabilities.

This project is intended to help you bridge the gap between what you know about a paper-based and a computer-based accounting information system. The benefit to you is an enhanced ability to contribute immediately to a business in a computer environment as an accountant, auditor, or user of information generated by such a system.

As you go through the different learning steps, you will be asked to identify where and how data are processed, perform inquiries into the system to analyze the data, make customized changes to the system, establish and test controls, and prepare reports to support business decision-making. You will learn these tasks by following the guidance provided as you proceed through the materials.

Three important characteristics of accounting software are (1) its ability to generate multiple-use information without entering information more than once, (2) the incorporation of shortcut methods to enter data, and (3) the embedding of internal controls in the software to detect and prevent errors. There are hundreds of examples of all three characteristics in most accounting software. An example of the first is the automatic update of other records when a sales invoice is prepared, such as the sales journal, general ledger, and accounts receivable master files. An example of the second is the automatic inclusion of a customer's name and address when the customer's identification number is entered in the system. An example of the third is the rejection of an accounting transaction where relevant information, such as the customer's name, is not entered for a transaction.

There is a wide variety of accounting software available for companies to purchase and modify for their company's needs. Examples of accounting software for medium-size and larger-size companies are Microsoft Dynamics® GP and Sage 500 ERP. The software used in this project, *QuickBooks*, is widely used by small businesses and is commercially available from Intuit. The reason for selecting *QuickBooks* for this project is that, although the software is relatively easy to use, it also includes several internal controls that can be implemented to improve controls over accessing and processing accounting information.

Key Activities Included in the Project

Accountants perform several types of activities when they use accounting software such as *QuickBooks* to keep accounting records for companies. You will do many of these activities during the project. These activities are introduced in Chapters 1 and 2 and are dealt with more extensively in later assignments. Following are key activities you will be doing in the project:

- **Install the *QuickBooks* software and back up data.** You will learn this in the E-materials (*QuickBooks* E-material) that you download from the Armond Dalton Resources website. *Note:* **Installing the *QuickBooks* software is not applicable to students doing the project in a computer laboratory where the software is already installed on the individual machines in the lab or on a network.**

- **Open the *QuickBooks* program and open a company.** You must be able to access *QuickBooks* and the company for which you will be performing activities.

- **Perform maintenance.** An important characteristic of accounting software is the automatic performance of many mechanical activities by the computer. To permit the computer to do these activities, maintenance is done to provide an adequate database of information. For example, *QuickBooks* permits a user to enter a customer's identification number and the software automatically includes the customer's name and address on a sales invoice.

- **Process transactions, including all information needed for record keeping.** Accountants spend most of their time processing transactions. This is a major emphasis of the project. An example of processing transactions is entering data to bill customers for shipped goods, preparing the sales invoice, and recording the sale and related accounts receivable, including updating subsidiary records for accounts receivable and inventory.

- **Inquire about and analyze recorded information.** Management, employees, and outside users frequently need information about data in the system. For example, if a customer calls about an apparent incorrect billing, it is important to respond quickly. *QuickBooks* provides access to data in a variety of ways to permit many different types of inquiries.

- **Review and print reports.** Users of accounting information need reports in a proper format with an adequate level of detail for their decision making. Examples include an aged trial balance for accounts receivable, an income statement, and a report of sales by sales person or product. *QuickBooks* permits many different reports to be prepared and printed and allows tailoring to meet users' needs.

2 Chapter

FAMILIARIZATION

Introduction

This chapter illustrates *QuickBooks* features and provides practice using the program so that you are able to complete assignments for the project. Do not skip this chapter.

The discussion assumes that you have a working knowledge of Windows. If you need additional guidance for using Windows, consult your Windows user manual.

There are five primary activities in *QuickBooks*, all introduced in this chapter and dealt with extensively in later chapters. In most cases there is more than one way to access each of these activities, also covered in this chapter.

1. **Open the *QuickBooks* Program and Open a Company**—Used as is implied by the title.
2. **Maintenance**—Used to add, change, or delete information about such things as customers, vendors, employees, and general ledger accounts (chart of accounts).
3. **Processing Information Tasks**—Used to perform the tasks needed to process and record transactions such as processing a sales transaction or weekly payroll.
4. **Obtaining Information**—Used to obtain information already included in *QuickBooks* about such things as a customer address or a recorded sales transaction.
5. **Reports**—Used to view and print a wide variety of reports, such as an aged trial balance and a comparative income statement.

Open the *Quickbooks* Program and Open a Company

Starting now, you will use the computer to perform tasks with *QuickBooks*. In this chapter, the instructions are reasonably detailed to make certain that you understand how to use *QuickBooks* correctly and efficiently.

Depending on which version of Windows you are using, certain window functions and the appearance of the windows may differ slightly. However, the *QuickBooks* elements and functions are identical between different versions of the Windows operating system.

Do not be concerned about making mistakes while performing familiarization activities in this chapter. It is important that you practice each activity and learn by doing.

If, at any time, you decide that you want to start the chapter again, you may do so by restoring the Rock Castle Construction and Larry's Landscaping & Garden Supply datasets following the instructions in the E-materials. You may want to do so if you believe that you do not understand the material in the chapter.

To begin using *QuickBooks*, complete the following steps.

▶ *Click the "QuickBooks Premier–Accountant Edition 2018" icon on your desktop. It may take a couple of minutes to load.*

▶ *If you do not have the program icon on your desktop, click Start → QuickBooks → QuickBooks Premier–Accountant Edition 2018 to get the opening QuickBooks screen.*

Depending on whether or not this is the first time you've installed *QuickBooks*, various windows may or may not open. None of them are critical to the installation process, so the windows are not illustrated here. Following are brief descriptions of these windows, as well as the steps you should follow if they open.

▶ *If you receive a message at any time that tells you there is a product update available, make sure you have Internet access and then follow the instructions for installing the product update.*

▶ *If the "How QuickBooks uses your Internet Connection" window opens, click OK to close it.*

▶ *If the "QuickBooks Setup" window opens that says "Let's get your business set up quickly," click Other Options → Open Existing File to open the Open or Restore Company window. Then close the Open or Restore Company window.*

▶ *If at any time the Accountant Center window opens, remove the checkbox in the lower-left corner next to "Show Accountant Center when opening a company file" and close the window.*

Eventually, the No Company Open window will appear on your screen.

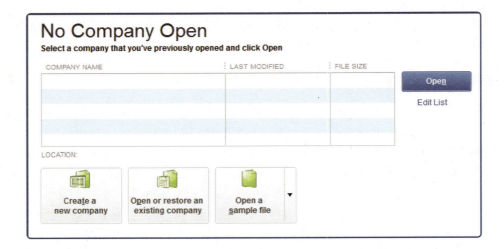

▶ *Click the Open a sample file button and select the option that says Sample product-based business.*

A window may open that says that it needs to update your company file.

▶ *Click Yes.*

Next, a window opens that informs you that the file is for practice only and that the date is set to 12/15/2022 for this sample company.

▶ *Click OK to open Rock Castle Construction, a product-based company that you will use extensively in the next four chapters. If the Accountant Center window opens, uncheck the "Show Accountant Center when opening a company file" checkbox.*

When Rock Castle Construction opens, the bar at the top of the window includes the company name.

▶ It is common in *QuickBooks* for the Live Community and Help window to open automatically on the screen. We suggest that you close that window. You can get help by using Help on the Menu Bar any time you want it.

▶ After the software is installed and during subsequent sessions using the software, various messages may appear from Intuit. Examples include windows that ask you to participate in a survey or whether or not you want to purchase Intuit's forms and documents. Whenever these types of windows open while you are using the software, check the box that says "Do not display this screen again" (if it is available) and then close the window. These materials do not list all of the possible pop-up windows that may appear. Follow the preceding instructions for each pop-up window so that it does not open again.

▶ *Click File → Close Company to close the Rock Castle Construction window. QuickBooks will again display the No Company Open window indicating that no company is open. However, this time the sample_product-based business is listed in the window.*

▶ *Open Rock Castle Construction again.*

Open Another Company

Whenever you open *QuickBooks,* you can open a previously opened company from the list in the No Company Open window. If you wish to open a different company, close that company and select a different company from the list.

You can also open a previously opened company without closing the existing company.

▶ *Select File to open a drop-down list shown at the top of the following page. (Only the top few items are shown here.) Next to Open Previous Company there is a small arrow. It is called an expansion arrow, which indicates the availability of an additional list.* **Note:** *Depending on the work you have already done on this project, your window might show fewer or more previous companies open.*

▶ *Hold the cursor over Open Previous Company to show an expansion list, which includes the names of companies that are available in QuickBooks software, listed in the order they were last opened. You can now open any company in the expansion list by clicking on the company name.*

▶ *Move up one menu option and select Open or Restore Company to open the Open or Restore Company window. Then click the Next button.*

▶ *Select sample_service-based business to open Sample Larry's Landscaping & Garden Supply.*

▶ *Practice opening and closing QuickBooks and opening and closing companies until you feel comfortable doing so.*

Quickbooks Navigation Overview

▶ *Open Rock Castle Construction. The window that is first displayed when you open a company is usually the Home Page, which is shown below.*

▶ *If the Home Page isn't displayed, click the Home icon shown near the top-left portion of the window.*

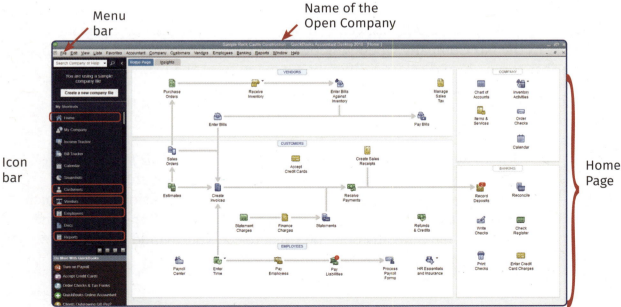

QuickBooks Window

Following is a description of key parts of the *QuickBooks* window with the Home Page open.

Name of the Open Company—In this case, it is Rock Castle Construction.

Menu Bar—The Menu Bar is organized like most Microsoft Windows applications, but adapted to *QuickBooks*. With a few exceptions discussed later these materials use only the File menu. You used the File menu in the previous section.

Icon Bar—The Icon Bar is located on the left side of the Home Page. The first icon is the Home Page icon. The other Menu Bar icons that you will use are Customers, Vendors, Employees, and Reports, all circled in the preceding illustration. *Note:* You may need to expand the icon list area if you cannot see it.

Home Page—The Home Page is the main part of the window and includes all information below the Menu Bar. The Home Page is critical for these materials because it is the primary way you will access all maintenance and processing activities.

The middle section of the Home Page includes three buttons that correspond to Centers on the Icon Bar. The three icons are for vendors, customers, and employees.

▶ *Click the Customers button in the Home Page shown below.*

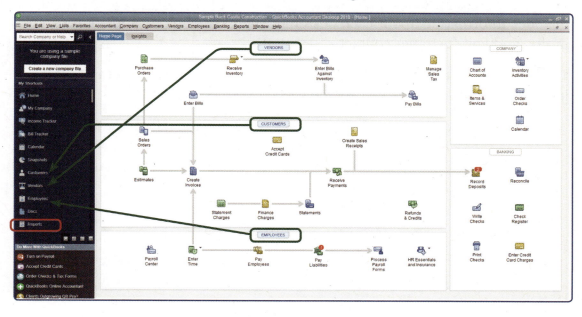

▶ *Observe the window and close.*

▶ *Click the Customers icon (* ▨ Customers *) on the Icon Bar.* It opens the same window. Either of these approaches to open the Customer Center is acceptable, depending on your preference. The same is true for the Vendor Center and the Employee Center, but there is no button on the Home Page for the Report Center so you will use the Icon Bar to access that center.

▶ *Return to the Home Page by clicking the Home icon (* ⌂ Home *) or closing the Customer Center window.*

▶ *Practice opening and closing centers using the center buttons on the Home Page and the Icon Bar icons including the Reports icon (* ▨ Reports *) until you feel comfortable doing so. Close all windows except the Home Page.*

> Use care in closing windows to avoid closing the *QuickBooks* main window. In most cases, the open window has a close icon identical to the main window.

Maintenance

As stated earlier in the chapter introduction, maintenance is used to add, change, or delete information about such things as customers, vendors, and general ledger accounts (chart of accounts). Maintenance is critical to computerized accounting systems and is discussed in detail later.

It is useful to think of maintenance as developing and providing information for master files in an accounting system. One major master file is for customers, which includes information about each customer such as the name and address, credit limits, and key customer contact.

The five key master files in *QuickBooks* are customers, vendors, employees, inventory (Items & Services), and chart of accounts.

▶ *Click the Vendors button (* VENDORS *) in the middle of the Home Page to open the Vendor Center window, the top portion which is shown on the following page.* One tab and one icon on the window deal with maintenance. (Both are circled in the following example.)

► *Click the Vendor's tab if it is not already highlighted.* A list of vendors is included on the left side of the screen. These vendors were previously added during maintenance.

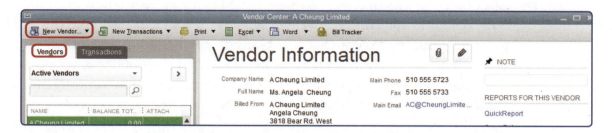

► *Double-click on Hamlin Metal to access the Edit Vendor window shown below. If a New Feature window opens, select "Do not display this message in the future" click OK, then close the window.* The Address Info tab is open and includes information about the company.

▶ *Click the Additional Info tab to see additional information entered during maintenance.*

▶ *Click OK to return to the Vendor Center.*

▶ *Click the New Vendor icon (* 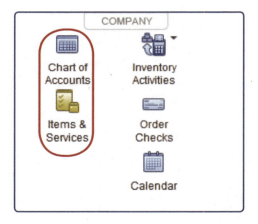 *) at the top of the window and select New Vendor to open the New Vendor window. It looks exactly like the previous window except that it is blank in the headings and other boxes. The only exception is the 12/15/2022 in the date box, which is the default date for Rock Castle Construction. For a new vendor, the QuickBooks operator will complete this window for all three tabs.*

▶ *Return to the Home Page and practice opening and closing maintenance tabs for existing and new customers, employees, and vendors, including opening the Edit windows until you feel comfortable doing so.*

▶ *Close all windows except for the Home Page.*

▶ *Click Chart of Accounts, then the Items & Services icons on the right portion of the Home Page partially shown below for practice. Both are accessed through the Home Page instead of the vertical icons or the Icon Bar icons, and neither of them has multiple tabs.*

Processing Information Tasks

As stated in the introduction, processing information tasks is used for processing and recording transactions. It is the primary activity in computerized accounting systems and the main focus of Chapters 5 through 8.

There are three primary information processing categories in *QuickBooks* that are consistent with the three most common cycles in accounting systems as follows:

1. Sales and cash receipts (Customers in *QuickBooks*)
2. Purchases and cash disbursements (Vendors in *QuickBooks*)
3. Payroll (Employees in *QuickBooks*)

▶ *Open the Home Page for Rock Castle Construction if it is not already open.* The tasks to process and record transactions are indicated by icons connected by arrows in the Vendors, Customers, and Employees portion of the Home Page as shown on each circled portion of the Home Page window below.

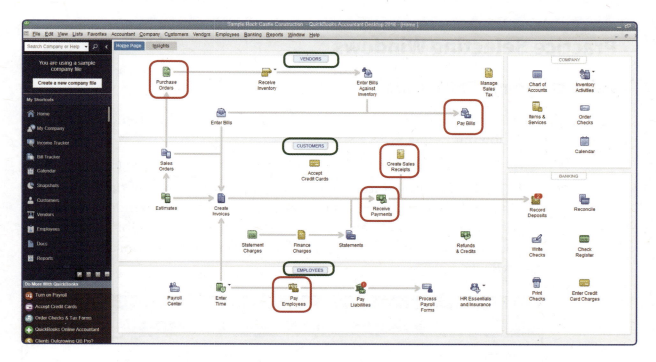

▶ *Move and hold your cursor over each icon in the Vendors portion of the Home Page without clicking on it. Doing so allows you to read what is accomplished when the activity indicated by the icon is completed.* The connecting lines show the direction of flows for processing information for the purchase process. The Vendors portion of the window flows from Purchase Orders through Receive Inventory to Enter Bills to Pay Bills. The Customers and Employees portions of the Home Page are similar to the one for Vendors, but obviously the activities are different.

▶ *Move your cursor over each icon in the Customers portion of the Home Page without clicking on it to learn what activity each icon is used to accomplish.*

Again, the icons include connecting lines showing the direction of flow of the sales process. The process starts with Estimates (if used), Create Invoices, and continues through Finance Charges to Receive Payments and Record Deposits.

▶ *Move your cursor over each icon in the Employees portion of the Home Page in the same way you did for vendors and customers.*

Review the flow of tasks for each of the three sections on the Home Page. Each starts with the beginning of an accounting process and continues through to the end of that process. As you will see later, the processes include the automatic recording of transactions and updating a variety of records. Completing the assignments in later chapters will help you better understand these processes.

Practice Selecting Windows

Using the Home Page Icons

To practice accessing *QuickBooks* windows using the icons on the Home Page complete the following.

▶ *Return to the Home Page for Rock Castle Construction if it is not already open.*

▶ *Click the Create Invoices icon (* Create Invoices *) in the Customers portion of the Home Page to open the Create Invoices window shown below. The window is in the format of a sales invoice and is used for billing customers for shipments of goods or providing services, and for processing sales transactions in QuickBooks.*

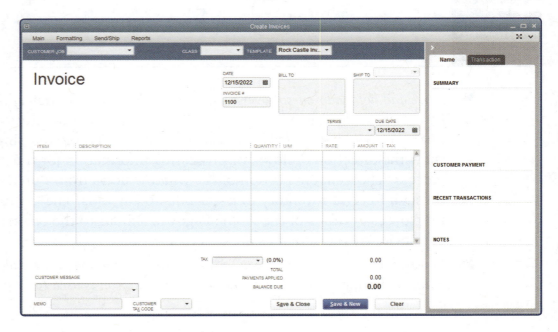

▶ *Close the Create Invoices window and then click the Purchase Orders icon (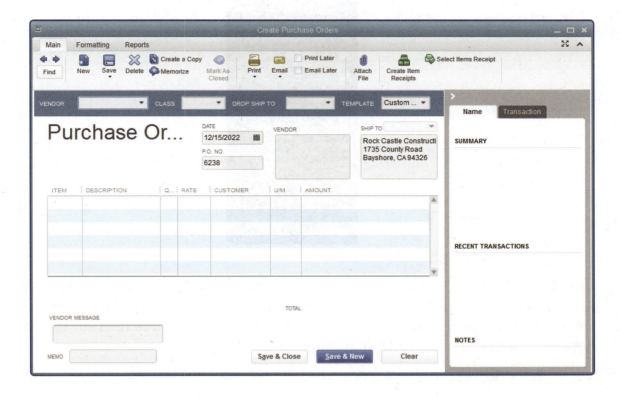) in the Vendors portion of the Home Page to open the Create Purchase Orders window shown below.*

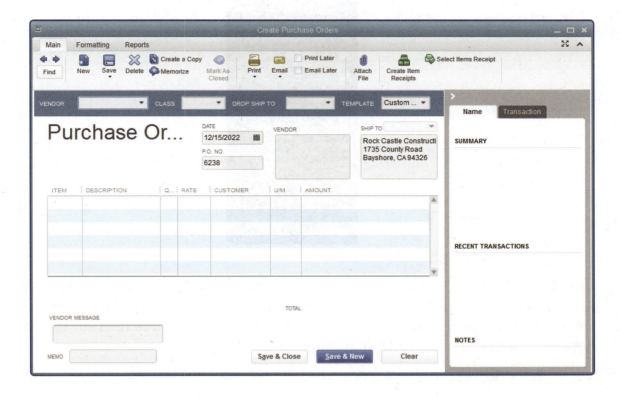

The document in the window is different, but the concept is the same as for the Create Invoices window you opened previously.

▶ *Practice clicking icons in each of the Vendors, Customers, and Employees portion of the Home Page until you feel comfortable doing so.*

▶ *Close all windows except the Home Page.* If there is a message asking if you want to record any transaction now, click No.

> The remainder of this book and the Reference book include instructions for recording transactions and performing maintenance and other tasks using *QuickBooks*. To limit repetition in the project, the materials do not instruct you to close each window when you have finished recording a transaction or completing maintenance or other tasks. When you are done with each window, close the window.

You will now practice opening windows using three different approaches that accomplish the same thing you just did.

Using the Centers on the Icon Bar

An alternative to using Home Page icons is to use the Icon Bar shown above. *Note: If you cannot see all of the Icon bar icons, you may need to expand the list by clicking and dragging the bottom border to expand it. The three relevant Icon bar icons are Customers, Vendors, and Employees.*

▶ *Click the Customers icon on the Icon Bar to show the Customer Center window partially shown below.*

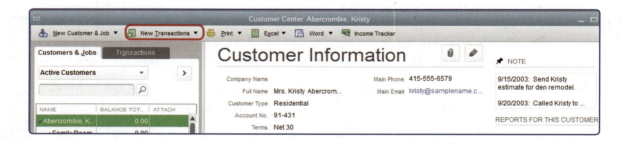

▶ *Click the New Transactions icon (* New Transactions ▼ *), circled in the preceding window) to drop down the list shown below.* The list includes almost the same names as the ones on the icons in the Customers portion of the Home Page.

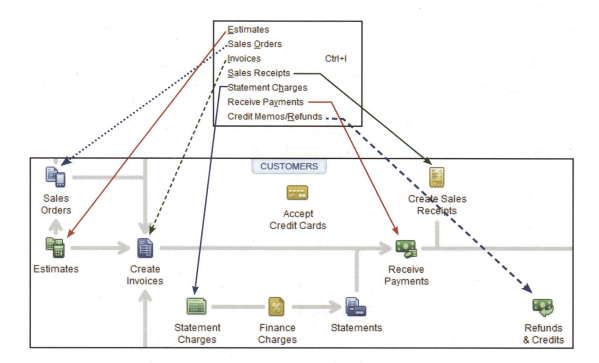

▶ *Click Invoices to open the same Create Invoices window that you opened using the Create Invoices icon on the Home Page.*

▶ *Return to the Home Page and click the Vendors icon (Vendors) on the Icon Bar → click on New Transactions → select Purchase Orders to open the same blank Create Purchase Orders window that you opened using the Purchase Orders icon on the Home Page.*

Using the Center Buttons on the Home Page CUSTOMERS VENDORS EMPLOYEES

Another alternative to using Home Page icons or Icon Bar icons is to use the center buttons on the Home Page. The three relevant buttons are Customers, Vendors, and Employees.

▶ *Return to the Home Page and click the Customers button on the Home Page to open the Customer Center, which is identical to the one you opened using the Customers icon in the Icon Bar.*

▶ *Practice using the center Home Page buttons for Customers, Vendors, and Employees until you feel comfortable opening a variety of windows.*

Using the Menu Bar

File   Edit   View   Lists   Favorites   Accountant   Company   Customers   Vendors   Employees   Banking   Reports   Window   Help

A third alternative to using Home Page icons, Icon Bar icons, or center buttons on the Home Page is to use the Menu Bar. The three relevant Menu Bar items are Customers, Vendors, and Employees.

▶ *Click Customers on the Menu Bar to open the drop-down list shown below.*

▶ *Click Create Invoices (circled in the previous list) to open the same Create Invoices window that you have now opened several times.*

▶ *Practice using the Menu Bar drop-down lists for Customers, Vendors, and Employees until you feel comfortable opening a variety of windows.*

These materials typically use the Home Page icons to open *QuickBooks* windows for processing and maintenance tasks, not the other alternatives discussed. The reason is to help you better understand the processes followed in maintaining, processing, and recording transactions. When you use the icons on the Home Page you can visualize the activities that have taken place before the item you are dealing with and what will take place next to complete all activities in a cycle.

You will use the Reports icon on the Icon Bar because reports are not included as icons on the Home Page. You will also use the Menu Bar for other activities that are not included on the Home Page, such as making general journal entries and preparing bank reconciliations.

A Typical *QuickBooks* Window

You will now explore a typical *QuickBooks* window and practice using some common features of the software.

▶ *Open Rock Castle Construction if it is not already open.*
▶ *Click the Home Page icon → Create Invoices to open the Create Invoices window, if it is not already open.* The window is now blank except for information in three boxes. The Date and Due Date boxes include 12/15/2022 and the Invoice # box includes 1100. This data is referred to as default information.

Default Information

Default settings are established through original company setup or subsequent maintenance activities. The availability of this default information reduces the time needed to enter repetitive transaction information and the likelihood of mistakes. Default settings are one of the most important benefits of computerized accounting software. The Create Invoices window on your screen includes only two items of default information. As you proceed, additional default information is automatically added.

The Create Invoices window on your screen should look like the one on page 2-21, except it will not include the circled letters. This window is an example of a typical window for processing transactions using *QuickBooks*. You will use this window to practice using common *QuickBooks* features. The window includes many description fields, entry boxes that are completed with default information, empty entry boxes, and various buttons.

To begin you will select a customer using a drop-down list.

▶ *Click the Customer: Job drop-down list arrow.* A listing of all customers for Rock Castle Construction is shown.

▶ *Locate the up/down scrolling arrows on the right side of the Customer: Job list. Use the down arrow to examine the listing.* Observe that almost every customer includes a name and subheading under the name. The subheading is the name of a job for that customer. For example, for the customer Ecker Designs there are two jobs, Office Expansion and Office Repairs.

Some companies in this project include both customers and jobs, while others have only customers. A common example of a company that uses jobs is a construction company that has one or more projects for the same customer. Sample Rock Construction is a construction company. Retail and wholesale companies have customers but do not typically use jobs.

▶ *Select the job Utility Shed for customer Johnson, Gordon as shown below.*

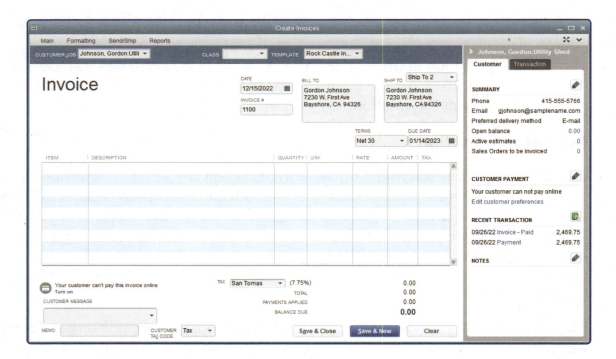

Description Fields

Description fields include labels that describe the information that must be typed, selected, or accepted to complete the window. For example, Customer: Job, Bill To, Date, and Invoice # are all description field labels in the Create Invoices window on your screen (shown on page 2-21). Each description field corresponds to an entry box where information exists or is to be entered.

Entry Boxes

Entry boxes are typically located directly below or to the right of the description fields but are sometimes located to the left or elsewhere. For simplicity, the term "box" is used throughout the remainder of the project, instead of "entry box."

In the Create Invoices window, many of the boxes now include considerably more default information than existed when you first opened the window. By selecting a customer, *QuickBooks* added additional information that has already been established during maintenance for this customer.

The [Tab] key on your keyboard is used to move to the next box in a window, to skip through boxes that you want to leave blank, to accept default data, or to move to the next box after you have entered data. You may also use your touchpad or mouse to skip around anywhere on the entry screen that you like. These project materials assume you are using the [Tab] key to move consecutively through the entry boxes.

To move the cursor backward through the boxes that already include information, use the mouse to move the cursor to a previous box or press [Shift] [Tab].

For now, you are only practicing/learning the use of [Tab] in *QuickBooks*, without concern for the contents of the boxes.

> ▶ *Make sure the cursor is on Johnson, Gordon: Utility Shed in the Customer: Job box. Press [Tab] three or four times and watch the cursor on the screen as it moves from one box to another.* When the cursor goes through a box that already includes information, that information is being accepted without change.
>
> ▶ *Next, use the touchpad or mouse to go back to the Customer: Job box.*

▶ *Press [Enter] once.* The following message appears.

Note: You should not use [Enter] to advance to the next box.

▶ *Click OK.*

▶ *Click the drop-down list arrow next to Customer: Job and change the customer to Smith, Lee: Patio.* The default information changes for the different customer, which is what you expect.

▶ *Change customers again to any one you choose.* There is now new information in the default boxes for the customer you selected.

In most *QuickBooks* windows used in this project, there are boxes in which you do not need to enter information. Either the default information is correct or the information is not applicable to the transaction. The Reference book instructions for recording each type of transaction in *QuickBooks* focus on boxes in which you need to perform some type of activity, such as typing information or selecting an option from a list of available choices. If a box is not discussed, you should not do anything with that box. For example, if the document number box is not mentioned in the instructions for a specific *QuickBooks* window, the default entry is correct and you can therefore press the [Tab] key to go to the next box.

Entering Information Into Boxes

There are different ways to enter information into the boxes of a window or show that information. The circled letters **A** through **D** in the Create Invoices window shown below correspond with the discussion in the following sections.

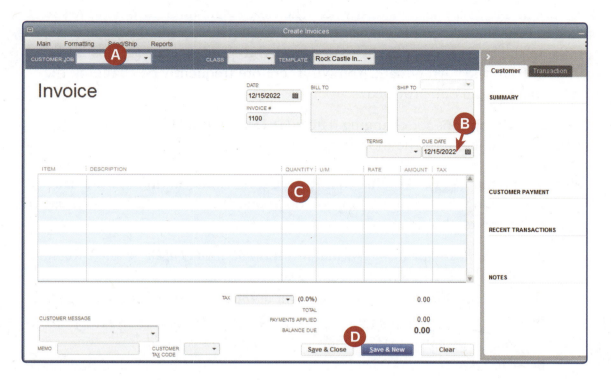

A Drop-Down List Arrow

Drop-down list arrows are identifiable by an adjacent down pointing arrow. You have already used drop-down list arrows several times. Clicking a drop-down list arrow allows users to see a list of available choices on the drop-down list. An example of a drop-down list arrow in the Create Invoices window is to the right of the Customer: Job box. Clicking on it drops down the complete list of company customers. You can then scroll up or down that list and select the customer you want by clicking on that line, then selecting it as you did earlier for Smith, Lee: Patio.

> ▶ *Open a brand new Create Invoices window by clicking the Clear button in the bottom-right corner. Then click in the Item box until the drop-down list arrow appears → click the arrow → select Wood Door: Exterior. Observe that the following boxes are automatically updated: Description, Rate, Amount, and Tax. This is another example of a default entry.*

B   Date Box   [🗓️]

Many *QuickBooks* windows include date boxes for entering the date of transactions or other date information such as the invoice date in the Create Invoices window. In many cases for windows where data is being processed, the default entry in the date box is the current date. That is not the case for maintenance windows for such things as the date an employee was hired in employee records. In these materials, unlike in the business world, the date provided to process transactions is always different from the current date. You will therefore frequently be changing the default date to the one provided.

There are two options for changing dates: (1) Use the calendar button, or (2) type the date as numbers for the month, day, and year.

Calendar Button (🗓️). Assume that Rock Castle Construction made a sale to Robson, Darci for the Robson Clinic job on September 16, 2022. You will use the calendar button to change the date.

> ▶ *Open the Create Invoices window if it is not already open* → *select Robson, Darci: Robson Clinic in the Customer: Job box.*
>
> ▶ *Click the calendar button to the right of the Date box to open the calendar.*
>
> ▶ *Change the month and year to September 2022 by clicking on the left arrow until September - 2022 is shown at the top of the calendar.*
>
> ▶ *Click day 16 in the calendar.* The date in the Date box should now read 09/16/2022.

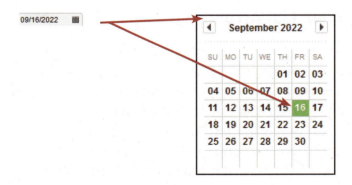

> ▶ *Delete the transaction by clicking the Clear button in the bottom-right portion of the window.*

▶ *Click on Customer: Job box → select Robson, Darci: Robson Clinic again.*

▶ *Highlight the entire date in the DATE box.*

▶ *Type the date using one or two numbers for the month, two for the day, and two or four for the year. Type 09/16/2022 → [Tab] to enter the same date that you entered above.*

There are other options to enter the same date. These include the following:

- 9/16/22
- 9.16.22
- 9-16-22
- 091622
- 09162022

You can also highlight any one or two numbers and change just those. You should use whichever method you prefer.

▶ *Practice entering dates using the calendar button or typing the date using different options until you feel comfortable doing so.*

C **Text Boxes**

Text boxes are boxes in which you enter information by typing. *QuickBooks* has several text boxes in most windows. One example in the Create Invoices window is Quantity for the quantity billed.

▶ *Open the Create Invoices window if it is not already open → select Duncan, Dave for the Customer: Job box → [Tab] to the Item box.*

▶ *Enter 9/20/22 in the Date box.*

▶ *Select Wood Door Exterior → [Tab], [Tab], then type 3 in the Quantity box for the shipment of the three exterior doors → [Tab]. Observe that QuickBooks automatically extends the unit selling price of $120 times the quantity and adds the 7.75% tax when you typed the quantity in the Quantity box. Your window should show a total of $387.90 for the invoice.*

Now assume that customer Dave Duncan decided to have the doors shipped to 1445 Forest St. (same city and state).

▶ *Use the mouse to enter the second line of the Ship To box.*

▶ *Type the new address in the Ship To address box. Your window should now look like the one that follows at the top of the next page.*

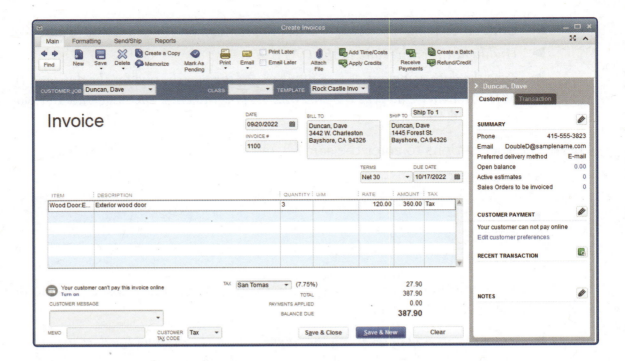

▶ *Click the Save & Close button at the bottom of the window to save the transaction.* The following window appears.

▶ *Click the Yes button.* That will change the address on Duncan's customer maintenance record.

▶ *Click on the Save & Close button if the window does not close automatically.*

D **Buttons**   Save & Close   Save & New   Clear

You have already used buttons several times. There are dozens of them in *QuickBooks*. You will practice selecting three buttons.

▶ *Open the Create Invoices window again.* The three buttons on the bottom of the window are common to all transaction windows.

▶ *Select Balak, Mike in the Customer: Job box.*

▶ *Click the Save & Close button.* A warning window opens indicating that you cannot record a blank transaction.

▶ *Click the OK button to return to the Create Invoices window.*

▶ *Click the Clear button.*

Radio Button (⦾   ⦿). A radio button is used where there are two or more options available, but only one option can be selected. *QuickBooks* has only a few radio button options. To practice using the radio button, complete the following:

▶ *Return to the Home Page and click the Pay Bills icon (⬛ Pay Bills) in the Vendors potion of the Home Page to open the Pay Bills window.* There are two sets of radio buttons on the window.

▶ *Click the Assign check number radio button at the bottom of the window to fill that radio button.* Observe that the To be printed radio button is now unmarked.

▶ *Click the To be printed radio button to change it back to the original setting.*

Tabs

Three of the five maintenance activities and some other windows use tabs to allow an easy way to include information while keeping the amount of information in each window manageable. You have already used tabs for the Vendors Center.

▶ *Return to the Home Page and click the center Customers button on the Home Page to open the Customer Center, partially shown at the top of the following page.* There are two tabs in the Customer Center: Customers & Jobs and Transactions. The Customers & Jobs tab is now open, indicated by the bold and highlighted text of the tab. A listing of all customers is included below the tab.

▶ *Click the name Allard, Robert on the list.* Observe that the left portion of the partial window shown below now includes the heading Customer Information.

▶ *Click the Edit Customer button on the right side of the window (✎) to open the Edit Customer window shown below. If the window on your screen is not the same as that shown here, it may be maximized. If so, click the Restore Down button (▫) on the menu row (File, Edit, View, etc.), but not the entire QuickBooks window. The Address Info tab is open for the window and includes information that was entered through maintenance. Almost all boxes include information.*

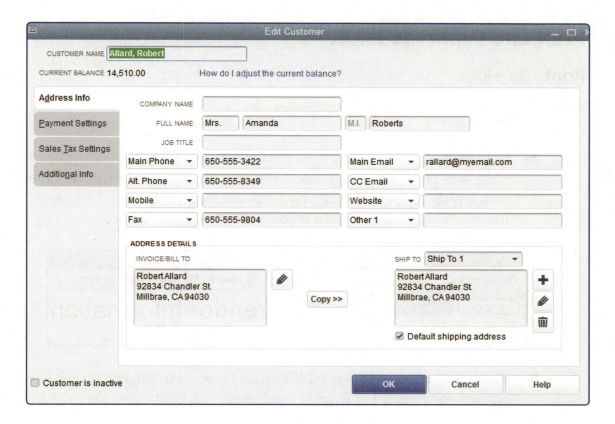

▶ *Click the Payment Settings tab.* Two boxes in this window have default information (Payment Terms and Preferred Delivery Method).

▶ *Click the Sales Tax Settings tab.* The Tax Code and Tax Item boxes contain default information.

▶ *Click the Additional Info tab.* The boxes are empty except for the Customer Type box.

▶ *Click the Cancel button to return to the Customer Center.*

Print and Export to Excel

Print    Print ▾

A wide variety of information in *QuickBooks* can be printed. Only one example is shown.

▶ *Close the Customer Center. From the Home Page, select the Vendor Center. Make sure the Vendors tab is open. You may have to click the Restore Down button if the top part of your screen does not look like the one shown next. The top of the window is shown below with two icons circled.*

▶ *Click the Print icon drop-down list arrow to show three items → click Vendor List to open a small List Reports window → click OK to open the Print Reports window shown below. If the List Reports window opens, select "Do not display this message in the future" and then click OK.* **Note:** *You may have different selections showing in this window, depending on what printer(s) you have available for your computer.*

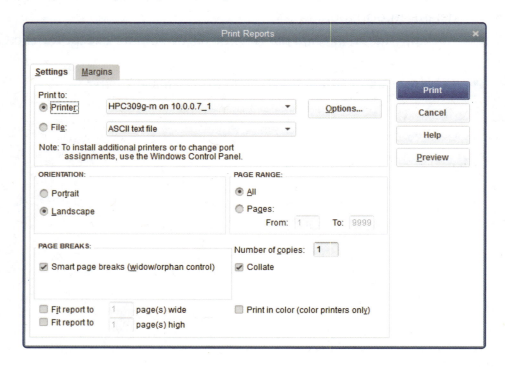

▶ *Click the Preview button (* Preview *) to show the Vendors List that will be printed if you print the list. Click Next page to see the rest of the list.* It is always desirable to preview the printing to avoid printing information that you do not want to print or prefer to be in a different format.

▶ *Click the Close button to return to the Print Reports window and then close the Print Reports window to return to the Vendor Center.*

Export To Excel   🔲 Excel ▼

It is often useful to export information from *QuickBooks,* especially when the data available includes more information than the user wants. For example if the user wants a list of vendor payables in excess of $1,000 that information can be provided by first exporting to Excel.

▶ *Make sure the Vendor Center is open → click once on Bayshore Water on the Vendors tab list.* **Note:** Do not click twice because that will open the Edit Vendor window.

▶ *Click the Excel drop-down list arrow to show three items → click Export Transactions to open the Export window shown below.*

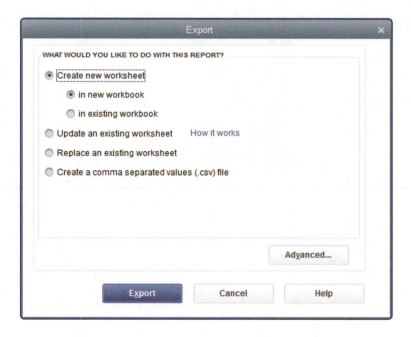

▶ *Click the Export button (* Export *) to export to Excel the Excel worksheet with a list of transactions for Bayshore Water.* You can now change the Excel worksheet, including formats to meet your needs.

▶ *Close the Excel worksheet without saving it.*

▶ *Return to the Home Page and click the Reports icon to open the Report Center with the Standard tab open to Company & Financial. Double-click on the report named "Profit & Loss Standard" to open the following report.*

The report on your screen may contain slightly different numbers depending on what you have done in the project so far. Do not be concerned about the content of the report.

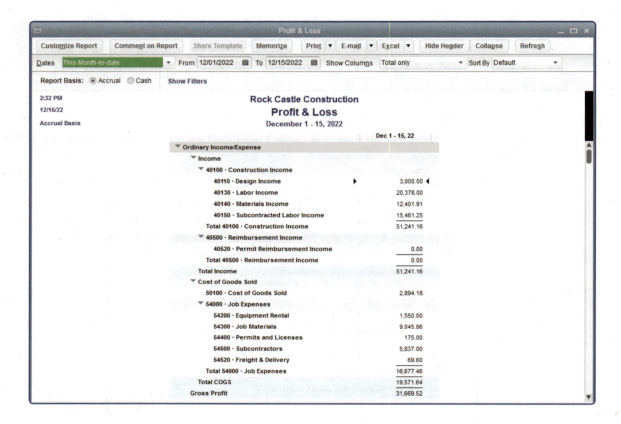

QuickBooks Help Menu

QuickBooks has an extensive Help menu that you can use if you need further information to complete activities in the project. Because the Help menu is easy to use and is likely similar to other Help menus you have used, detailed instructions are not included here.

The Help menu includes a QuickBooks Desktop Help option.

▶ *Click Help → QuickBooks Desktop Help option to open the following window.*

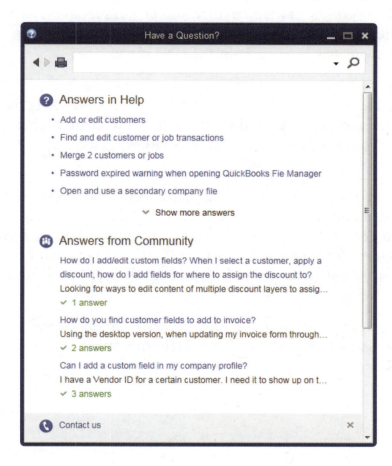

You can type terms, phrases, or entire questions in the "Have a Question?" box to find additional guidance for specific topics. *QuickBooks* even provides step-by-step guidance in several areas.

▶ *Close all Help menus to return to the Home Page.*

Chapter Summary

After completing Chapter 2, you have now learned:

- ✔ how to open and close companies.
- ✔ how to work with various Centers (Customer, Vendor, Employee, Report).
- ✔ how to work with the features that are available on the Home Page, Menu bar, and Icon bar.
- ✔ an overview of maintenance and processing information tasks.
- ✔ how to print information and export to Excel.
- ✔ how to access features in the Help menu.

You should now save your work by making periodic backups of Rock Castle Construction and Larry's Landscaping & Garden Supply using the instructions in the E-Materials you downloaded from the Armond Dalton Resources website. Be sure to use descriptive file names, such as "Rock Castle after Ch 2" or something similar.

Before starting the homework for Chapter 2, you should restore both Rock Castle Construction and Larry's Landscaping & Garden Supply using the initial backups you downloaded from the Armond Dalton Resources website. Restore both companies before proceeding to the Chapter 2 homework assigned by your instructor either online or in the Student Problems & Cases book (consult your instructor).

If you cannot recall how to make a periodic backup or restore a backup, refer to the E-materials located at www.armonddaltonresources.com.

OVERVIEW OF MAINTENANCE, PROCESSING INFORMATION, AND INTERNAL CONTROLS

Introduction

This chapter provides an overview of maintenance, processing information, and internal controls in *QuickBooks*. You need to understand each of these because they are used extensively in the remainder of these materials.

As discussed in Chapter 2, there are four related types of information or activities included in *QuickBooks*. The first two included below are discussed in this chapter and the last two in the next chapter. In addition, this chapter includes processing controls over processing information.

1. **Maintenance**—Used to add, change, and delete information about such things as customers, vendors, and general ledger accounts. Maintenance is critical to computerized accounting systems. It was introduced in the last chapter and is discussed in detail next.
2. **Processing Information Tasks**—Used to perform the tasks needed to process and record transactions. It is the primary activity in computerized accounting systems and was also introduced in the last chapter.
3. **Obtaining Information**—Used to obtain a wide variety of information in *QuickBooks* that has already been entered through maintenance and processing information. Examples include a list of all customers and a list of all sales transactions for a month. *QuickBooks* permits using alternative methods to access this information.
4. **Reports**—Used to view and print a wide variety of reports such as an aged trial balance and a comparative income statement.

Maintenance

Maintenance means establishing the information in the system that permits the automatic and convenient processing of transactions and preparing reports. Examples include setting up general ledger accounts, customer accounts, and employee payroll information. After such information databases are set up in maintenance, *QuickBooks* automatically completes many parts of data entry in the processing windows, such as completing the customer address, terms, and the general ledger sales account for processing sales to a customer.

You will use five maintenance windows to perform maintenance for the five areas introduced in Chapter 2: customers, vendors, inventory items, employees, and chart of accounts. For each maintenance window there are three possible maintenance tasks: add, change, or delete information.

The maintenance part for setting up a new company in *QuickBooks* is time consuming. You will do maintenance to set up a simple company in a later chapter. For now, you are introduced to maintenance to understand what has already been entered into the system.

An important internal control in maintenance is the inability to add, change, or delete any transaction data by doing any type of maintenance task. For example, it is impossible to add, change, or delete sales transactions by doing customer maintenance.

Examine Existing Maintenance

You will now view customer maintenance that has already been done for Rock Castle Construction.

▶ *Open Rock Castle Construction.*

▶ *Click on Home Page → Customers → Customers & Jobs tab, if not already open, to access the list partially shown below.*

Customer Center Window → Customers Jobs Tab

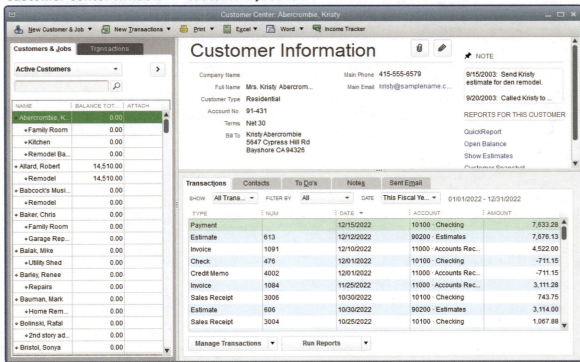

▶ *In the first box at the top of the Customers & Jobs tab section of the window, click the drop-down list arrow and select Customers with Open Balances, then click once on Campbell, Heather. This shows only those accounts with open balances, partially shown below. The window shows all customers with open balances on the left side and shows Heather Campbell's information on the right side.*

Customer Center Window → Customers & Jobs—Customers with Open Balances

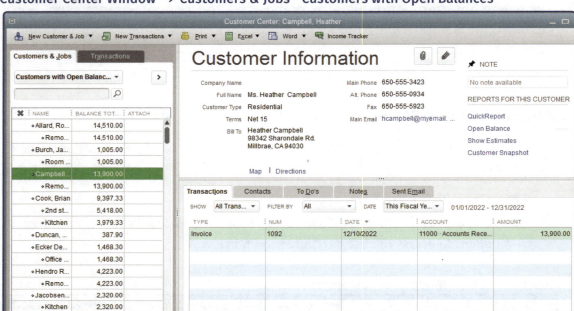

▶ *Double-click Campbell, Heather to open the Edit Customer window shown at the top of the following page. If you cannot see her full name in the Customers & Jobs tab section of the previous window, you may have to widen the Name column.*

Edit Customer Window—Address Info Tab

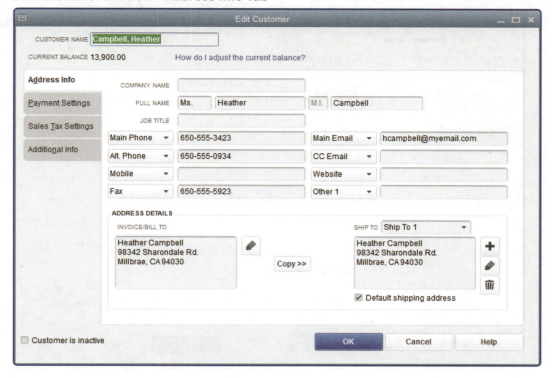

All of the information in the Edit Customer window was previously entered through maintenance. The top part of the window includes the customer name and the current balance. The Address Info tab is already open. Notice that the Company Name is blank for Campbell, Heather. It is common to have empty boxes in a window for the information that is not relevant to that customer. The information completed includes the customer name, invoice/bill to and ship to addresses, and several boxes for contact information.

▶ *Click the Additional Info tab.* Tabs are common in maintenance windows to permit including a considerable amount of information. The same window is edited, but the information in each tab differs. Notice how several boxes are complete in this tab but just as many are blank.

Edit Customer Window—Additional Info Tab

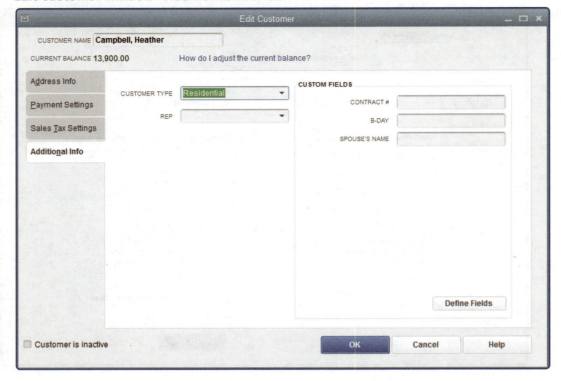

► *Click the drop-down list arrow next to the Customer Type box.* The customer in this case can be residential or commercial.

► *Click the drop-down list arrow next to the blank Rep box.* The only thing shown is <Add New>.

► *Click <Add New> to open the New Sales Rep window.* This permits the *QuickBooks* operator to add a sales representative at any time. Most drop-down list arrows in maintenance permit adding a new first item or another item to an existing list by including <Add New>.

New Sales Rep Window

► *Close the New Sales Rep window and click the Payment Settings tab.* In this case no information is included.

Edit Customer Window—Payment Settings Tab

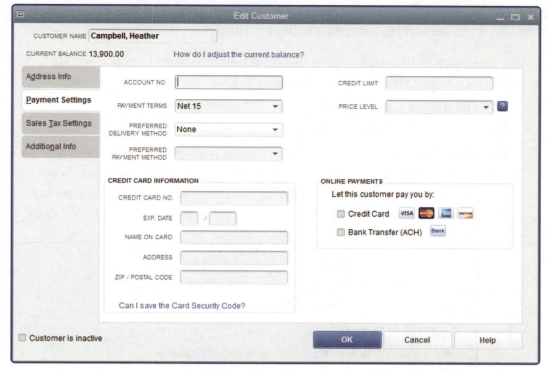

▶ *Close the Edit Customer window for Campbell, Heather and reopen it.* Observe that the default open tab is Address Info even though it was not the last open tab.

▶ *Close the Edit Customer window for Campbell, Heather.*

▶ *Close the Customer Center window to return to the Home Page.*

Based on this discussion you should reach two conclusions:

1. A large amount of information can be entered into each type of maintenance window.
2. The information in the maintenance windows was originally entered through setup maintenance unless it was changed during editing.

Performing Maintenance For a New Account

Next, you will practice adding a new record, which is one of the three types of maintenance tasks listed earlier. For now you will only add a new customer record. You will do additional maintenance in Chapter 5.

When a company obtains a new customer, a new record must be created in *QuickBooks*. Sufficient information must be included when the new customer information is added to permit effective and efficient transaction processing and useful information for management. For example, if management wants to know the sales territory and sales rep for a given sale at some future time, that information needs to be entered during maintenance.

During this part of practice, you will add a new customer record for Rock Castle Construction, using the New Customer window.

▶ *Open Rock Castle Construction if it is not already open.*

▶ *Open the Customer Center.*

▶ *Add a new customer record for Wooden Nursery using the information in the box that follows and the instructions listed after the box, but do not save the new record until you are told to at the end of this section. Not all boxes in the New Customer window and related tabs are applicable to Wooden Nursery.*

- **Customer Name:** Wooden Nursery
- **Opening Balance:** 0
- **As of:** 8/16/2022

Address Info Tab

- **Company Name:** Wooden Nursery
- **Main Phone:** 650-701-1000
- **Main E-mail:** wooden@woodennursery.com
- **Website:** www.woodennursery.com
- **Fax:** 650-701-7254
- **Invoice/Bill To:** Wooden Nursery
  Fred Wooden
  561 Lancaster
  Holland, CA 94036
- **Ship To:** Same Address (Note: click on Copy >>)

Payment Settings Tab

- **Credit Limit:** 30,000.00
- **Payment Terms:** 2% 10, Net 30
- **Price Level:** Commercial

Sales Tax Settings Tab

- **Tax Code:** Tax (Taxable Sales)
- **Tax item:** E. Bayshore/County (not East Bayshore option)

Additional Info Tab

- **Customer Type:** Commercial

▶ *Click the Home Page → Customers button.* The Customer Center displays the list of customers under the Customers & Jobs tab.

▶ *Click on the arrow next to the New Customer & Job button on the top of the window. Click New Customer to open the New Customer window.*

▶ *Complete all relevant boxes for the heading and the four tabs using the preceding information. Some boxes may not be applicable to the new customer.*

▶ *Review the contents of the New Customer window for completeness and accuracy.*

The diagram below and those on the following two pages show the completed New Customer windows, including the four tabs you completed for Wooden Nursery.

Completed New Customer Window—Address Info Tab

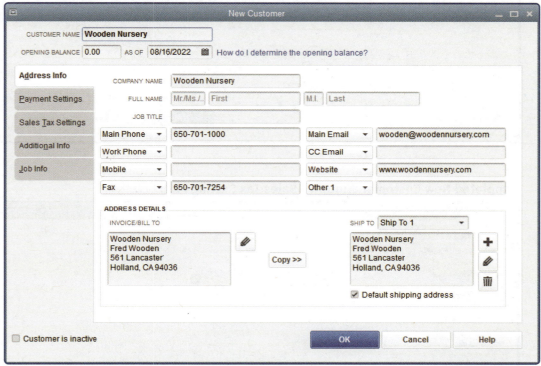

Completed New Customer Window—Payment Settings Tab

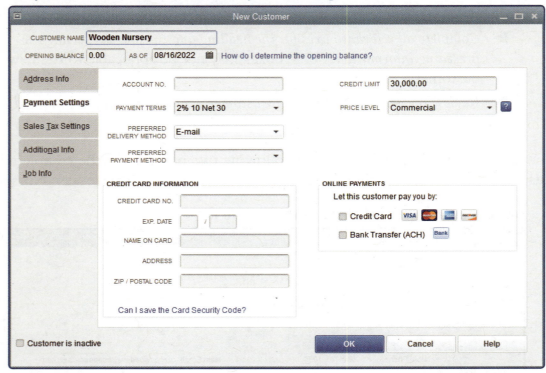

Completed New Customer Window—Sales Tax Settings Tab

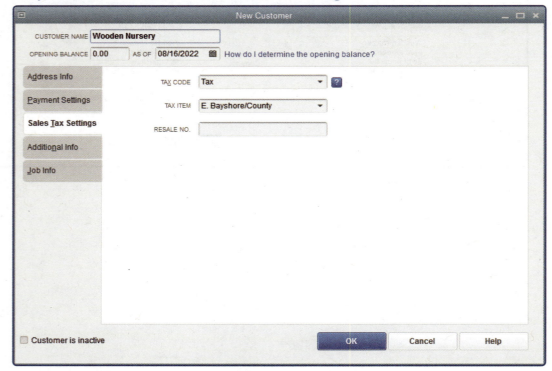

Completed New Customer Window—Additional Info Tab

> ► *If your data entry windows are consistent with the preceding diagrams, click the OK button in the New Customer window to save the information. If there are errors, correct them before saving the record.*

Performing Maintenance To Delete an Account

When you delete any account in maintenance, the account is eliminated from *QuickBooks* permanently. You must therefore delete only accounts that are not likely to be used in the future.

Fortunately, *QuickBooks* has built-in controls to prevent users from deleting accounts that include financial data. For example, it would cause considerable difficulties if a customer with an account receivable balance was eliminated from the records. In that case, the customer would not be billed for the amount owed and the accounts receivable trial balance would not agree to the general ledger.

> ► *Click the Customers & Jobs tab in the Customer Center to display the list of customers, if they are not already displayed.* Observe that customer Ecker Designs on the list has a balance of $1,468.30.

► *Right-click Office Repairs for Ecker Designs on the list of customers → Delete Customer: Job on the list of options provided to open the QuickBooks Message window below. QuickBooks* identifies both the problem and a solution. Read that information.

► *Click the Cancel button and the customer will not be deleted.* That is the approach that is followed in this project.

► *Select "All Customers" in the first drop-down list of the Customers & Jobs tab if not already selected.* Observe that Carr's Pie Shop on the list of customers has a zero balance outstanding.

► *Right-click Remodel under Carr's Pie Shop on the list of customers → Delete Customer: Job on the list of options provided to open the same QuickBooks Message as the preceding one.* Even though the balance is zero, *QuickBooks* will not permit the account to be deleted because it is used in at least one transaction this accounting period.

► *Click the Cancel button and the customer will not be deleted.* The account can be deleted at the beginning of the next year after the current year's accounts are closed.

► *Right-click Wooden Nursery on the list of customers → Delete Customer:Job on the list of options provided to open the Delete Customer:Job window.*

► *Click OK to delete the account.*

► *Examine the list of customers and attempt to locate Wooden Nursery.* It should no longer be on the list.

► *Close all windows and return to the Home Page.*

Processing Information Tasks

After an accountant is satisfied with maintenance for all five windows, information processing activities can begin. Additional maintenance, such as adding a new customer, can be and often is done during these activities.

Information processing activities are the most time-consuming part of any accounting system. For most companies, there are both a wide variety and a large volume of processing activities, usually done every day, but certainly every week and month. Although these activities are technically processing activities, it is easier to think of them as processing accounting transactions. For example, getting a customer order is not a sales transaction, but the purpose is to sell goods, which is a sales transaction. For the remainder of the book we refer to these activities as processing transactions.

Turn to the table of contents for Chapters 6 through 8 and observe that most of the topics involve processing transactions. You will begin processing transactions in this chapter, and then learn and practice all types of transactions included in this book in Chapters 6 through 8.

As an introduction to transaction processing, complete the following steps to process a sales transaction.

▶ *Open Rock Castle Construction if it is not already open.*
▶ *Click the Home Page → Create Invoices icon to open the Create Invoices window.*
▶ *Close the Create Invoices window.*

To learn an alternative method of opening the Create Invoices window, complete the following steps.

▶ *Click on the Customers button in the middle of the Home Page to open the Customer Center window → click on the drop-down list arrow next to New Transactions near the top of this window → select Invoices to open the same Create Invoices window as the preceding one.*
▶ *Close the Create Invoices window.*

To learn a third method of opening the Create Invoices window, complete the following steps.

▶ *Click the Customers button on the Menu Bar to open a drop-down list → select Create Invoices to open the same Create Invoices window as the one shown previously.*

All three methods are equally acceptable and are applicable for opening most windows in *QuickBooks*. We will use the first method whenever a window can be accessed using an icon on the Home Page.

The Create Invoices window is now used to help you understand processing a sales transaction.

▶ *Select the Customers button from the middle of the Home Page → click the Customers & Jobs tab if not already open → click once on Remodel under Babcock's Music Shop to highlight that line → click the drop-down button next to New Transactions and select Invoices.*

▶ *Press [Tab] and observe that the cursor goes to Class.* Observe the large amount of information that has already been included as default information for Babcock's Music Shop. Assume you have no interest in Class or Print Preview and you are satisfied with the Rock Castle Invoice format in the next box.

▶ *Press [Tab] until you are in the DATE box → Change the date to 12/20/2022, using the Date button.* Observe that the TERMS are Net 15 and the DUE DATE was automatically changed to 01/04/2023.

▶ *Press [Tab] to reach the SHIP TO box and use the drop-down list to select Ship To 1.*

▶ *Press [Tab] until you enter the TERMS box and observe that the default terms are Net 15 and the DUE Date is now 15 days later than the invoice date you just entered. Change TERMS to Net 30.* Observe the automatic change in the DUE DATE.

▶ *Press [Tab] until your cursor is in the blank box under the ITEM heading. Select Light Pine under the Cabinets heading.* Observe that the unit cost of $1,799.00 is included in the RATE column and the total amount of $1,799.00 is included in the AMOUNT column. The San Tomas tax jurisdiction is included along with the tax rate of 7.75% and the tax amount. The total amount due is also included.

▶ *Press [Tab] until you enter the QUANTITY column, and type 4 for the quantity invoiced.* Observe that the amount changed as well as the tax and total.

▶ *Press [Tab] until the cursor is on the second row under the Item heading. Select Framing (Framing labor) from the drop-down list and type 8 to represent 8 hours in the Quantity box. In the Rate box select the commercial rate of $49.50.* Observe that much of the information in the window resulted from you having to make only a small number of data entries. The sources of most of the other default information are maintenance and the retrieval and calculation abilities of *QuickBooks*.

What is the increase in accounts receivable from Babcock's Music Shop for this transaction? It is $8,149.69, which is shown as the Balance Due at the bottom of the window. What is the accounting entry for the transaction? It includes a debit to accounts receivable of $8,149.69, a credit to sales tax payable of $557.69, and a credit to sales revenue for the difference. It also includes a debit to cost of goods sold and a credit to inventory for the cost of the four light pine kitchen cabinets. You will learn to access that information in the next chapter.

Do not save this transaction yet, but you can see by examining the buttons at the bottom of the window that it is easy to do so when you are ready to record the transaction.

Correcting a Transaction Before It Is Recorded

Transaction entry errors can be corrected by clicking the box containing the error and reentering the correct information.

> ▶ *Change the date to December 15, 2022.*
> ▶ *Change the first Item to Wood Door: Interior.*
> ▶ *Change the Quantity for the first item in the invoice window to 5.*
> ▶ *Change the Rate category in the second line to Industrial, with a rate of $46.75. The total amount of the sale at the bottom of the screen should now be $761.90.*
> ▶ *Click the Save & Close button to record the sales invoice and return to the Home Page and click Yes when asked if you want to change the default credit terms for this customer.*

Now you will determine if the transaction was included in the *QuickBooks* records.

> ▶ *Click the Customers button on the Home Page to go to the Customer Center. Click the Transactions tab. If Invoices is not the highlighted item, make it so now. The top of the window now shows Customer Center: Invoices. Is the preceding sales transaction included? Yes, unless you made an error. Close the Customer Center: Invoices window.*

Deleting or Voiding a Transaction

There is a difference between deleting and voiding transactions even though they have the same effect on all totals, including those on the financial statements. When you delete a transaction, all evidence of the transaction is eliminated from the records. When you void a transaction, *QuickBooks* retains the transaction information in the system, but records the amount as zero.

A *QuickBooks* user should void a transaction if it is impractical to prepare the same document again. This is common when a company uses pre-numbered documents and discovers an error after the document was printed, or if the document was mutilated or destroyed during the printing process. Good internal controls require that the sequence of all pre-numbered documents be accounted for in the relevant transactions journal. This is the reason for including the voided items as part of the audit trail. For example, assume the company prepared and printed pre-numbered payroll Check No. 864 to an employee but the recorded number of hours worked was incorrect or the printer jammed and destroyed the check in the printing process. In either case a new check will have to be prepared. The *QuickBooks* payroll preparer will void the original check, preserving the audit trail, and prepare a new one.

A *QuickBooks* user should delete a transaction if there is no need to preserve an audit trail. For example, if the transaction errors can be corrected before printing pre-numbered documents or if the documents are not pre-numbered, then the user can simply delete the current attempt and reenter the transaction. In the previous example, assume that internal controls require an independent review of the payroll preparation before the checks have been printed. Errors discovered during this process can be corrected using *QuickBooks*, and printed afterward. In the preceding example, if the error was caught before Check No. 864 was printed, the preparer can delete the transaction and prepare a new transaction for Check No. 864. Even in this case, it is equally acceptable to void the transaction and prepare a new one instead of deleting the transaction.

You will now practice both deleting and voiding transactions for Rock Castle Construction.

▶ *Open the Customer Center. It should open to Customer Center: Invoices window. The Transactions tab and Invoices on the drop-down list will have been already selected because that was the setting when you last closed the Customer Center above.*

▶ *Double-click Invoice No. 1093 for the Lew Plumbing sale on 12/12/2022 to open the Create Invoices for that transaction.*

▶ *Click the arrow beneath the Delete button to reveal two choices: Delete and Void → click Void.*

▶ *Click the Save & Close button at the bottom of the window, and click Yes when asked if you want to save your changes.*

▶ *Return to the Customer Center: Invoices and examine the line for Invoice No. 1093.* The customer name is still there, but it now has a zero invoice total.

▶ *Double-click the invoice number to open the invoice.* Observe that the invoice is now marked "Void" in the Memo box. This is the method used by *QuickBooks* to deal with voided transactions.

▶ *Close the Create Invoices window, returning to the Invoices List. Double-click Invoice No. 1089 for the Mike Violette sale on 12/05/2022 to open the Create Invoices window for that transaction.*

▶ *Click the arrow beneath the Delete button to reveal two choices: Delete and Void → click Delete.*

▶ *Click OK when the Delete Transaction window appears. Also click OK if a second window appears.*

▶ *Return to the Customer Center: Invoices and examine the line for Invoice No. 1089. The transaction is no longer there.* This is the method used by *QuickBooks* to deal with deleted transactions.

▶ *Close all windows to return to the Home Page.*

▶ *Click the Reports icon from the Icon Bar → Accountant & Taxes in the left column → double-click on the Voided/Deleted Transactions Summary report under Account Activity to open the report.* You can also use the Run button (green arrow) beneath the report to open it. Observe that both the Deleted and Voided transactions you just completed are included. This is a useful internal control.

▶ *Return to the Report Center → select Sales in the left column → double-click on the Sales by Customer Detail report (or click the Run button beneath the report) to open the report. Scroll down to Lew Plumbing and observe that the transaction is still included, but now with a zero invoice amount. Now scroll down the report and observe that invoice No. 1089 to Mike Violette is not included.* Again, you can see the difference between a voided and deleted transaction.

Quick Add

Assume the *QuickBooks* user is processing sales transactions for a day and one of them is a credit sale to a new customer. When the company expects additional transactions with the customer in the future, it is appropriate to follow the maintenance process that was shown earlier. The maintenance is likely to be done as a part of the transaction recording process. This is often called "maintenance on the fly."

If the company does not expect additional transactions with that customer, which is reasonably common, it is appropriate to include less information when doing the maintenance procedures than is normally followed. This is referred to as Quick Add, which is shown next. Assume that a one-time new customer, Abbot Nursery, is purchasing inventory for cash.

▶ *Click on Create Sales Receipts icon under Customers on the Home Page to open the Enter Sales Receipts window.* You will enter only the minimum information that permits you to practice Quick Add. Observe in the window before you enter anything that the only default information included is for the DATE, SALES NO., and Tax.

▶ *Type Abbot Nursery in the Customer: Job box → press [Tab] once to open the following window:*

▶ *Click the Quick Add button. Now observe that the only additional information that was added was the name in SOLD TO.*

▶ *Change the date to 12/13/2022 → click the Check button → Type 16214 in the CHECK NO. box → press [Tab] until the ITEM box becomes highlighted.*

▶ *Select Wood Door: Exterior for the ITEM → Type 3 for the QTY.*

▶ *Using the Tax drop-down list arrow at the bottom of the window, select E. Bayshore/County (not the East Bayshore option). The Invoice Total should be $388.98.*

▶ *Click the Print button. The Information Changed window below appears.*

▶ *Click Yes.* Do not print the document, but return to the Enter Sales Receipts window and click Save & Close.

▶ *Click the Customers button on the Home Page → Customers & Jobs tab.* Is Abbot Nursery included on the list? Yes, and the customer information section of the window shows the cash sale of $388.98.

▶ *Click the Reports icon from the Icon Bar → select Sales → select Sales by Customer Detail.* Observe that the cash sale to Abbot Nursery was recorded.

▶ *Click Abbot Nursery to open the Edit Customer window.* Observe that there is no information included other than the name. It is easy to add additional information to the record later if the customer becomes a repeat customer.

Maintenance on the fly is commonly done for all five of the categories of maintenance discussed in the first part of the chapter, but Quick Add usually applies only to customers and vendors. Why? Answer: It is common to encounter new customers, vendors, employees, inventory, and general ledger accounts while processing transactions, therefore Maintenance on the fly applies to all five categories. Information about all new employees, inventory items, and general ledger accounts is normally needed for at least the rest of the current accounting period. For one-time sales to customers and one-time purchases from vendors there is no need for continuing information; therefore, Quick Add only generally applies to these two.

Internal Controls

There are many internal controls included in the *QuickBooks* software to minimize the likelihood of errors and fraud. In this chapter, a few controls are introduced to help understand these internal controls.

Two important categories of controls are those applicable for processing transactions and those that restrict access to the software. The next section discusses controls applicable to processing transactions.

Controls Over Processing Transactions

QuickBooks includes many controls to minimize errors and reduce the time it takes to enter data. To keep the amount of time needed to complete the chapter reasonable, only a sample of processing controls is included. The controls discussed include: automatic entry controls, automatic calculation and posting controls, complete data controls, valid data controls, and exceeded limits controls.

Automatic Entry Controls

Automatic entry controls automatically enter information after receiving a cue from the person entering the data. These controls reduce data entry time and reduce the likelihood of data entry errors.

▶ *Open Rock Castle Construction if it is not already open. Open the Home Page and then open the Create Invoices window.*

▶ *Click the drop-down list arrow for Customer: Job and select Family Room for Chris Baker. Then in the ITEM box select Wood Door: Interior using the drop-down list arrow. Some information is automatically entered as a result of these two entries. The window below shows these.* Note that you may have a different document number on your window, depending on what you have done in the project so far. Ignore the document number.

Completed Create Invoices Window

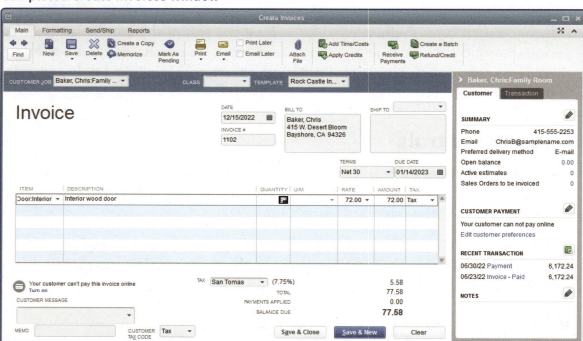

Almost every window in *QuickBooks* includes automatic entry controls.

Automatic Calculation and Posting Controls

Automatic calculation and posting controls automatically calculate or post information after the person entering data has entered other information or clicked certain buttons, usually Print or Save. These controls reduce the time required to process and record transactions, as well as decrease the likelihood of errors. An example of an automatic calculation control is the calculation of sales invoice information.

▶ *Using the Create Invoices for Family Room for Chris Baker in the previous steps, type 4 in the Quantity box and press [Tab]. Observe that several calculations are automatically made in the Create Invoices window. These are shown in the window that follows.*

Create Invoices Window

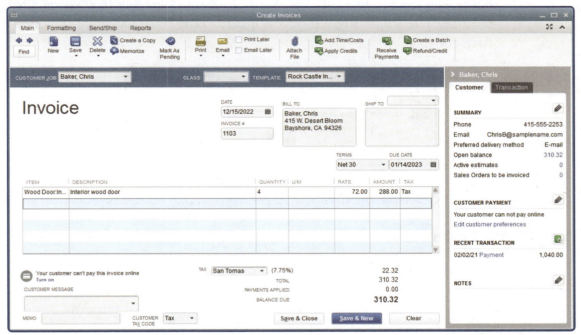

▶ *Close the window and do not save the transaction.*

In addition to the calculations made in a transaction processing window, *QuickBooks* also automatically updates considerable other information when the transaction is saved. For example, for a sales transaction like the preceding one, the information updated includes the following: various sales listings, customer balance listing, aged receivables, general ledger trial balance, income statement, and balance sheet.

Automatic posting controls are used throughout *QuickBooks*. Transaction information is automatically posted to subsidiary records, the general ledger, and other records through the use of the Print and Save buttons in various windows after transaction information is entered. Internal controls are among the most important reasons companies use accounting software.

Complete Data Controls

Complete data controls inform the person entering data into the system that information needed to process the transaction has been omitted. When these controls are not met, the software provides an error message in the window informing the user that certain information is missing.

▶ *Click Create Invoices on the Home Page to open the Create Invoices window.*

▶ *Select Wood Door: Interior in the ITEM box using the drop-down list arrow, but leave the Customer: Job box blank. Type 4 in the Quantity box → click Save & Close. The following window appears. This is an example of complete data control.*

▶ *Click OK → close the Create Invoices window and do not save the transaction.*

Valid Data Controls

Valid data controls minimize the likelihood of entering incorrect data into the system. Similar to complete data controls, *QuickBooks* provides an error message or does not recognize the data when invalid data is entered into a window.

▶ *Click Create Invoices on the Home Page to open the Create Invoices window.*

▶ *Click the drop-down list arrow for Customer: Job and select Family Room for Chris Baker. Then in the ITEM box select Wood Door: Interior using the drop-down list arrow → type the letter h in the Quantity box → press [Tab]. The following window appears.*

This field is numeric and will not accept alphanumeric data.

▶ *Click OK and change the h to 3.*

▶ *While still in the Create Invoices window, change the date in the DATE box to 12/35/2022 → press [Tab]. The following window appears.*

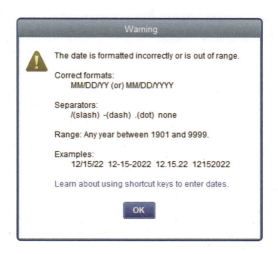

▶ *Click OK → close the Create Invoices window and do not save the transaction.*

These are just two examples of valid data controls in *QuickBooks*.

Exceeded Limits Controls

Exceeded limits controls inform the person entering data that the transaction is not authorized due to such things as a sale exceeding a customer's credit limit or a sale when there is insufficient inventory on hand to fill the order. The following is an example of this control.

▶ *Click Create Invoices on the Home Page to open the Create Invoices window.*
▶ *Click the drop-down list arrow for Customer: Job and select Family Room for Chris Baker. Then in the ITEM box select Door Frame using the drop-down list arrow → type 30 under QUANT → press [Tab]. The following window appears.*

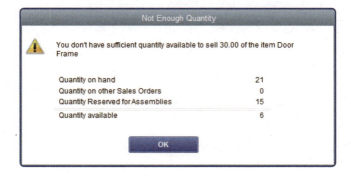

You will therefore need to reduce the units in the Quantity box to be equal to or less than the quantity on hand.

▶ *Click OK → close the Create Invoices window and do not save the transaction.*

Password Control

An important internal control in computer systems is limiting who has access to the software. Restricting access reduces the likelihood of both errors and fraud. *QuickBooks* permits restricting access to the software by the use of a password. After a password is initially set up in *QuickBooks*, the software can be accessed only by entering the password whenever *QuickBooks* is opened.

Chapter Summary

After completing Chapter 3, you have now learned:

- ✔ an overview of maintenance activities.
- ✔ how to process data and correct data entry.
- ✔ some of the internal control features available in the software.

You should now save your work by making periodic backups of Rock Castle Construction and Larry's Landscaping & Garden Supply using the instructions in the E-Materials you downloaded from the Armond Dalton Resources website. Be sure to use descriptive file names, such as "Rock Castle after Ch 3" or something similar.

Before starting the homework for Chapter 3, you should restore both Rock Castle Construction and Larry's Landscaping & Garden Supply using the initial backups you downloaded from the Armond Dalton Resources website. Restore both companies before proceeding to the Chapter 3 homework assigned by your instructor either online or in the Student Problems & Cases book (consult your instructor).

If you cannot recall how to make a periodic backup or restore a backup, refer to the E-materials located at www.armonddaltonresources.com.

OBTAINING INFORMATION FROM *QUICKBOOKS*, INCLUDING REPORTS

Introduction

The maintenance and processing information tasks that were discussed in Chapters 2 and 3 are the inputs into *QuickBooks* that create several outputs. These include such things as sales invoices, checks, updated master files, and lists of transactions. A considerable amount of these outputs are relevant and needed by management and other users to make business decisions. This chapter provides an overview of ways this information can be obtained from *QuickBooks*, including how to access transactions and other data to correct errors or make changes in the data. Following is a repetition of Chapter 3 of the two categories of obtaining information:

1. **Obtaining Information**—Used to obtain a wide variety of information in *QuickBooks* that has already been entered through maintenance and processing information. Examples are a list of all customers and all sales transactions for a month. *QuickBooks* permits using alternative methods to access this information.

2. **Reports**—Used to view and print a wide variety of reports such as an aged trial balance and a comparative income statement.

Obtaining Information

A typical company's accounting information in *QuickBooks* includes a wide variety of data that can be easily accessed. There are several alternative ways to access the information, depending on personal preference and the type of information you want to access. The four main ways are:

1. **Maintenance**—Accessed starting from the Home Page, then through the various centers such as the Vendor Center. Maintenance is used primarily to obtain information about master files including customers, vendors, employees, inventory, and the chart of accounts. You were introduced to using maintenance to obtain information about customers earlier in Chapter 3. This will be covered in greater detail in Chapter 5.

2. **Lists**—Accessed through Lists in the Menu Bar. Lists includes several category titles that can be accessed to obtain a wide variety of information.

3. **Company Snapshot Accessed through the Icon Bar**—A wide variety of detailed information and summaries can be accessed through the Company Snapshot button.

4. **Drill Down**—Accessed by double-clicking on lines in a large number of lists and other records. Users can access detailed information about transactions and other records by using the drill-down feature.

You have already practiced obtaining information using Maintenance in Chapter 3. The other three methods are covered in this chapter.

Lists

Customers, vendors, employees, and management often want information about recorded transactions or balances and need to determine whether information has been recorded correctly. Lists, such as a list of customers, are one way to get that information. These lists for vendors, customers, and employees and their related transactions are accessed through Centers. Because chart of accounts and inventory items do not have related transactions lists, they are accessed directly from the Home Page. First we deal with Centers.

Centers can be accessed either from the buttons in the center of the Home Page or from the Icon Bar.

> ▶ *Open Rock Castle Construction if it is not already open.*
> ▶ *Click the Vendors icon on the Icon Bar to open the Vendor Center.*
> ▶ *If it is not already the default tab, click the Vendors tab to open a list of Rock Castle's vendors, including account balances. Your screen should look similar to the following illustration.*

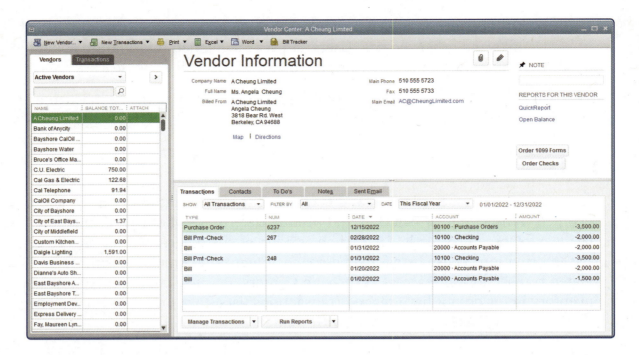

▶ *Click the Transactions tab to open a list of options shown in the following window.*

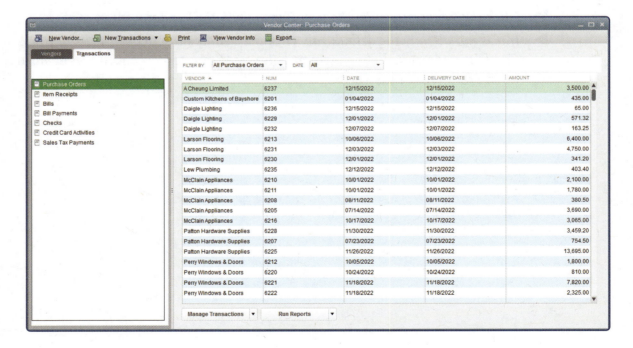

▶ *Close all windows to return to the Home Page.*

▶ *Click the Vendors button on the Home Page to open the Vendor Center → click the Vendors tab to open a list of all vendors, including account balances.*

▶ *Click the Transactions tab to open a list of options that are identical to the preceding window.* These few steps show that either approach works. The same is true for customers and employees. The chapters in this book typically start with the Home Page, but either approach for accessing the Center is acceptable.

▶ *Click Bills in the list of options shown in the Transactions tab. Observe that the top of the window now includes Vendor Center: Bills.* This same type of information is provided for all options for the three centers.

The vendor bills listing to the right of the options includes all recorded purchase transactions for a selected period. The information can be shown in a variety of ways.

It is often useful to reorder the information in a window to make it easier to view and analyze. Any list can be sorted on any column heading, reordering it either alphabetically, numerically, or by date. It is done by clicking the column heading for the column you want sorted. *QuickBooks* sorts the list on the column you click without changing the details of any record. It is similar to the Sort command in Excel, except that it is done automatically by *QuickBooks* after you click the column heading. Clicking the column heading a second time reverses the ascending or descending order of the sort that was just done.

▶ *Click the Open Balance heading at the top of the Open Balance column.* Observe how the Bills List is sorted by the amount of the account balance with the order of the list changing to ascending order from the smallest amount due to the largest amount due. (You may need to scroll the list to see this.)

▶ *Click the Open Balance heading again.* Observe how the order of the list reverses to descending order from the largest amount due to the smallest amount due.

▶ *Click once and then again on each of the other column headings and observe the change in the order of the list for that column.*

▶ *Click the drop-down list arrow in the Filter By box near the top of the Vendor Center: Bills window → click Open Bills to list only the vendors that are still owed money. Then click All Bills on the drop-down list and observe that there are more items listed.* Similar information is also available for customer transactions but not for employee transactions.

▶ *Click the drop-down list arrow in the Date box near the top of the Vendor Center: Bills window.* Observe that there are many date options available for users to select.

▶ *Click successively on two or three of the options and observe how the lists change.* These date options are also available for both customer and employee transactions.

Hiding Information From View

Lists windows contain considerable information that can be hidden from view. This is done to avoid excessive information being displayed in the window. Deleting a column has no effect on the *QuickBooks* data and any column in the window can be added back to the list at any time.

▶ *Return to the Home Page and open the Customer Center: Invoices list, using the method you just learned for opening Vendor Center: Invoices.*

▶ *Click View in the Menu Bar → select Customize Transaction List Columns on the drop-down list of options to show the following Customize Columns window.*

The Chosen Columns box includes the columns currently included in the Customer Center: Invoices window in the order they are in the window, from left to right. The Available Columns box list includes all other columns that can be added to the window.

▶ *Click Terms in the Available Columns box list to highlight that item and then click the Add button (* Add > *) in the center of the window to add it to the bottom of the Chosen Columns box.* Note that you can also double-click on an item to add it to the Chosen Columns list.

▶ *Click Aging in the Chosen Columns list to highlight that item and then click the Remove button (* < Remove *) in the center of the window. Click OK to accept and* *save the two changes you made.* Doing so closes the Customize Columns window. The Customer Center: Invoices window now includes the changes that you made.

You can change the width of columns in the window in the same way that you do in Excel. Move the cursor to the line next to column name until it turns into a cross and move the column to the width you want. Double-clicking on the line adjusts the size automatically.

▶ *Click View in the Menu Bar → select Customize Transaction List Columns on the drop-down list again → click Default and click OK.* The two changes you just made are removed and the settings are now returned to the default settings.

▶ *Open the Customize Columns list window again (View → Customize Transaction List Columns) and then click any item in Chosen Columns. Click Remove repeatedly.* When you try deleting the last item, you should get the following message:

> "This column can't be removed because it's currently the only column displayed."

▶ *Click OK, and then click the Default button to return the columns to the original settings.*

It is also possible to change the order of the columns in a Lists window by using the Move Up (Move Up) and Move Down (Move Down) buttons.

▶ *In the Customize Columns window, select Customer under Chosen Columns. Click the Move Down button five times to move Customer to the next to the last item. Select Amount and then click the Move Up button four times to move it to the top of the list.*

▶ *To practice using Customize Columns, add, remove, and reorder items until you have only the following items included in the Chosen Columns list, from top to bottom: Customer, Type, Terms, Amount, and Open Balance.*

▶ *Click OK to see the resulting Customer Center: Invoices list. Are they in the correct order?* If not, return to the Customize Columns window and make the necessary changes.

▶ *Open the Customize Columns window again → click Default and click OK.* The settings are now returned to the default settings.

Lists are equally useful for both Chart of Accounts and Items & Services (Inventory), but these two are opened directly from the Home Page with one available list for each.

▶ *Return to the Home Page and click the Items & Services icon in the Company section of the Home Page to open the Item List.*

▶ *Click View on the Menu Bar → click Customize Columns → scroll down the Available Columns list and observe the large number of options that are available for inventory items for customization.*

To summarize, the List function is one of the most valuable sources of information in *QuickBooks*. You have already seen the large number of lists available and how there are different ways to access lists. You have also seen how you can change the order of the data in a list by clicking on a column heading. One characteristic of the List function is the ability to add and delete columns from lists. It is the only source of information in *QuickBooks* that includes this feature.

Exporting to Excel

You have already learned in the last section dealing with Lists that you can sort information in Lists windows to provide the information in different formats. You also learned that you can add, delete, or change the order of columns in a Lists window. *QuickBooks*, however, does not permit removing from view individual or certain types of transactions, such as say those under $50.

Exporting to Excel is useful to further analyze, delete, or rearrange information and to prepare reports of the revised information. Exporting information is easy and can be done with almost all *QuickBooks* data, including data in Reports, Maintenance, and Lists. You will now learn how to export to Excel.

▶ *If not open already, reopen the Customer Center: Invoices window for Rock Castle Construction (click Customers → Transactions tab → Invoices).*

▶ *Use the Customize Columns window to include only the following columns in the order listed: Amount, Customer, and Date.*

▶ *Select All Invoices in the Filter By box and then select This Fiscal Year-to-date in the Date box.*

▶ *Click the Export icon at the top of the Customer Center: Invoices window to open the Export window shown on the following page.*

▶ *The default selection is typically "Create new worksheet." If the default is not as shown in the preceding window, make sure that the "Create a new worksheet" radio button is selected.*

▶ *Click Export.* All data in the Customer Center: Invoices window is sent to an Excel worksheet, which will be displayed on your computer screen. *If it is not displayed, click the Excel icon on the bottom of your screen.*

▶ *Change the data on the Excel worksheet in the following ways:*

- *Arrange the amounts in descending order.*
- *Delete all accounts with a balance of less than $7,000.00.*
- *Type a new header including the following three lines: Rock Castle Construction, Customer Invoice List, and 1/1/22 to 12/15/22.*
- *Bold the new header and all column headings.*
- *Total the amount column.*

The report on your screen should look similar to the following report. Note that the illustration includes just the bottom portion of the spreadsheet. Do not be concerned if your report has slightly different balances.

▶ *Close Excel without saving the file.*

Rock Castle Construction

Customer Invoice List

1/1/2011 to 12/15/2022

Amount	Customer	Date
20,300.00	Rahn, Jennifer:Remodel	06/17/2022
17,270.00	Sage, Robert:Remodel	05/08/2022
15,870.00	Tony's Barber Shop:Remodel	08/20/2022
15,435.00	Pretell Real Estate:155 Wilks Blvd.	11/20/2022
14,900.00	Carr's Pie Shop:Remodel	03/15/2022
14,560.00	Davies, Aaron:Remodel	02/16/2022
14,538.54	Memeo, Jeanette:2nd story addition	08/13/2022
14,510.00	Allard, Robert:Remodel	09/12/2022
14,085.30	Cook, Brian:2nd story addition	10/30/2022
13,900.00	Larsen's Pet Shop:Remodel	01/16/2022
13,900.00	Samuels Art Supplies:Remodel	07/20/2022
13,900.00	Mackey's Nursery and Garden Supply:Greenhouse Addition	10/20/2022
13,900.00	Campbell, Heather:Remodel	12/10/2022
12,754.14	Melton, Johnny:Dental office	11/20/2022
12,530.00	Babcock's Music Shop:Remodel	04/14/2022
12,420.98	Robson, Darci:Robson Clinic	12/15/2022
11,481.80	Luke, Noelani:Kitchen	07/14/2022
8,656.25	Natiello, Ernesto:Kitchen	12/14/2022
8,361.67	Smith, Lee:Patio	08/27/2022
8,305.95	Teschner, Anton:Sun Room	11/28/2022
7,764.78	Teichman, Tim:Kitchen	10/15/2022
279,344.41		

Company Snapshot

Company Snapshot provides access to much of the same information that is available in the various centers.

▶ *Open Rock Castle Construction to the Home Page. Click Company from the Menu Bar and then click Company Snapshot to access the following window. Do not be concerned if your window contents differ slightly.*

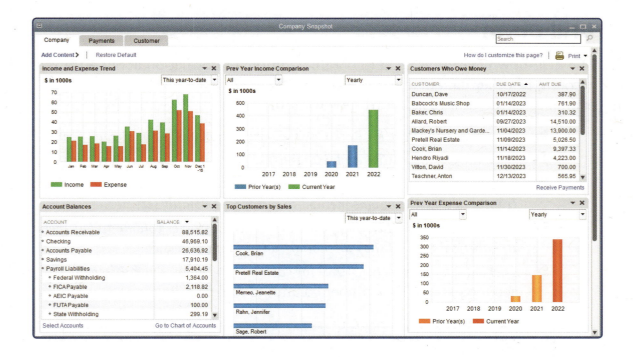

Observe on the Company Snapshot window that there are many sections of data. You may have to maximize the window and scroll down to see all of the available areas. This same information is available elsewhere by accessing information from the Home Page, for example:

- **Account Balances** — Available using the Chart of Accounts icon.
- **Customers Who Owe Money** — Available using the Customer Center: Invoices window (Customers icon → Transactions tab → Invoices).

The Income and Expense Trend graph and the remaining portion of the window are helpful to some users but are not used in these materials. As a result, the Company Snapshot window is not essential and is therefore not used in these materials.

▶ *Click View in the Menu Bar – Customize Icon Bar → select Snapshots on the list under the Icon Bar Content → click Delete and click OK to remove the Snapshots icon from the Icon Bar.* Assuming you deleted the icon, it should now be gone.

▶ *Close the Company Snapshot window.*

It is acceptable to leave any icon on the Icon Bar or remove it if it is not used. The default settings on the Icon Bar are established by *QuickBooks* and are set for each company. Therefore, if you want one or more icons removed from the Icon Bar for all companies, you must do so separately for each company.

Drill Down

The drill-down feature is used to obtain additional information about a transaction, account, or other data in a list or report. Following are just a few of many possible examples of the *QuickBooks'* drill-down feature.

▶ *Open the Home Page for Rock Castle Construction if it is not already open.*

▶ *Open the Customer Center: Invoices list window if it is not already open.*

▶ *If the layout of the window is not the default layout, use the View → Customize Transaction List Columns method to restore the default settings.*

▶ *Click the Customer heading at the top of the Customer column to put them in alphabetical order if they are not already in that order.*

▶ *Click once on Invoice No. 1069 for customer Jacobsen, Doug:Kitchen. You may need to expand the number column to find that invoice.*

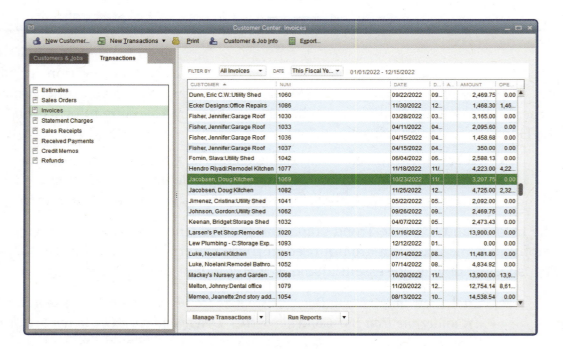

Observe that the entire row of data for the transaction is now highlighted.

▶ *Click once on other rows under any of the column headings and observe that it doesn't matter where you click the row, the entire row becomes highlighted.*

▶ *Double-click the line for Invoice No. 1069 for Jacobsen, Doug:Kitchen, opening the Create Invoices window. The window shown is the invoice that was prepared when the customer was billed October 23, 2022.*

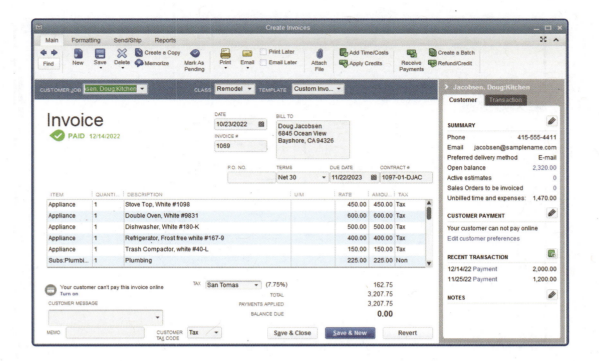

If Jacobsen or management wants information about the invoice, it is easy for the accountant at Rock Castle Construction to provide the information by starting with the Customer Center: Invoices and drilling down to the sales invoice of interest.

▶ *Close the Create Invoices window and the Customer Center: Invoices window to return to the Home Page.*

▶ *Click the Items & Services icon to open the Item List window.*

▶ *Double-click Cabinets: Light Pine (Light pine kitchen cabinet wall unit in the Description column) to open the Edit Item window for that inventory item.* Like for sales transactions, *QuickBooks* can provide useful information for management or outsiders. This same drill-down approach is applicable to all *QuickBooks* lists.

▶ *Close all windows except the Home Page.*

▶ *Click the Reports icon and then click the Sales selection to open the Report Center window shown on the following page.*

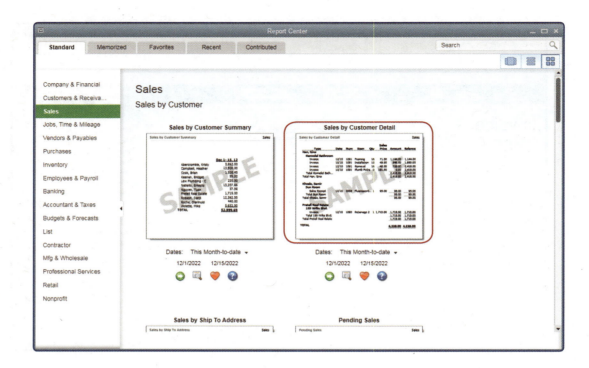

▶ *Double-click Sales by Customer Detail (circled) to open the Sales by Customer Detail window. Select This Fiscal Year-to-date in the Dates' drop-down list. Do not be concerned if your report contains slightly different amounts.*

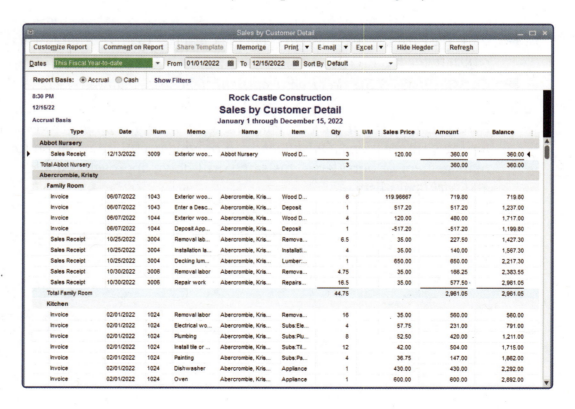

➤ *Scroll down to Jacobsen, Doug:Kitchen and observe that there are several lines with the number 1069. Move your cursor over any portion of any of the invoices. Observe that a magnifying glass with a Z in the center appears. This has the same effect as the highlight you used above.*

➤ *Double-click any line on the Jacobsen, Doug:Kitchen entries to open the same Create Invoices window that you opened earlier.*

➤ *Close the Create Invoices and the Sales by Customer Detail window to return to the Report Center window.*

➤ *Click Accountant & Taxes on the left column → double-click General Ledger to open the General Ledger window → select This Fiscal Year-to-date in the Dates drop-down list → scroll to account 40130, Labor Income, and double-click Num 1069. You will open sales invoice 1069 for Jacobsen, Doug:Kitchen.*

➤ *Close all windows to return to the Home Page.*

➤ *Return to the Customer Center and open the Received Payments list in the Transactions tab.*

➤ *Double-click Check No. 9384 from Aaron Davies to open the Receive Payments window shown below.*

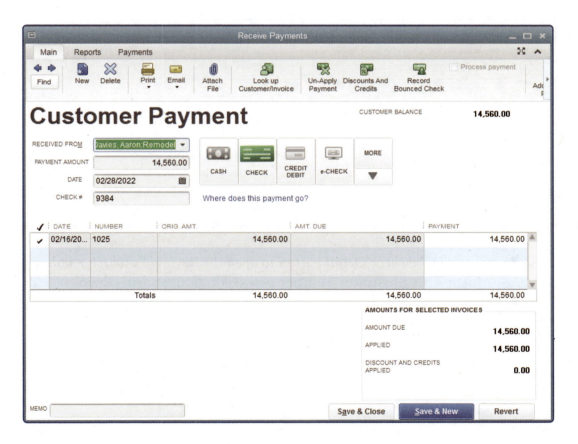

➤ *Double-click anywhere on the line that includes the amount of the payment. This opens the customer invoice that resulted in the payment on Check No. 9384. This an example of a double drill down, a highly useful feature in QuickBooks.*

➤ *Close all windows to return to the Home Page.*

➤ *Click the Employees button to open the Employee Center window → select the Transactions tab and select Paychecks from the list.*

➤ *Double-click Check No. 10073 to Dan T. Miller to open the Paycheck – Checking window shown below.*

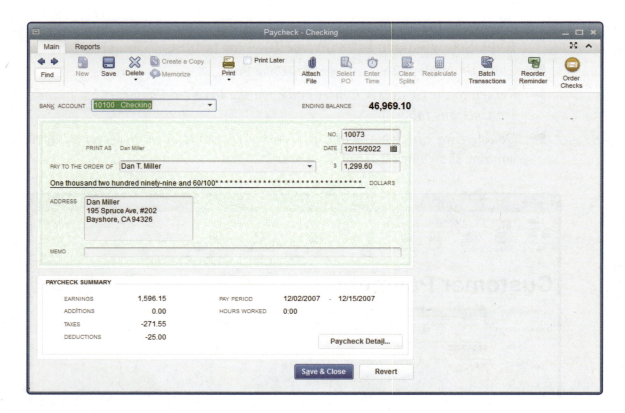

➤ *Click the Paycheck Detail button to open the Review Paycheck window for Dan Miller.* This illustrates another way to do double drill down.

➤ *Practice the drill-down feature for different lists and reports using the methods you learned in this section until you feel comfortable obtaining information this way.*

Custom Transaction Detail Reports

QuickBooks has a unique feature that enables users to obtain a wide variety of detailed information about transactions in a company's database. This section provides practice obtaining that information.

▶ *Make sure Rock Castle Construction is open. Click Reports on the Menu Bar (not the Reports icon) → Custom Reports → Transaction Detail → select This Fiscal Quarter in the Dates box to open the window shown below.*

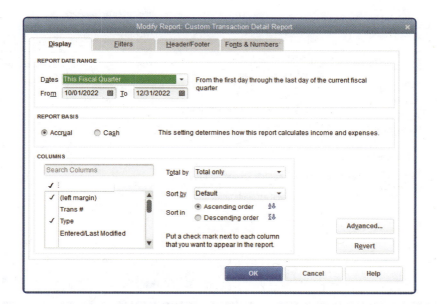

▶ *Click the Filters tab to open the window that appears below.*

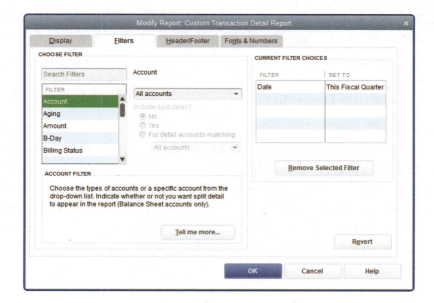

Account is highlighted in the Filter box and All accounts is shown in the Account filter box.

▶ *Scroll down the Account drop-down list and observe the large number of options available. Click All current assets and the window now appears as shown below.*

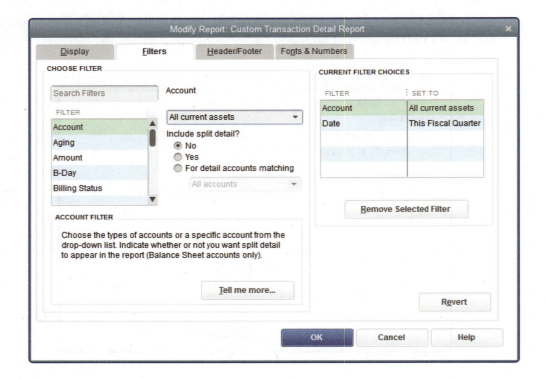

Notice that the window under Current Filter Choices now includes both the account you selected and also the date you selected earlier.

▶ *Click OK.* Now observe that the list does not include all current assets for the period, but all transactions for the period that affected any current assets. This is consistent with the title of the Menu Bar item — Custom Transaction Detail Report.

▶ *Close the report window.*

▶ *Open the Custom Transaction Detail Report again (Reports menu item → Custom Reports → Transaction Detail), but set the Dates to Last week → click the Filters tab → scroll down the list in the Filter box and again see the large number of options available → click Amount in the Choose Filter box to open the window shown on the following page. Amount now appears directly to the right of the Filter box and the information in the window also changes.*

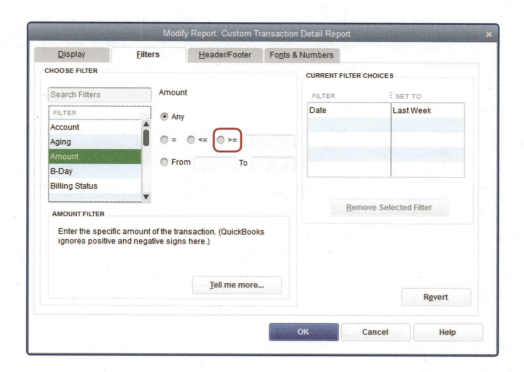

▶ *Click the red circled radio button shown in the preceding window and type 2000 in the box to the right → press the Tab key. Click Transaction Type in the Choose Filter box and select Invoice under Transaction Type.* Your window should now look like the one below.

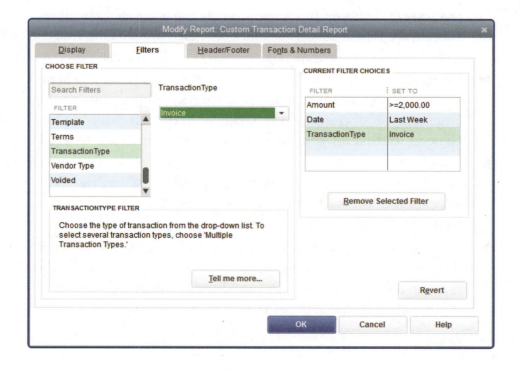

Examine the Current Filter Choices box and observe that you will be filtering for three things. (You can remove any of those filter choices by highlighting the line and clicking the Remove Select Filter button, but do not do that.)

▶ *Click OK*. The result should look like the window below. Using this feature allows you to obtain a large amount of information about transactions.

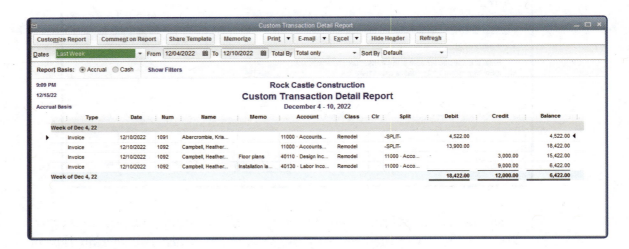

▶ *Close the report and then open the Modify Report: Custom Transaction Detail Report window again (Reports menu → Custom Reports → Transaction Detail) and leave the Dates as This Month-to-date and again click the Filters tab → select Transaction Type in the Choose Filter box and Journal on the Transaction Type drop-down list → click OK.* There should be no transactions in the window.

▶ *Do not close the report, but change the setting to This Fiscal Year in the Dates box at the top of the report.* The list of journal transactions now includes a large number of transactions. Observe in the left column that the Type is General Journal for general journal entries, which are covered in Chapter 8.

▶ *Double-click one of the lines for Num 1108 to open the Make General Journal Entries window.* You can now see that the entry was the monthly allocation of prepaid insurance to insurance expense. This is another example of drill down.

▶ *Practice using the Custom Transaction Detail Report window until you are comfortable obtaining transaction information.* The most important part is using alternative filters in the Filter box and the related information immediately to the right of that box.

Reports

Reports are the output of the accounting data and are typically all that information users see. Some of the reports are only for management within the company, such as an aged trial balance, whereas other reports are useful to both management and outsiders, such as banks and unions. These include the income statement and balance sheet. There is a wide variety of reports available in *QuickBooks*. First, there is a brief introduction to a few of the most important reports and then a followup with practice in accessing various reports.

General Ledger Balances (Trial Balance)

The general ledger balances are the accounts in an organization's chart of accounts that have balances at any designated point in time. They are the basis for preparing financial statements and the result of all transaction processing to the point where the trial balance is prepared.

A trial balance is a report that lists the general ledger accounts, including account balances on a given date.

Subsidiary Ledger Balances (Master Files)

A subsidiary ledger maintains detailed records for certain general ledger accounts. They are frequently referred to as master files. An example is accounts receivable. The accounts receivable general ledger records all transactions affecting accounts receivable and includes the total balance on a specific date. The subsidiary ledger for accounts receivable does the same thing for each customer making up the accounts in accounts receivable. The subsidiary ledger includes the balance due from each customer on a specific date. The total of all customer balances equals the general ledger balance for accounts receivable.

There is a variety of information included in *QuickBooks* for each of the following four subsidiary ledger accounts:

- Accounts Receivable
- Accounts Payable
- Payroll
- Inventory

Financial Statements

Three financial statements that are commonly created for most companies are the balance sheet, income statement, and statement of cash flows. Companies typically print monthly, quarterly, and annual financial statements.

Accessing Reports

There are several ways to access reports. The only method discussed is the one introduced earlier.

▶ *Make sure Rock Castle Construction is open and that all windows are closed except the Home Page.*

▶ *Click the Reports icon on the Icon Bar → then select Company & Financial (if not already selected) to open the window partially shown below.*

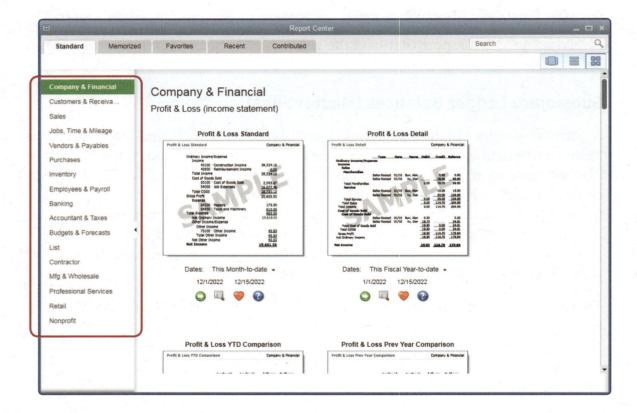

The above circled listing on the left includes categories of available reports. These are used to access the reports included in *QuickBooks*.

▶ *Click Customers & Receivables on the left to display all Customers & Receivables related reports on the right.*

▶ *Click Vendor & Payables in the Reports listing and observe how a new set of reports is included in the main section on the right.*

▶ *Click other titles in the Reports listing until you feel comfortable with selecting reports.*

▶ *Click Customers & Receivables again to open the window shown below.*

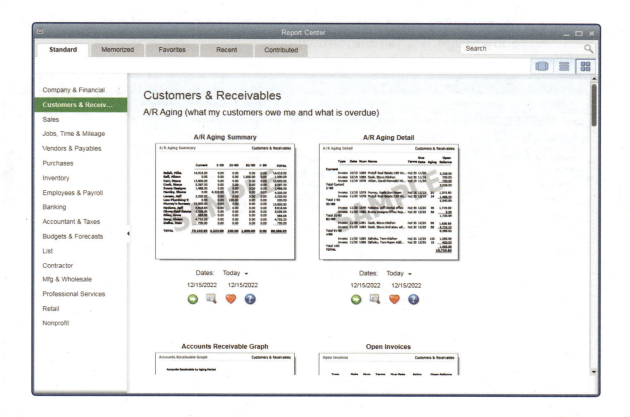

The Customers & Receivables window has several categories related to accounts receivable starting with A/R Aging, which includes seven reports. The right side of the screen includes a brief illustration/explanation of each report to help the user decide which report best meets his or her needs.

▶ *Move your cursor over the Open Invoices (due but not received) report, which opens more information about this report near the bottom of the report description.*

▶ *Click the Info button at the bottom of the report illustration to open an information window about the report.* In the lower part of the information window is a description of what the report tells you.

▶ *Close the information window to return to the Customers & Receivables Reports window.*

▶ Scroll down and double-click the Customer Balance Detail report under the Customer Balance category to see a listing of customer balances, including the detailed transactions that make up each customer balance due.

▶ Close the Customer Balance Detail window.

▶ Click Company & Financial on the Reports listing → double-click Profit & Loss Standard to access the Profit & Loss report for December 1–15, 2022, the top portion of which is shown below. Do not be concerned if your window shows slightly different numbers.

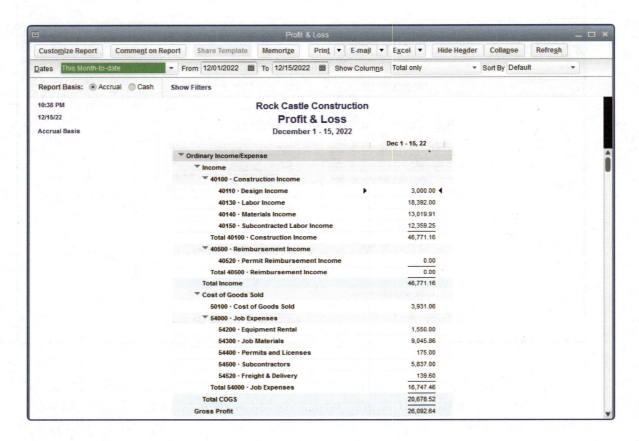

Selecting Report Dates and Other Options

QuickBooks provides users the option to select an accounting period or date for each report and to also select other display options. This section includes discussion of three options:

- **Dates:** Permits selecting different accounting periods or ending dates for a report.

- **Columns:** Permits selecting different totals or subtotals for a report.
- **Expand/Collapse:** Permits a different appearance when there are subtotals.

Dates

▶ *With the Profit & Loss report open, click the drop-down list arrow next to the Dates box and observe the many options for accounting periods available to users → select This Fiscal Year-to-date to open a year-to-date income statement.* Again, your screen may contain different amounts, but do not be concerned about that.

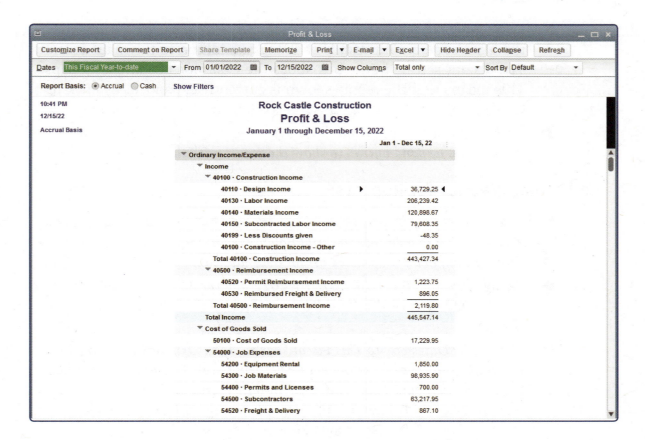

Whenever selecting any of the options to change a report window, develop the habit of also clicking the Refresh button (Refresh). In most cases, the window refreshes automatically, but not always.

▶ *Select different date options to observe the changes in the income statement.*

Columns

▶ *Click the drop-down list arrow next to the Show Columns box and observe the many options for different information available to users → select Quarter and observe the new quarterly as well as the previous total information provided.*

▶ *Select Employee in the drop-down list in the Show Columns box.* The information provided is the payroll expense for each employee.

▶ *Select different items in the Show Columns drop-down list to see the large amount of information available to users when this box is selected.*

Expand/Collapse

▶ *Select Month in the drop-down list in the Show Column box.* Notice that there are sub-account details showing in addition to the main account. The default setting in the box above the Columns heading is Collapse. This means that the report shown is currently expanded. By clicking the Collapse button (Collapse), the sub-accounts are no longer visible.

▶ *Click the Collapse button and it changes to an Expand button (Expand). Click on the Expand button to show the report with the sub-accounts again.*

Additional Practice for Reports

▶ *Return to the main Report Center window → click Accountant & Taxes in the Reports list → double-click the Trial Balance report under Account Activity → select Custom in the drop-down list for the Dates box.*

▶ *Change the From Box date to 07/01/2022 and the To box date to 08/01/2022 → click the Refresh button.* Notice that the amounts all changed.

▶ *Change the Dates box to This Fiscal Year-to-date. Drill down on account 15000, Furniture and Equipment by hovering over the dollar amount for this account. (Make sure This Fiscal Year-to-date is still in the Dates box.)* What were the two additions to this account during the year? The answer is two purchases from Kershaw Computer Services.

▶ *Drill down on the $6,500 transaction to examine the original Enter Bills transaction.*

▶ *Close all windows to return to the Report Center.*

▶ *Open the Journal report in the Accountants & Taxes Reports under Account Activity → select Today in the Dates box.* This is a useful window to determine if you have correctly recorded a transaction that you recorded on that day. For this project you often record a transaction on a date other than the current date, but you want to examine the transaction in the journal to make sure it was correctly recorded. One way to do that is to select any period in the Dates box that includes the date you are processing data, and then scroll to the appropriate transaction. An alternative that makes it easier to find the transaction is to change the dates to include only the date you are interested in.

Assume you processed a transaction to Sergeant Insurance on 12/07/2022 and want to determine if it was correctly recorded.

► *While you are in the Journal report, select Custom for the Dates box →* *12/07/2022 for the From box → 12/07/2022 for the To box → Refresh.* It is now easy to find and review the transaction because there were only four transactions recorded on that date.

► *Select This Fiscal Year-to-date for the Dates box in the Journal report → select Date in the Sort By box → click the AZ button to the right of Sort By. Now scroll down to the 12/07/2022 date to find Sergeant Insurance a different way.* You will be locating transactions you have just recorded several times in later chapters. There are alternative ways to locate them.

Chapter Summary

After completing Chapter 4, you have now learned:

- ✔ how to obtain information from your accounting system using lists and drill down.
- ✔ how to export information to Excel.
- ✔ what information is available in Company Snapshot.
- ✔ how to modify information obtained through Custom Transaction Detail Reports.
- ✔ how to generate reports.

You should now save your work by making periodic backups of Rock Castle Construction and Larry's Landscaping & Garden Supply using the instructions in the E-Materials you downloaded from the Armond Dalton Resources website. Be sure to use descriptive file names, such as "Rock Castle after Ch 4" or something similar.

Before starting the homework for Chapter 4, you should restore both Rock Castle Construction and Larry's Landscaping & Garden Supply using the initial backups you downloaded from the Armond Dalton Resources website. Restore both companies before proceeding to the Chapter 4 homework assigned by your instructor either online or in the Student Problems & Cases book (consult your instructor).

If you cannot recall how to make a periodic backup or restore a backup, refer to the E-materials located at www.armonddaltonresources.com.

PRACTICE— MAINTENANCE ACTIVITIES

Introduction

Maintenance activities were introduced in Chapter 2 and further expanded in Chapter 3. This chapter's material provides a review, a more detailed study of maintenance, and introduces use of the Reference book. Additional practice is provided.

Maintenance Overview

When a transaction is recorded in a *QuickBooks* window, one of the most useful features of the program is the default information stored in the system. For example, after a customer name is selected in the Create Invoices window for a credit sale, the software automatically completes many areas of the window, such as the customer's name, address, and the discount terms for sales to that customer.

Default information is stored in the system through maintenance. As discussed in Chapter 3, there are five maintenance windows, and for each of the five there are three possible maintenance tasks that can be performed with the window. These are as follows:

Five Maintenance Windows
1. Vendors
2. Customers & Jobs
3. Inventory Items & Services
4. Employees
5. Chart of Accounts

Three Types of Maintenance Tasks
1. Add a new record
2. View and change information in an existing record
3. Delete a record

Maintenance tasks have no effect on transactions already recorded, but certain maintenance tasks affect the amounts recorded in subsequent transactions. For example, changing an inventory item's cost and selling price has no effect on previously recorded sales transactions, but it will affect the amount of sales revenue and cost of goods sold posted to the general ledger for future sales of the item.

Reference Material

Before beginning the practice section for maintenance activities, you should read the introduction on pages 3 and 4 of the Reference book. Read the section Suggested Way to Use the Reference Book carefully. Most of the Reference book is used for processing transactions, but it is also useful for maintenance activities. You will deal with processing transactions in Chapters 6 through 8.

Instructions for using each of the five maintenance windows to perform the three types of maintenance tasks are on pages 58 through 85 of the Reference book. The reference material for maintenance is less detailed than for the other reference sections because there is a wider variety of information that may or may not be entered or changed in each window.

Read and understand the overview material in the Perform Maintenance Activities section on pages 58 and 59 of the Reference book before practicing maintenance.

In this section, you will practice working with each of the five types of maintenance windows shown in the Reference book. All maintenance tasks are processed through one or more of the maintenance windows listed previously.

Practice Tasks

As discussed previously and in the Reference book, maintenance windows are used to (1) add a new record, (2) view and change information in an existing record, and (3) delete a record.

Maintenance Practice Task #1 — Add a New Vendor Record

When a company makes a purchase from a new vendor, a new record must be created in *QuickBooks*. During this part of the practice section, you are to add a new vendor record for the sample company, Jackson Supply Company, using the New Vendor window.

Jackson Supply Company has a new vendor, XYZ Warehouse, from which it purchases inventory.

▶ *Open Jackson Supply Company and add a new vendor record for XYZ Warehouse using the Reference book instructions on page 64 and the information in the box that follows, but do not save the new record yet.* ***Note:*** All default information is correct for the vendor unless otherwise noted. Also remember that not all boxes in the New Vendor window are applicable to this vendor.

- ■ **Vendor Name:**   XYZ Warehouse
- ■ **Opening Balance:**   0
- ■ **As of:**   2/18/2019

Units Address Info Tab
- ■ **Company Name:**   XYZ Warehouse
- ■ **Main Phone:**   (201) 235-0039
- ■ **Main Email:**   xyzw@xyzwarehouse.com
- ■ **Website:**   www.xyzwarehouse.com
- ■ **Fax:**   (201) 234-9002
- ■ **Addresses: (Billed From   XYZ Warehouse
  & Shipped From —   500 Westland Park Dr.
  use Copy button)**   Upper Saddle River, NJ 07458

Payment Settings Tab
- ■ **Account No.:**   XYZ0001
- ■ **Terms:**   2% 10, Net 30
- ■ **Print on Check as:**   XYZ Warehouse

Account Settings Tab
- ■ **First Prefill:**   10400 Inventory
- ■ **Second Prefill:**   30700 Purchases Discounts
- ■ **Third Prefill:**   30800 Freight costs

Maintenance Practice Transaction #1. Add a New Vendor Record

Maintenance Practice Transaction #1. Add a New Vendor Record *(continued)*

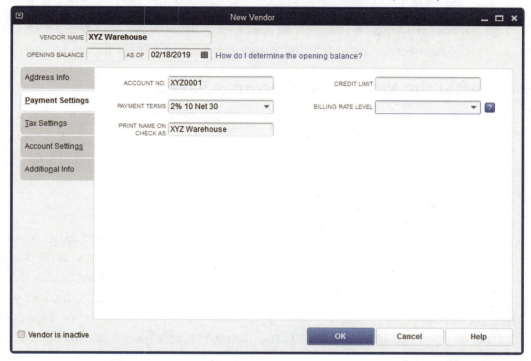

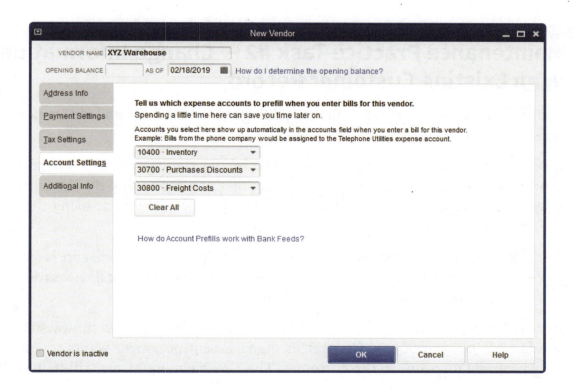

The preceding diagrams show the completed tabs of the New Vendor window.

▶ *If the windows on your screen are consistent with the diagrams, click OK to save the new vendor record. If there are errors, correct them before saving the record.*

Review After the New Vendor Record is Saved

▶ *Click Vendors in the Home Page → Vendors tab → right-click vendor XYZ Warehouse, which should now appear in the list of Active Vendors → select Edit Vendor to open the Edit Vendor window. Review the information for the new vendor and correct the information if it is wrong.*

The purpose of the Maintenance procedure you just completed is to review the addition of a new vendor and verify that the information you added was correct. You should follow this procedure for any new maintenance activity or change of maintenance information to determine if the information was entered and saved correctly. *Note:* This subsequent review is not repeated in this or later chapters for new maintenance or change in maintenance information.

Maintenance Practice Task #2 — Change Information in an Existing Customer Record

Adding a customer record is similar to adding a vendor record and is therefore not repeated here.

Often, it is necessary to change information in an existing customer's record. Examples include changing a customer's address and modifying the default general ledger accounts for a customer. Edits to existing records are made using the Edit Customer window.

▶ *Change the customer record for Sunway Suites using the Reference book instructions on page 61 and the following information, but do not save the revised record yet.*

Sunway Suites, an existing customer, added a new website of www.sunwaysuites.com. In addition, Jackson Supply Company's management no longer wants to extend an early payment discount to the customer. The new payment terms will be Net 30.

Maintenance Practice Transaction #2. Change Information in an Existing Customer Record

▶ *If the windows on your screen are consistent with the preceding diagrams, click OK to save the changes to the customer record.*

Maintenance Practice Task #3 — Delete a Customer Record

It is often desirable to remove an existing customer record from *QuickBooks*. An example is a customer to which the company no longer sells goods or services. Deletion is possible only if customer accounts have no active transactions or account balances.

▶ *Delete Traver's Bed & Breakfast Customer ID using the Reference book instructions on page 61.*

Maintenance Practice Transaction #3. Delete a Customer Record

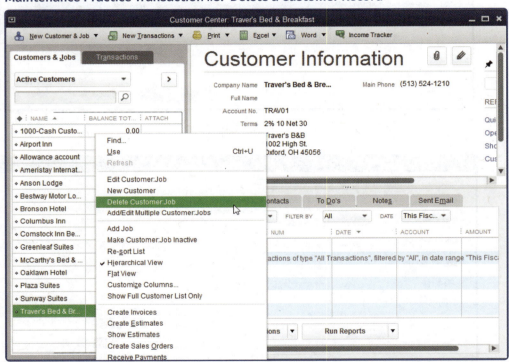

Review After the Customer is Deleted

▶ *Click Customers on the Home Page → Customers & Jobs tab, if it is not already open. As shown on the next page, notice that Traver's Bed & Breakfast no longer exists in the list of customers. If the customer is still listed, select it again and repeat the steps to delete the record for Traver's Bed & Breakfast.*

Maintenance Practice Transaction #3. Delete a Customer Record *(continued)*

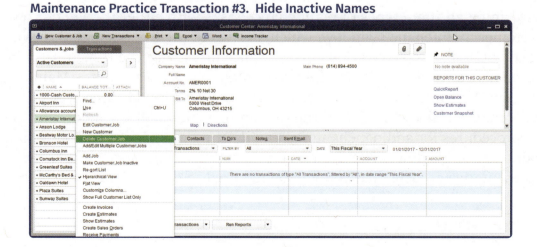

> You should follow this same review procedure for any deletion maintenance activity to determine if the information was deleted. This subsequent review is not repeated in this or subsequent chapters for deletion maintenance activities.

Hide Inactive Names

Jackson Supply Company is planning to no longer do business with an existing customer, Ameristay International. Practice deleting the customer record using the Customer Maintenance window as follows:

▶ *Attempt to delete Ameristay International's customer record using the Reference book instructions on page 61.*

Maintenance Practice Transaction #3. Hide Inactive Names

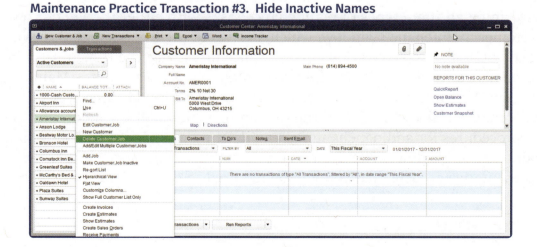

A warning message appears in the QuickBooks Message window saying that you cannot delete this record because it has a balance or it is used in at least one transaction. Transactions and balances from an existing customer must be cleared before the software will allow that customer to be deleted.

Observe in the QuickBooks Message window that a solution is provided. Hide the inactive customer so they don't appear on the list of customers.

▶ *Click the Make Inactive button and click Yes if you receive a message saying the customer has an outstanding balance.* Ameristay International's name should no longer be included in the Customers & Jobs window because the View box indicates only Active Customers are shown.

▶ *Click the drop-down list arrow next to the first box in the Customers & Jobs tab, and then click All Customers. Ameristay International is now included, this time with an X in the left margin.*

After completing the first three maintenance practice tasks, you have practiced performing all three types of maintenance tasks with either the Vendor or Customer maintenance window, which are two of the five maintenance windows. Next, you will complete various maintenance tasks with the three remaining maintenance windows: Chart of Account Maintenance, Employees Maintenance, and Item & Services Maintenance. You will not practice all three types of maintenance tasks with each of the remaining maintenance windows because the procedures for adding, changing, and deleting records are similar among maintenance windows. The five remaining maintenance practice tasks are representative of tasks that you will be required to complete in later chapters.

Maintenance Practice Task #4 — Add a General Ledger Account

The next maintenance task is to add a new general ledger account for Jackson Supply Company.

▶ *Add a new general ledger account record using the Reference book instructions on page 82 and the information that follows, but do not save the new record yet.*

- ■ **Account Type:**   Other Current Liability
- ■ **Number:**   20210
- ■ **Account Name:**   401K Deductions Payable
- ■ **Description:**   Liability for employee 401K deductions from gross pay
- ■ **Tax-Line Mapping:** B/S-Liabs/Eq: Other current liabilities

Maintenance Practice Transaction #4. Add New Account

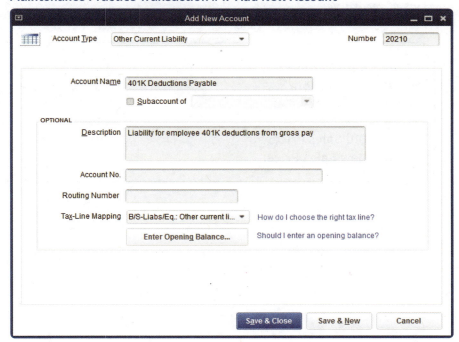

The preceding diagram includes the completed Add New Account window for the new general ledger account. Do not be concerned if the general ledger account numbers and balances on your screen differ slightly from those shown.

▶ *If the window on your screen is consistent with the diagram, save the new record. If there are errors, correct them before saving the record.*

Changing or deleting an existing general ledger account is relatively simple when you follow the instructions on page 83 of the Reference book. No practice exercises are considered necessary.

Maintenance Practice Task #5 — Change an Inventory Item's Cost and Selling Price

The next maintenance task is to edit an inventory item's record for changes in the item's cost and selling price.

▶ *Change the inventory item record for Item No. 103 using the Reference book instructions on page 69 and the following information, but do not save the revised record yet.*

■ Item Number:	103
■ Description:	Washcloths – 100 pack
■ Old Cost:	$85.00
■ New Cost:	$87.00
■ Old Sales Price:	$110.50
■ New Sales Price:	$113.00

Maintenance Practice Transaction #5. Change an Inventory Item's Cost and Selling Price

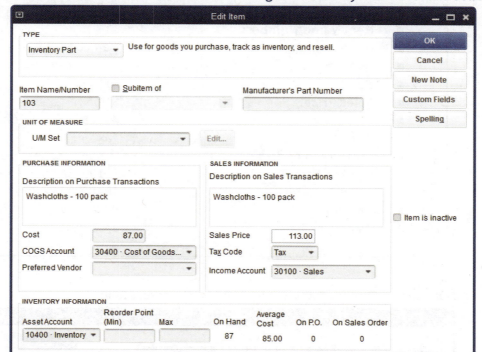

The preceding diagram includes the revised Edit Item maintenance window for Item No. 103. Ignore the contents of the On Hand box; your window may contain a different quantity.

▶ *If the window on your screen is consistent with the diagram, click the OK button to save the revised inventory item record. If there are errors, correct them before saving the record.*

Maintenance Practice Task #6 — Add an Inventory Item

The next maintenance task is to add a new inventory item's record.

▶ *Add an inventory item record for Item No. 118 using the Reference book instructions on page 68 and the following information, but do not save the new record yet.*

■ **Type:**	Inventory Part
■ **Item Name/Number:**	118
■ **Description on Purchase Transactions:**	Hair conditioner – box of 50
■ **Description on Sales Transactions:**	Hair conditioner – box of 50
■ **Cost:**	$11.75
■ **Sales Price:**	$14.50
■ **COGS Account:**	50000 Cost of Goods Sold
■ **Tax Code:**	Tax
■ **Preferred Vendor:**	Omni Incorporated
■ **Income Account:**	30100 Sales
■ **Asset Account:**	10400 Inventory
■ **On Hand:**	0.00
■ **As of:**	2/15/2019

Maintenance Practice Transaction #6. Add an Inventory Item

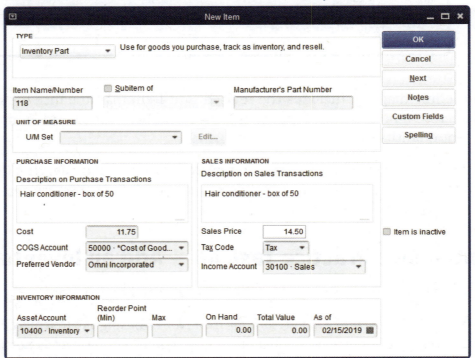

The preceding diagram includes the New Item maintenance window for Item No. 118.

▶ *If the window on your screen is consistent with the diagram, click the OK button to save the new item record. If there are errors, correct them before saving the record.*

Maintenance Practice Task #7 — Increase an Employee's Pay Rate

The sample company, Jackson Supply Company, increased the hourly and overtime pay rates for Mark Phelps, an employee.

▶ *Change Mr. Phelps' employee record to reflect his increased hourly pay rate, using the Reference book instructions on page 75 and the following information, but do not save the information yet.*

> - **Employee Name:**   Mark C Phelps
>
> **Payroll Info Tab**
> - **Old hourly rates:** $14.00 hourly, $21.00 overtime
> - **New hourly rates:** $15.00 hourly, $22.50 overtime

Maintenance Practice Transaction #7. Increase an Employee's Pay Rate

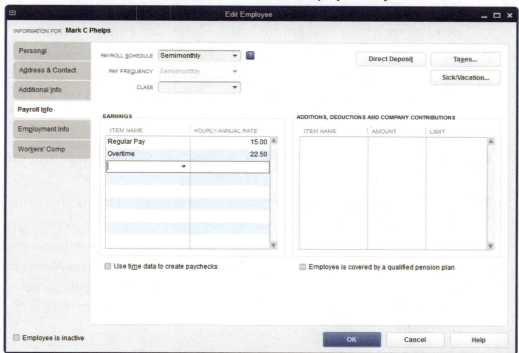

The preceding diagram includes the revised Edit Employee Payroll and Compensation Info tab window for Mr. Phelps.

▶ *If the window on your screen is consistent with the diagram, click the OK button to save the hourly pay rate increase. If there are errors, correct them before saving.*

Maintenance Practice Task #8 — Perform Maintenance "On the Fly"

It is common for the person entering sales invoice information in the Create Invoices window to find no name in the Customer drop-down list because it is a new customer. The same situation occurs when entering information for transactions related to the other four maintenance activities. *QuickBooks* solves this by allowing the person to add the new customer or other maintenance activity while the transaction is being processed. This is often called maintenance "on the fly."

▶ *Click Create Invoices on the Home Page → click the drop-down list arrow next to Customer: Job box. Observe that the top item is <Add New>.*

▶ *Click <Add New> to access the New Customer window.* This is the same window in which you entered data earlier for a new customer.

▶ *Type ABCD in the Customer Name box and click OK. You have returned to the Create Invoices window and can now continue entering data.* (Normally, in practice, the person entering data enters information for all tabs in the New Customer window but, to avoid repetition of what you did earlier, only the name is included.)

▶ *Click the drop-down list arrow in the Item box.* Observe that the top item is <Add New>.

▶ *Click <Add New> to access the New Item window.* This is the same window in which you entered data earlier for a new inventory item.

▶ *Leave the default setting for Type as Service.* (A new inventory item would have been added during purchases.) *Type 137 in the Item Name/Number box, 30400 in the Account box, then click OK to return to the Create Invoices window again.* (Again, in practice, the person entering data completes the entire window.)

▶ *Close the Create Invoices window and do not record the transaction.*

▶ *Delete both customer ABCD and Item 137 from the maintenance activity records.* You will not be using them again.

The same approach you just completed is equally applicable to new vendors, employees, and account numbers.

No illustrative windows are included for Maintenance "on the fly."

Chapter Summary

After completing Chapter 5, you have now learned how to perform detailed maintenance activities for customers, vendors, inventory, employees, and general ledger accounts. You have practiced the five types of maintenance activities and the three types of changes for each activity. This knowledge is essential for the remaining chapters in the book.

You should back up your data files for Jackson Supply Company using the instructions in the E-materials. Be sure use a descriptive file name, such as "Jackson after Ch 5" or something similar.

If you do not feel comfortable with your ability to complete each of these purchases and cash disbursements activities, you can practice further in one of two ways.

1. You can do the entire chapter again by following the instructions in the E-materials to restore the Jackson Supply Company dataset from the initial backup you downloaded and extracted to your QB Backup Files folder in the E-materials. Doing so will return all data to its original form and permit you to practice all procedures again.

2. You can do additional practice by completing any or most of the sections in this chapter again. You can continue to use Jackson Supply Company.

There are no problems for Chapter 5. There will be questions dealing with maintenance in subsequent chapters as you learn to process transactions. Next, you will move on to learn and practice processing a wide variety of transactions and doing other activities in the next three chapters, starting with purchases and cash disbursements cycle activities in Chapter 6.

This page is intentionally blank.

PRACTICE — PURCHASES AND CASH DISBURSEMENTS CYCLE ACTIVITIES

Introduction

In Chapters 6 through 8, you will learn how to record transactions and perform other activities commonly done using *QuickBooks*. You will use Jackson Supply Company to complete these chapters. For each of the practice exercises summarized in Chapters 6 through 8 you will be given information about a transaction or other activity and Reference book pages that provide a summary to help you complete the practice exercise. You are already familiar with the Reference book from the last chapter, but for the next three chapters more detail is provided. Windows showing the relevant information are provided to help your learning.

Chapters 6 through 8 are organized by the transaction cycle approach. The transaction cycle approach means that related transactions and account balances are included in the same cycle. For example, sales transactions, collections on accounts receivable, sales returns, and write-offs, as well as the account balances associated with those transactions, are included in the sales and cash receipts cycle. Systems designers develop accounting systems using cycles. Similarly, auditors perform audits following a cycle approach.

The following three cycles are included in the next three chapters:

- Chapter 6 — Purchases and cash disbursements cycle
- Chapter 7 — Sales and cash receipts cycle
- Chapter 8 — Payroll cycle and other activities

Before beginning the practice section, you should reread the introduction on pages 3 and 4 of the Reference book. Read the section Suggested Way to Use the Reference Book especially carefully. You should do all of the practice exercises in the order listed in Chapters 6 through 8. The knowledge you gain in these practice chapters will be applied to subsequent chapters.

The transaction cycle included in Chapter 6 is the purchases and cash disbursements. The beginning process for operating a business that sells inventory for a profit is to purchase inventory and turn it into saleable commodities.

▶ *Open Jackson Supply Company if it is not already open.*

▶ *Open the Home Page if it is not already open.* Observe in the top-left corner of the flow diagram that the first step is Purchase Orders, which is used to order inventory or services. Following to the right is the receipt of inventory, entering bills, and then paying those bills. You will follow that process in this chapter and learn how to complete each major task in the cycle using *QuickBooks*. You will be dealing with all of the processes on the Home Page in the next three chapters. Even though the flow diagram shows purchase orders as the first step, the process is ongoing, with purchases & cash disbursements, sales & cash receipts, and payroll working simultaneously and continuously.

Be sure to carefully read the following considerations before you begin processing transactions to avoid making mistakes.

Several considerations are relevant for Chapters 6 through 8.

- Throughout Chapters 6 through 8, 2019 has been adopted as the business year for purposes of demonstrating transaction entry and report generation. You will need to pay close attention to dates to ensure that you enter the correct 2019 date. (Of course, when companies process information using *QuickBooks* in real time, the current date usually is the proper default date for the entry window.)
- You will be given a specific date for processing each practice transaction. Unless otherwise noted, use this same date throughout the entire transaction.
- Before starting to process transactions for Chapters 6 through 8, you should make sure that the "Use the last entered date as default" radio button is selected in the General Preferences window (Edit → Preferences → General). Then make sure that each time you open Jackson Supply Company and record a practice transaction you use the correct date for that transaction. As long as the company remains open, that same date will be used as the default date for other transactions. Even if you forget to change the date to another date in February 2019, at least the transaction will be recorded in the correct month. When you close Jackson Supply Company and open it back up again, the date defaults to 12/15/2019. You will have to be careful to enter the correct date for the first transaction you record when opening the company again.
- The only company that is used to process practice transactions is Jackson Supply Company. The problem material for Chapters 6 through 8 often uses a different company.
- Recall from Chapter 5 that if a box is not mentioned in the Reference book, you do not have to do anything with that box.
- You can make most corrections prior to saving a transaction by clicking the appropriate boxes and correcting the errors.

- Do not be concerned about making mistakes during any of the practice sections until you get to the problems at the end of each chapter. These practice sections are for your benefit only and any errors that you make will not affect the graded assignments.
- If, at any time, you decide that you want to start the chapter over, you may do so by restoring the Jackson Supply Company dataset using the instructions in Chapter 1. You may want to do so if you believe that you do not understand the material in the chapter.

For the purchases and cash disbursements cycle, the following activities are included in this chapter:

- Prepare a purchase order
- Receive goods on a purchase order
- Purchase inventory without a purchase order — no payment made at time of purchase
- Purchase non-inventory items or services without a purchase order — no payment made at time of purchase
- Pay a vendor's outstanding invoice
- Purchase inventory without a purchase order — payment made at time of purchase
- Purchase non-inventory items or services without a purchase order — payment made at time of purchase
- Return inventory from a purchase

Prepare a Purchase Order

Reference Material

A purchase order is prepared through the Create Purchase Orders window, an example of which is shown on page 7 of the Reference book. Read and understand the Prepare a Purchase Order overview on page 6 of the Reference book before processing the transaction. Then follow the instructions on pages 6 and 7 as you complete the practice section.

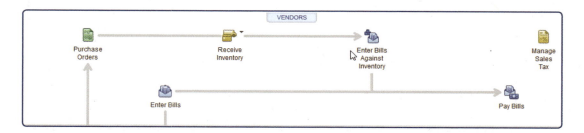

Practice Transaction #1. Prepare a Purchase Order For Inventory Items

▶ *Process the first purchase order using the following information, but do not save it yet.*

- **Vendor:** Omni Incorporated
- **Date:** February 11, 2019
- **PO No.:** 5876
- **Inventory items ordered:**

Item	Description	Qty.	Rate
114	Soap – box of 50	20	$21.15
117	Hand Lotion – 50 pack	10	9.00

- **Purchase Order Total (check figure):** $513.00

Practice Transaction #1. Create Purchase Orders Window

The preceding window shows the Purchase Orders entry window with Purchase Order No. 5876 entered.

▶ *If your Create Purchase Orders window is consistent with the one shown, click Save & Close. If there are errors, correct them before saving.*

Purchase Order Review After the Transaction Is Saved

▶ *Click Vendors on the Home Page → Transactions tab – Purchase Orders → double-click Purchase Order 5876. Review the information for Purchase Order No. 5876 on the same Create Purchase Orders window that you just prepared and correct the information if there are errors.*

Practice Transaction #2. Prepare a Purchase Order With a Different Shipping Location and With Item Cost Other than Standard

▶ *Process the second purchase order using the following information, but do not save it yet.* Because this transaction is for a shipment to a different location than the default one and the purchase price of the products is not the standard price charged by American Linen, both must be changed. This is a one-time change in both the price negotiated with American Linen and the shipping location, so answer No when asked if you want to make these changes permanent.

- **Vendor:** American Linen Supply
- **Date:** February 13, 2019
- **PO No.:** 5877
- **Shipping Address:** Jackson Supply Company
  1726 Carbon Lane
  Cincinnati, Ohio 43196 (no phone number)
- **Inventory items ordered:**

Item	Description	Qty.	Rate
105	Queen sheet set	40	$23.75
109	Standard comforter	10	31.00

- **Purchase Order Total (check figure):** $1,260.00

Practice Transaction #2. Create Purchase Orders Window

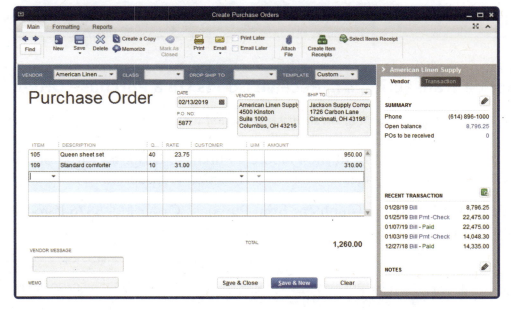

▶ *If your Create Purchase Orders window is consistent with the one shown, click Save & Close. If there are errors, correct them before saving. Click No when asked if you want to change the shipping address for future purchases.*

Purchase Order Review After the Transaction Is Saved

▶ *Click Vendors on the Home Page → Transactions tab – Purchase Orders → double-click Purchase Order 5877. Review the information for Purchase Order No. 5877 on the same Create Purchase Orders window that you just prepared and correct the information if there are errors.*

Receive Goods on a Purchase Order

Reference Material

The receipt of inventory and other goods when a purchase order has been prepared is processed through the Enter Bills window, an example of which is shown on page 9 of the Reference book. Read and understand the Receive Goods on a Purchase Order overview on page 8 of the Reference book before processing the transaction. Then follow the instructions on pages 8 and 9 of the Reference book as you complete the practice section.

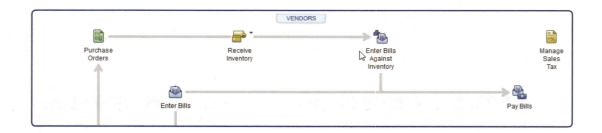

Practice Transaction #1. Receive Goods On a Purchase Order

▶ *Process a receipt of inventory transaction where a purchase order had been prepared using the information that follows, but do not save it yet.* You processed the purchase order in the Prepare a Purchase Order for inventory practice section on pages 6-5 and 6-6 (Practice Transaction #1).

On February 15, 2019, a shipment from Purchase Order No. 5876 is received from Omni Incorporated, along with an invoice. Other details of the transaction follow.

> - **Vendor:** Omni Incorporated
> - **Purchase Order No.:** 5876
> - **Date:** February 15, 2019
> - **Ref. No. (Vendor Invoice No.):** X21478
> - **Terms:** 2% 10 Net 30
> - **Goods received:**
>
> | Item | Description | Qty. | Rate |
> |------|-------------|------|------|
> | 114 | Soap – box of 50 | 20 | $21.15 |
> | 117 | Hand Lotion – 50 pack | 10 | 9.00 |
>
> - **Invoice Total (check figure):** $513.00

Practice Transaction #1. Enter Bills Window

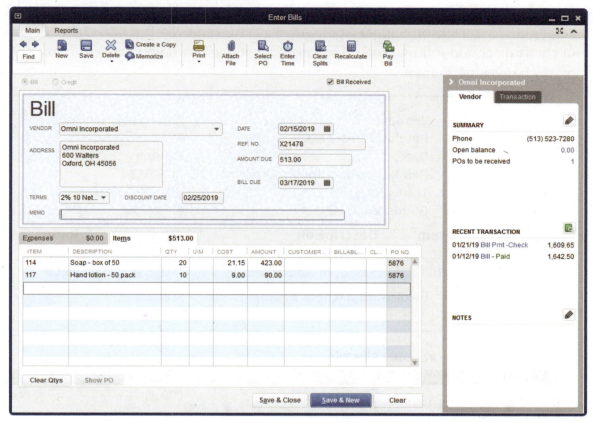

The preceding window shows the Enter Bills windows immediately before the Save & Close button is clicked.

▶ *If your Enter Bills window is consistent with the one shown, click Save & Close to save the transaction. If there are errors, correct them before saving.*

Transaction Review After the Transaction Is Saved

▶ *Click Vendors on the Home Page → Transactions tab – Bills → double-click Vendor Invoice X21478 (Num). Review the information for Vendor Invoice No. X21478 on the same Enter Bills window that you just prepared and correct the information if there are errors.*

Practice Transaction #2. Receive Goods On a Purchase Order

▶ *Process a receipt of inventory using the following information, but do not save it yet. You processed the purchase order in the Prepare a Purchase Order practice section on pages 6-6 and 6-7 (Practice Transaction #2).*

On February 18, 2019, a partial shipment from Purchase Order No. 5877 is received from American Linen Supply, along with an invoice. Due to special pricing, the selling price to Jackson Supply Company was reduced even further. Other details of the transaction follow. *Note:* Answer No when asked if you want to make these price changes permanent.

- Vendor: American Linen Supply
- Purchase Order No.: 5877
- Date: February 18, 2019
- Ref. No. (Vendor Invoice No.): ALS2663
- Terms: 2/10, Net 30
- Goods received:

Item	Description	Qty.	Rate
105	Queen sheet set	30	$21.50
109	Standard comforter	8	28.00

- Invoice Total (check figure): $869.00

Practice Transaction #2. Enter Bills Window

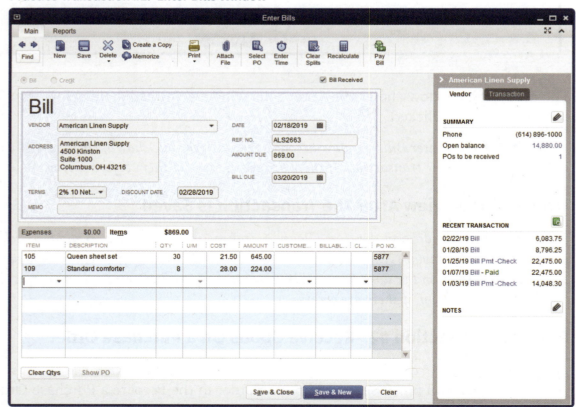

The preceding window includes the Enter Bills windows immediately *before* the Save & Close button is clicked.

▶ *If your Enter Bills window is consistent with the one shown, save the transaction. If there are errors, correct them before saving.*

Transaction Review

▶ *Click Vendors on the Home Page → Transactions tab – Bills → double-click Vendor Invoice ALS2663 (Num). Review the information for Vendor Invoice No. ALS2663 on the same Enter Bills window that you just prepared and correct the information if there are errors.*

Purchase Inventory Without a Purchase Order — No Payment Made At Time of Purchase

Reference Material

A purchase of inventory without a purchase order and without a corresponding cash disbursement is processed through the Enter Bills window, an example of which is shown on page 11 of the Reference book. Read and understand the Purchase Inventory Without a Purchase Order — No Payment Made At Time of Purchase overview on page 10 of the Reference book before processing the transaction. Then follow the instructions on pages 10 and 11 of the Reference book as you complete the practice section.

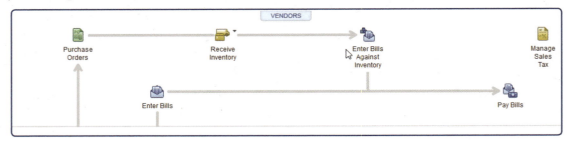

Practice Transaction. Purchase Inventory Without a Purchase Order — No Payment Made At Time of Purchase

▶ *Process a vendor's invoice using the following information, but do not save it yet.*

On February 22, 2019, Jackson Supply Company received an invoice from American Linen Supply for inventory delivered on the same day. Other details are in the box that follows.

- **Vendor:** American Linen Supply
- **Date:** February 22, 2019
- **Ref. No. (Vendor Invoice No.):** ALS2714
- **Terms:** 2/10, Net 30
- **Goods received:**

Item	Description	Qty.	Rate
103	Washcloths – 100 pack	50	$87.00
107	Pillows set of 2	15	22.25
112	Draperies	25	56.00

- **Invoice Total (check figure):** $6,083.75

Practice Transaction. Enter Bills Window

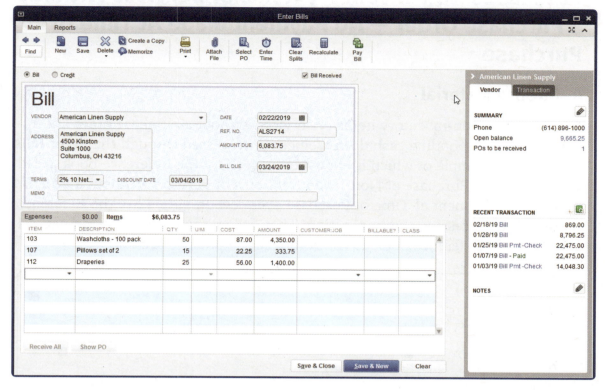

The preceding window shows the Enter Bills window with Invoice No. ALS2714 entered.

▶ *If your window is consistent with the one shown, save the transaction. If there are errors, correct them before saving.*

Transaction Review After the Transaction Is Saved

▶ *Click Vendors on the Home Page → Transactions tab – Bills → double-click Vendor Invoice ALS2714 (Num). Review the information for Vendor Invoice No. ALS2714 on the same Enter Bills window that you just prepared and correct the information if there are errors.*

Purchases Non-Inventory Items or Services Without a Purchase Order — No Payment Made At Time of Purchase

Reference Material

A purchase of non-inventory items or services without a purchase order and without a corresponding cash disbursement is processed through the Enter Bills window, an example of which is shown on page 13 of the Reference book. Read and understand the Purchase of Non-Inventory Items or Services Without a Purchase Order — No Payment at Time of Purchase overview on page 12 of the Reference book before processing the transaction. Then follow the instructions on pages 12 and 13 of the Reference book as you complete the practice section.

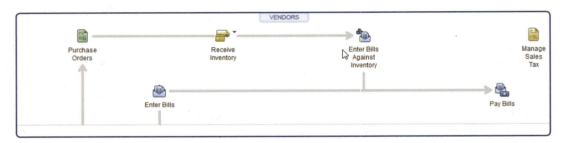

Practice Transaction #1. Purchase of Non-Inventory or Services Without a Purchase Order — No Payment Made At Time of Purchase

▶ *Process the following vendor's invoice using the following information, but do not save it yet.*

On February 25, 2019, Jackson Supply Company received an invoice from Standard Office Supplies for two color laser printers delivered on the same day. Other details are in the box that follows.

■ Vendor:	Standard Office Supplies	
■ Date:	February 25, 2019	
■ Ref. No. (Vendor Invoice No.):	26978	
■ Terms:	2/10, Net 30 (one-time change)	
■ Goods received:		

Account	Amount	Memo
10800 Fixed Assets	$1,550.90	XPL612C Color Laser Printer
40300 Office Supplies	126.15	20 reams of printer paper
■ Invoice Total (check figure):	$1,677.05	

Practice Transaction #1. Enter Bills Window

The window above shows the Enter Bills window with Invoice No. 26978 entered.

▶ *If your window is consistent with the one shown, save the transaction. If there are errors, correct them before saving. Click No when asked if you want to save the payment terms for future transactions.*

Transaction Review After the Transaction Is Saved

▶ *Click Vendors on the Home Page → Transactions tab – Bills → double-click Vendor Invoice 26978 (Num). Review the information for Vendor Invoice No. 26978 on the same Enter Bills window that you just prepared and correct the information if there are errors.*

Practice Transaction #2. Purchase of Non-Inventory or Services Without a Purchase Order—No Payment Made At Time of Purchase

▶ *Process the following vendor's invoice using the following information, but do not save it yet.*

On February 28, 2019, Jackson Supply Company received an invoice from Ohio Power & Light for electric and gas utilities for the month. Other details are in the box that follows.

- **Vendor:**   Ohio Power & Light
- **Date:**   February 28, 2019
- **Ref. No. (Vendor Invoice No.):**   487993241
- **Terms:**   Net 30
- **Services invoiced:**

Account	Amount	Memo
41600 Utilities	$482.71	Electric and gas utilities

- **Invoice Total (check figure):** $482.71

Practice Transaction #2. Enter Bills Window

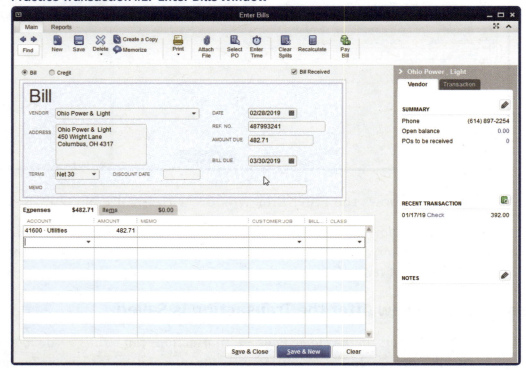

The preceding window shows the Enter Bills window with Invoice No. 487993241 entered.

▶ *If your window is consistent with the one shown, save the transaction. If there are errors, correct them before saving.*

Transaction Review After the Transaction Is Saved

▶ *Click Vendors on the Home Page → Transactions tab – Bills → double-click Vendor Invoice 487993241 (Num). Review the information for Vendor Invoice No. 487993241 on the same Enter Bills window that you just prepared and correct the information if there are errors.*

Pay a Vendor's Outstanding Invoice

Reference Material

Payment of a vendor's outstanding invoice is processed through the Pay Bills window, an example of which is shown on page 15 of the Reference book. Read and understand the Pay a Vendor's Outstanding Invoice overview on page 14 of the Reference book before processing the transaction. Then follow the instructions on pages 14 through 16 as you complete the practice section.

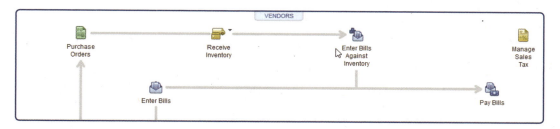

Practice Transaction #1. Full Payment of Vendor's Outstanding Invoice

▶ *Process a cash disbursement using the following information, but do not save it yet.*

On February 25, 2019, the company issued a check to American Linen Supply in payment of an outstanding invoice within the discount period. Recall that the invoice being paid was recorded earlier.

- **Vendor ID:** American Linen Supply
- **Check amount:** $748.72
- **Ref. No. (Invoice paid):** ALS2663, totaling $869.00
- **Amt. Due:** $869.00
- **Disc. Used:** $17.38
- **Amt. To Pay:** $851.62
- **Discount Account:** 30700 - Purchases Discounts
- **Check Number:** 513

Practice Transaction #1. Discount and Credits Window

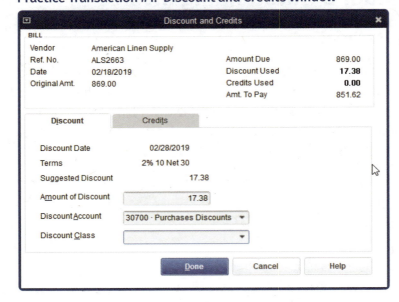

Practice Transaction #1. Pay Bills Window

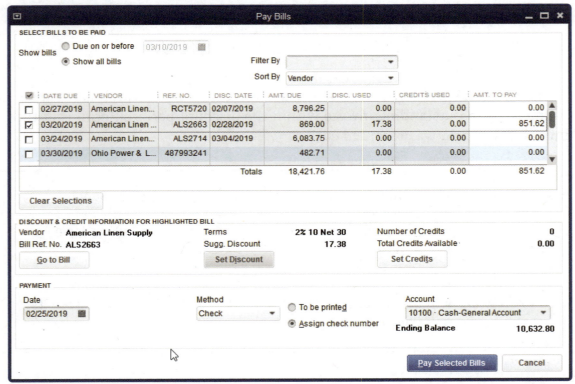

Practice Transaction #1. Assign Check Numbers Window

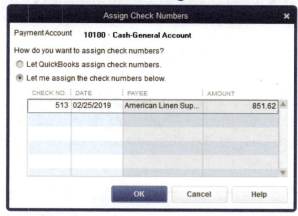

Practice Transaction #1. Payment Summary Window

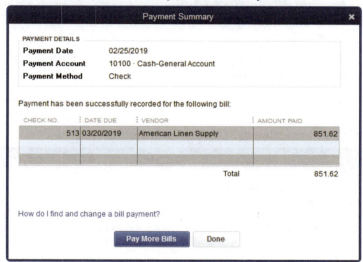

The preceding three windows show the Pay Bills, Assign Check Numbers, and Payment Summary windows with the practice transaction entered. Do not be concerned if your window shows some miscellaneous differences in the other transactions included in the window as long as the current transaction information is correct.

▶ *If your windows are consistent with the ones shown, save the transaction in the Pay Bills window. If there are errors, correct them.*

Transaction Review After the Transaction Is Saved

▶ *Click Vendors on the Home Page → Transactions tab – Bill Payments → double-click check number 513 (Num). Review the information for check number 513 on the same Bill Payments window that you just prepared and correct the information if there are errors.*

Practice Transaction #2. Partial Payment of Vendor's Outstanding Invoice

▶ *Process a cash disbursement using the following information, but do not save it yet.*

On February 20, 2019, the company issued a check to Omni Incorporated in partial payment of an outstanding invoice. There are no discounts on partial payments. *Note:* You will have to change the discount amount to 0 in the Discounts and Credits window.

■ Vendor ID:	Omni Incorporated
■ Check amount:	$300.00
■ Ref. No. (Invoice paid):	X21478, totaling $513.00
■ Amt. Due:	$513.00
■ Disc. Used:	None
■ Amt. To Pay:	$300.00
■ Check Number:	514

Practice Transaction #2. Discount and Credits Window

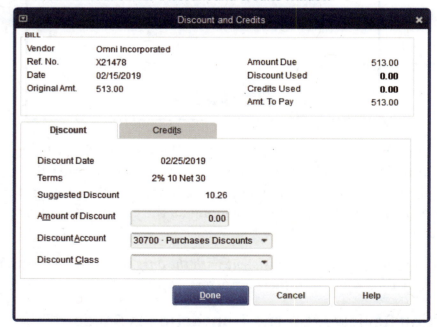

Practice Transaction #2. Pay Bills Window

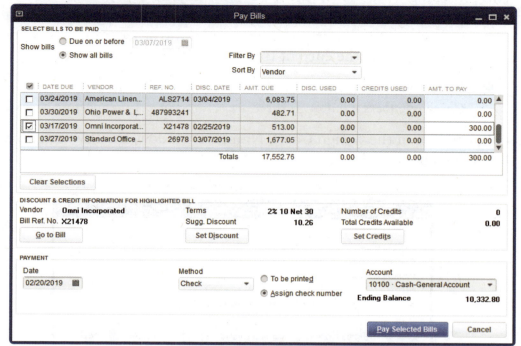

Practice Transaction #2. Assign Check Numbers Window

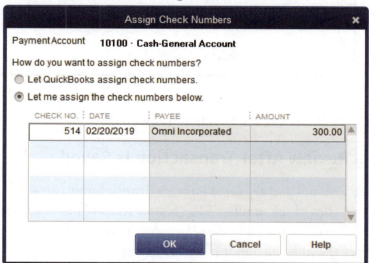

The two preceding windows show the Pay Bills and the Assign Check Numbers windows with the practice transaction entered. Do not be concerned if your windows show some miscellaneous differences in the other transactions included in the window as long as the current transaction information is correct.

▶ *If your windows are consistent with the ones shown, click OK in the Assign Check Numbers window. If there are errors, correct them before clicking OK.*

After clicking OK in the Assign Check Numbers window, the Payment Summary window shown below appears.

Practice Transaction #2. Payment Summary Window

▶ *Click Done to close the Payment Summary window.*

Transactions Review After Transaction Is Saved

▶ *Click Vendors on the Home Page → Transactions tab – Bill Payments → double-click check number 514 (Num). Review the information for check number 514 on the same Bill Payments window that you just prepared and correct the information if there are errors.*

Purchase Inventory Without a Purchase Order — Payment Made At Time of Purchase

Reference Material

A purchase of and payment for inventory without a purchase order is processed through the Write Checks window, an example of which is shown on page 19 of the Reference book. Read and understand the Purchase Inventory Without a Purchase Order — Payment Made at Time of Purchase overview on page 18 of the Reference book before processing the transaction. Then follow the instructions on pages 18 and 19 as you complete the practice section.

Practice Transaction. Purchase Inventory Without a Purchase Order — Payment Made At Time of Purchase

▶ *Process the following invoice and its corresponding payment using the following information, but do not save it yet.*

On February 27, 2019, Jackson Supply Company received an invoice from American Linen Supply for inventory delivered on the same day. Jackson Supply Company issued a check for payment at the time of delivery. Other details are in the box that follows.

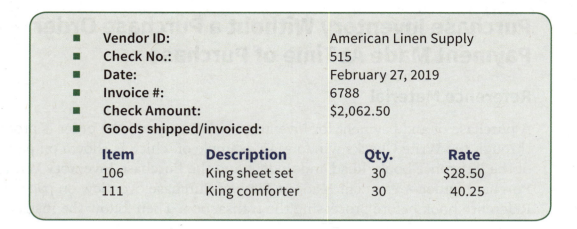

- **Vendor ID:** American Linen Supply
- **Check No.:** 515
- **Date:** February 27, 2019
- **Invoice #:** 6788
- **Check Amount:** $2,062.50
- **Goods shipped/invoiced:**

Item	Description	Qty.	Rate
106	King sheet set	30	$28.50
111	King comforter	30	40.25

Practice Transaction. Write Checks – Cash-General Account Window

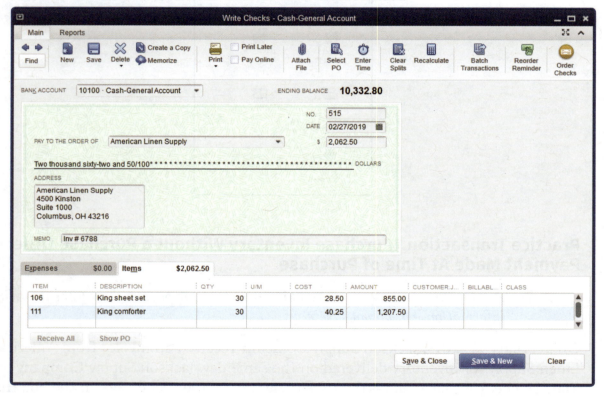

The preceding window shows the Write Checks – Cash-General Account window with the practice transaction entered. Do not be concerned if your screen shows a different Ending Balance (cash) near the top of the window.

▶ *If your window is consistent with the one shown, save the transaction. If there are errors, correct them before saving.*

Transaction Review After the Transaction Is Saved

▶ *Click Vendors on the Home Page → Transactions tab – Checks → double-click check number 515 (Num). Review the information for check number 515 on the same Write Checks – Cash-General Account window that you just prepared and correct the information if there are errors.*

Purchase Non-Inventory Items or Services Without a Purchase Order — Payment Made At Time of Purchase

Reference Material

A purchase of and payment for non-inventory items or services without a purchase order is processed through the Write Checks window, an example of which is shown on page 21 of the Reference book. Read and understand the Purchase of Non-Inventory Items or Services Without a Purchase Order — Payment Made at Time of Purchase overview on page 20 of the Reference book before processing the transaction. Then follow the instructions on pages 20 and 21 as you complete the practice section.

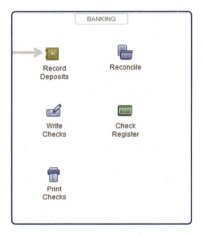

Practice Transaction. Purchase of Non-Inventory or Services Without a Purchase Order — Payment Made At Time of Purchase

▶ *Process the following invoice and its corresponding payment using the following information, but do not save it yet. Use Quick Add so that you do not have to enter anything into the system for the vendor except the name.*

On February 28 2019, Jackson Supply Company received an invoice from Hawkins Web Design for designing a web page for the company. Jackson Supply Company issued a check on the same day in full payment of the invoice. Other details about the invoice and payment follow.

- ■ **Vendor ID:**   Hawkins Web Design
- ■ **Check No.:**   516
- ■ **Invoice and Check Amount:** $875.00
- ■ **Invoice No.:**   115890
- ■ **General ledger account:**   41700 Professional Fees

Practice Transaction. Name Not Found Window

▶ *Select Quick Add.*

Practice Transaction. Select Name Type Window

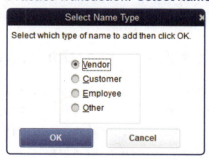

▶ *Click Ok.*

Practice Transaction. Write Checks–Cash-General Account Window

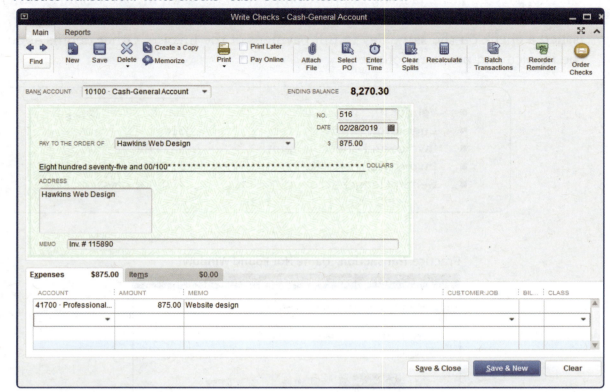

The preceding window shows the Write Checks – Cash-General Account window with the practice transaction entered. Do not be concerned if your window shows a different cash balance as long as the current transaction is correct.

▶ *If your window is consistent with the one shown, save the transaction. If there are errors, correct them before saving.*

Transaction Review After the Transaction Is Saved

▶ *Click Vendors on the Home Page → Transactions tab – Checks → double-click check number 516 (Num). Review the information for check number 516 on the same Write Checks – Cash-General Account window that you just prepared and correct the information if there are errors.*

Return Inventory From a Purchase

Reference Material

A purchase return is processed through the Enter Bills window, an example of which is shown on page 23 of the Reference book. Read and understand the Return Inventory from a Purchase overview on page 22 of the Reference book before processing the transaction. Then follow the instructions on pages 22 and 23 as you complete the practice section.

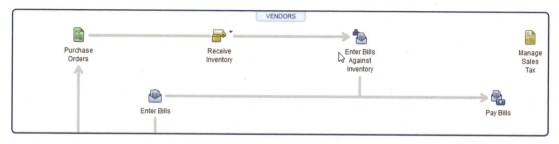

Practice Transaction. Return Inventory From a Purchase

▶ *Process the purchase return using the following information, but do not save it yet.*

On February 28, 2019, Jackson Supply Company returned inventory items that were originally purchased from American Linen Supply on February 23, 2017. Other details about the purchase return follow.

- **Vendor ID:**   American Linen Supply
- **Date:**   February 28, 2019
- **Ref. No.:**   445877 (debit memo number)
- **Vendor's Invoice No. for original purchase:**   ALS2714 (Memo box)
- **Items returned:**

Item	Description	Qty. Returned
103	Washcloths – 100 pack	10
112	Draperies	5

- **Credit Amount (check figure):** $1,150.00

Practice Transaction. Enter Bills

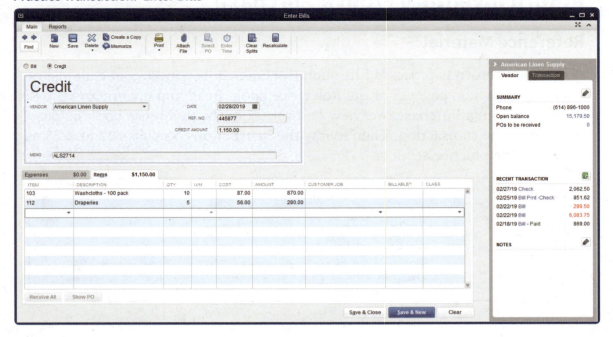

The preceding window shows the Enter Bills window with the purchase return entered.

▶ *If your window is consistent with the one shown, save the transaction. If there are errors, correct them before saving.*

Transaction Review

▶ *Click Vendors on the Home Page → Transactions tab – Bills → double-click Vendor Credit 445877 (Num). Review the information for the credit to Vendor Credit 445877 on the same Enter Bills window that you just prepared and correct the information if there are errors.*

Chapter Summary

After completing Chapter 6, back up your data files for Jackson Supply Company using the instructions in the E-materials you downloaded from the Armond Dalton Resources website. Be sure to use a descriptive file name, such as "Jackson after Ch 6."

You have now practiced processing transactions for the eight purchases and cash disbursements cycle activities. If you are satisfied with your understanding of the purchases and cash disbursements transactions in this chapter, you should now proceed to the Chapter 6 homework assigned by your instructor either online or in the Student Problems & Cases book (consult your instructor).

Because of practice exercises and chapter problems in previous chapters, each of you may have different transactions and balances in the company datasets used in the homework. To ensure consistent answers across everyone in the class, please restore both the Rock Castle Construction and Larry's Landscaping & Garden Supply datasets using the initial backups you downloaded from the Armond Dalton Resources website. **Restore both companies before proceeding to the Chapter 6 homework assigned by your instructor**.

If you do not feel comfortable with your ability to complete each of these purchases and cash disbursements activities, you can practice further in one of two ways.

1. You can do the entire chapter again by following the instructions in the E-materials to restore the Jackson Supply Company dataset from the initial backup you downloaded from the Armond Dalton Resources website. Doing so will return all data to its original form and permit you to practice all procedures again.
2. You can do additional practice by completing any or most of the sections in this chapter again. You can continue to use Jackson Supply Company.

Document numbers will, of course, be different if you redo sections without restoring the dataset. In addition, there are certain activities that you cannot do without first recording other information. For example, you cannot complete the Pay a Vendor's Outstanding Invoice section without first recording receipt of the goods in the Receive Inventory on a Purchase Order section. All practice sections that are dependent upon the completion of an earlier practice section are clearly identified in this chapter.

If you cannot recall how to make a periodic backup or restore a backup, refer to the E-materials located at www.armonddaltonresources.com.

This page is intentionally blank.

PRACTICE — SALES AND CASH RECEIPTS CYCLE ACTIVITIES

Introduction

Next you will learn about activities in the sales and cash receipts cycle. The approach is similar to those followed in the purchases and cash disbursements cycle. The following seven activities are included:

- Make a credit sale
- Collect an account receivable and make a deposit
- Make a cash sale
- Process a sales return or allowance (credit memo)
- Write off an uncollectible account receivable
- Receive a miscellaneous cash receipt
- Prepare a monthly accounts receivable statement

The same considerations included in Chapter 6 (pages 6-3 and 6-4) are equally relevant for Chapter 7. Carefully read these considerations again before you begin processing transactions in this chapter to avoid making mistakes.

Make a Credit Sale

Reference Material

A credit sale is processed through the Create Invoices–Accounts Receivable window, an example of which is shown on page 25 of the Reference book. Read and understand the Make a Credit Sale overview on page 24 of the Reference book before processing the transaction. Then follow the instructions on pages 24 and 25 as you complete the practice section.

The first practice transaction involves the sale of inventory items at the standard price level.

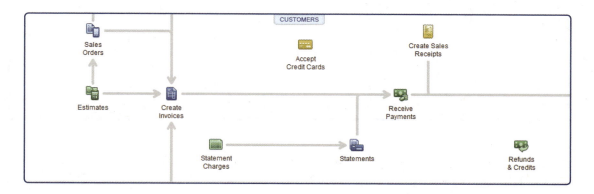

Practice Transaction #1. Sale of Inventory Items at Standard Price Level

▶ *Process a credit sale invoice using the information that follows, but do not save it.*

- **Customer:** Columbus Inn
- **Date:** February 1, 2019
- **Invoice No.:** 5128
- **PO No.:** 63921
- **Terms:** 2% 10, Net 30
- **Products sold:**

Qty.	Item	Description	Rate
20	112	Draperies	$72.50
10	116	Shower cap – 25	3.40

- **Ohio Sales Tax:** 5%
- **Invoice Total (check figure):** $1,558.20

Practice Transaction #1. Create Invoices–Accounts Receivable Window

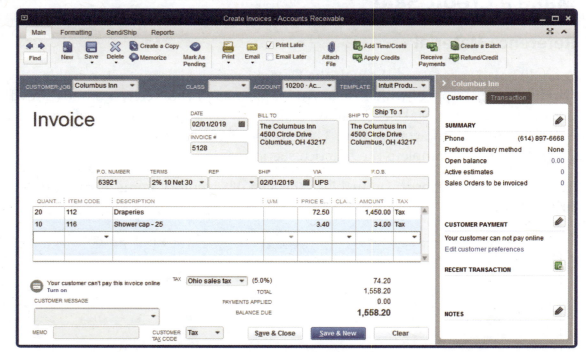

The previous diagram shows the Create Invoices–Accounts Receivable window with Practice Transaction #1 entered.

▶ *If your window is consistent with the diagram, save the transaction. If there are errors, correct them before saving.*

Transaction Review after the Transaction is Saved

▶ *Click Customers in the Home Center → Transactions tab → Invoices → double-click Invoice No. 5128. Review the information for Invoice No. 5128 on the same Create Invoices–Accounts Receivable window that you just prepared and correct the information if it is wrong.*

Practice Transaction #2. Sale of Inventory with Change in Price Level

The second practice transaction involves the sale of inventory items at a different price level for the inventory items sold.

▶ *Process another credit sale invoice using the information that follows, but do not save it yet.*

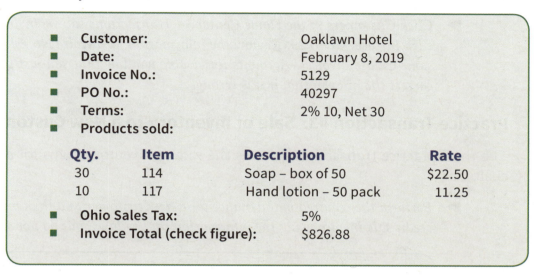

- **Customer:** Oaklawn Hotel
- **Date:** February 8, 2019
- **Invoice No.:** 5129
- **PO No.:** 40297
- **Terms:** 2% 10, Net 30
- **Products sold:**

Qty.	Item	Description	Rate
30	114	Soap – box of 50	$22.50
10	117	Hand lotion – 50 pack	11.25

- **Ohio Sales Tax:** 5%
- **Invoice Total (check figure):** $826.88

Practice Transaction #2. Create Invoices–Accounts Receivable Window

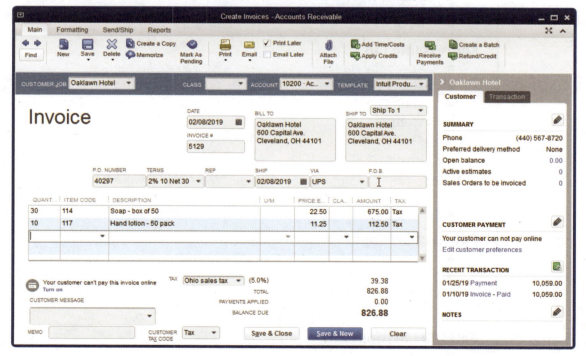

The previous window shows the Create Invoices–Accounts Receivable window with Practice Transaction #2 entered.

▶ *If your window is consistent with the diagram, save the transaction. If there are errors, correct them before saving.*

Transaction Review after the Transaction is Saved

▶ *Click Customers in the Home Center → Transactions tab → Invoices → double-click Invoice No. 5129. Review the information for Invoice No. 5129 on the same Create Invoices–Accounts Receivable window that you just prepared and correct the information if it is wrong.*

Practice Transaction #3. Sale of Inventory to a New Customer

The third practice transaction involves the sale of inventory items for a new customer.

▶ *Perform the required maintenance for a new customer and then process the credit sale invoice using the information that follows, but do not save it yet.*

New Customer Information
- Customer Name: New Place Motor Lodge
- Opening Balance: $0
- As of: 02/13/2019

Address Info tab
- Company Name: New Place Motor Lodge
- Main Phone: (404) 543-2117
- Main Email: newplace@newplacemotor.com
- Website: www.newplacemotorlodge.com
- Fax: (404) 543-7002
- Invoice/Bill To: New Place Motor Lodge
                        1622 New Place Road
                        Cleveland, OH 44101
- Ship To 1:   Same address

Payment Settings tab
- Account No.:   New01
- Credit Limit: $25,000
- Payment Terms: 1% 10, Net 30
- Preferred Delivery Method: Mail

Sales Tax Settings tab
- Tax Code: Tax
- Tax Item: Ohio sales tax

Chapter 7: Practice — Sales And Cash Receipts Cycle Activities

Practice Transaction #3. New Customer Window — Address Info Tab

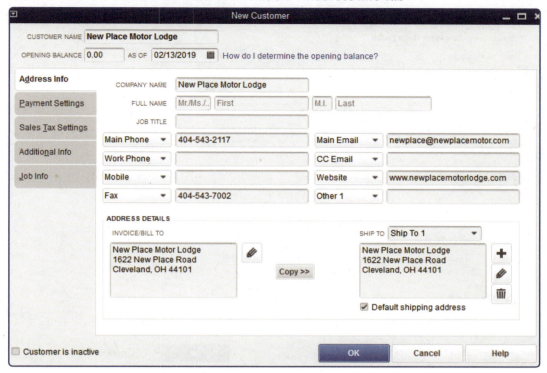

Practice Transaction #3. New Customer Window — Payment Settings Tab

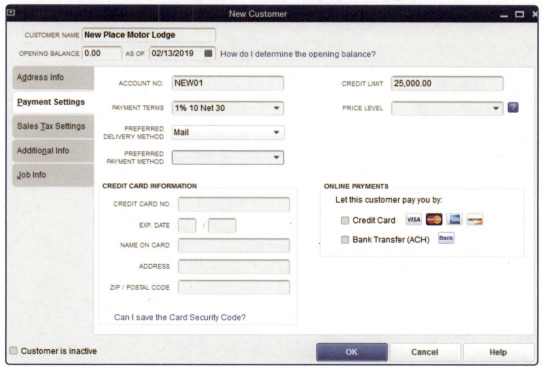

Practice Transaction #3. New Customer Window — Sales Tax Settings Tab

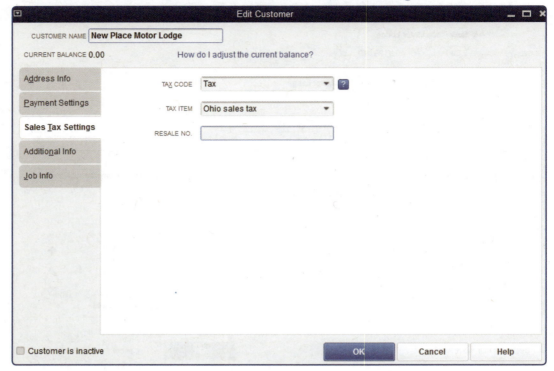

Sale information for Practice Transaction #3 follows:

- **Customer:** New Place Motor Lodge
- **Date:** February 13, 2019
- **Invoice No.:** 5130
- **P.O. No.:** P4487
- **Terms:** 1% 10, Net 30
- **Products sold:**

Qty.	Item	Description	Rate
4	109	Standard comforter	$44.50

- **Ohio Sales Tax:** 5%
- **Invoice Total (check figure):** $186.90

Practice Transaction #3. Create Invoices–Accounts Receivable Window

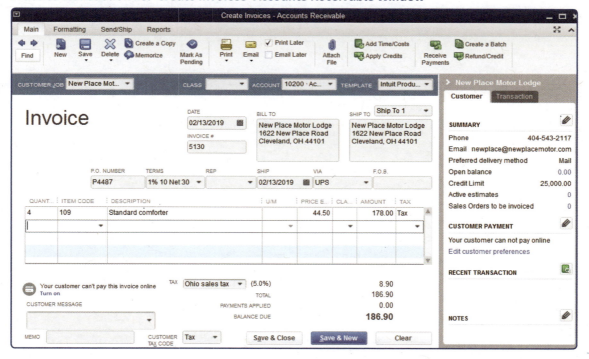

The preceding window shows the Create Invoices–Accounts Receivable window with Practice Transaction #3 entered.

▶ *If your window is consistent with the diagram, save the transaction. If there are errors, correct them before saving.*

Transaction Review after the Transaction is Saved

▶ *Click Customers in the Home Center → Transactions tab → Invoices → double-click Invoice No. 5130. Review the information for Invoice No. 5130 on the same Create Invoices–Accounts Receivable window that you just prepared and correct the information if it is wrong.*

Collect an Account Receivable and Make a Deposit

Reference Material

A collection of an account receivable is processed through the Receive Payments window, an example of which is shown on page 27 of the Reference book. Both processing and applying cash receipts are necessary for each collection. Read and understand the Collect an Account Receivable overview on page 26 of the Reference book before processing the transaction. Then follow the instructions on pages 26 through 28 of the Reference book as you complete the practice section. You will also be making a bank deposit using the Reference Book instructions on pages 30 through 32.

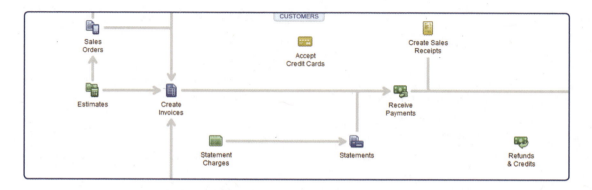

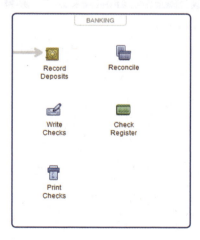

Practice Transaction #1. Collection Within the Discount Period

The first practice transaction illustrates an account receivable collection made within the early payment discount period.

▶ *Record Practice Transaction #1 using the following information, but do not save it yet.*

On February 1, 2019, Jackson Supply Company received a check from Greenleaf Suites in payment of Invoice No. 5127, which was dated January 24, 2019. Jackson Supply Company's payment terms for Greenleaf is 2% 10, Net 30.

- **Customer:** Greenleaf Suites
- **Date:** February 1, 2019
- **Invoice No.:** 5127
- **Terms:** 2% 10, Net 30
- **Amount:** $12,286.26
  (early payment discount = $250.74)
- **Discount Account:** 30300 Sales Discounts
- **Pmt. Method:** Check
- **Customer's Check:** 8421

Practice Transaction #1. Discount and Credits Window

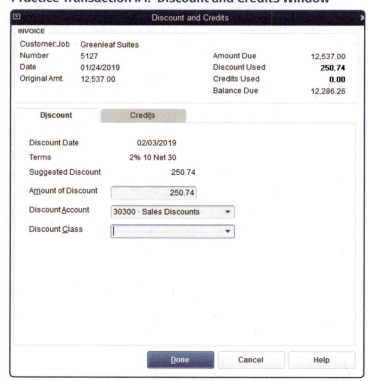

Practice Transaction #1. Receive Payments Window

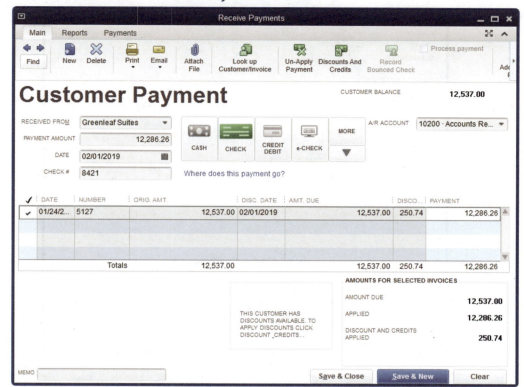

The preceding diagram shows the Receive Payments window with Practice Transaction #1 entered.

▶ *If your window is consistent with the diagram, save the transaction. If there are errors, correct them before saving.*

Transaction Review after the Transaction is Saved

▶ *Click Customers in the Home Center → Transactions tab → Received Payments → double-click Customer Payment for Check No. 8421. Review the information for Check No. 8421 on the same Receive Payments window that you just prepared and correct the information if it is wrong.*

Practice Transaction #1. Make a Bank Deposit

Record the bank deposit for the check received from Greenleaf Suites.

▶ *Process the bank deposit of $12,286.26 on 02/01/2019 in the general cash account (#10100 Cash - General Account) following the instructions on pages 30 through 32 of the Reference book, but do not save it yet.*

Practice Transaction #1. Payments to Deposit Window

Practice Transaction #1. Make Deposits Window

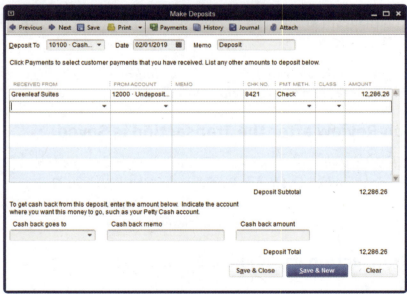

The previous diagram shows the Make Deposits window with the deposit completed.

▶ *If your window is consistent with the diagram, save it. If there are errors, correct them before saving it.*

Transaction Review After the Transaction Is Saved

▶ *Click Report Center → Banking → Deposit Detail → select This Fiscal Year in the Dates box → scroll to the last deposit, which is for Greenleaf Suites. If the deposit is for an amount other than $12,286.26, click on the transaction and make the correction.*

Practice Transaction #2. Partial Collection

The second practice transaction is a partial payment of an outstanding invoice.

▶ *Record the account receivable collection for Practice Transaction #2 using the information that follows, but do not post the transaction yet.*

On February 8, 2019, McCarthy's Bed & Breakfast sent a check in partial payment of Invoice No. 5126. The payment was outside the discount period.

■ **Customer:**	McCarthy's Bed & Breakfast
■ **Date:**	February 8, 2019
■ **Invoice No.:**	5126
■ **Amount:**	$3,000.00
■ **Pmt. Method:**	Check
■ **Customer's Check:**	7563

Practice Transaction #2. Receive Payments Window

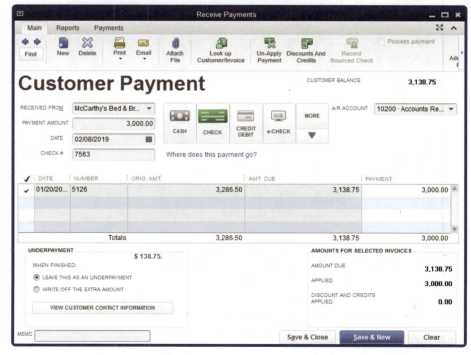

The preceding diagram shows the Receive Payments window with Practice Transaction #2 entered.

▶ *If your window is consistent with the diagram, save the transaction and close the window. If there are errors, go back and correct them before saving.*

Transaction Review After the Transaction Is Saved

▶ *Click Customers in the Home Center → Transactions tab → Received Payments → double-click Customer Payment for Check No. 7563. Review the information for Check No. 7563 on the same Receive Payments window that you just prepared and correct the information if it is wrong.*

Practice Transaction #2. Make a Bank Deposit

Record the bank deposit for the check received from McCarthy's Bed & Breakfast.

▶ *Process the bank deposit of $3,000 on 02/08/2019 in the general cash account (#10100 Cash - General Account) following the instructions on pages 30 through 32 of the Reference book, but do not save it yet.*

Practice Transaction #2. Payments to Deposit Window

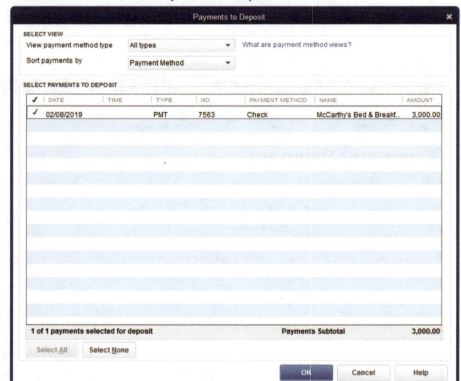

Practice Transaction #2. Make Deposits Window

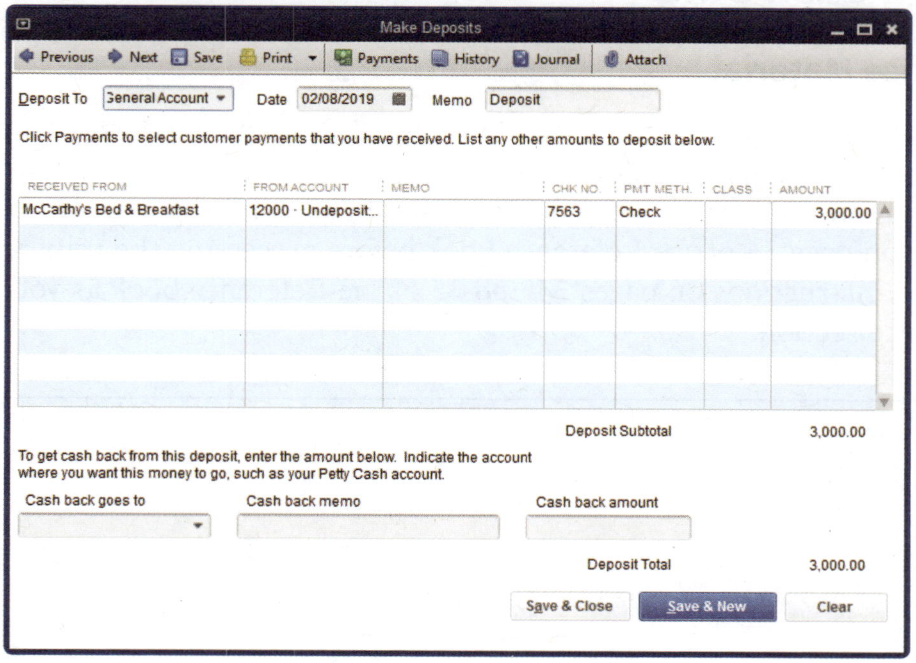

The previous diagram shows the Make Deposits window with the deposit completed.

▶ *If your window is consistent with the diagram, save it. If there are errors, correct them before saving it.*

Transaction Review After the Transaction Is Saved

▶ *Click Report Center → Banking → Deposit Detail → select This Fiscal Year in the Dates box → scroll to the last deposit for McCarthy's Bed & Breakfast. If the deposit is for an amount other than $3,000.00, click on the transaction and make the correction.*

Make a Cash Sale

Reference Material

A cash sale is processed through the Enter Sales Receipts window, an example of which is shown on page 35 of the Reference book. In this project, all payments received are in the form of a check. Read and understand the Make a Cash Sale overview on page 34 of the Reference book before processing the transaction. Then follow the instructions on pages 34 and 35 of the Reference book as you complete the practice section.

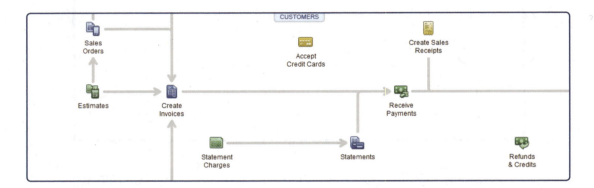

Practice Transaction. Make a Cash Sale Receipt

▶ *Process a cash sale invoice using the following information, but do not save it yet.*

- **Customer:** 1000–Cash Customer (10Cash Cust)
- **Payment Method:** Check
- **Date:** February 13, 2019
- **Sale No.:** CASH549
- **Sold To:** Rockview Inn
  6117 Green Blvd.
  Cleveland, OH 44101
- **Check No.:** 78645
- **Products sold:**

Item	Description	Qty.	Rate
105	Queen sheet set	15	$35.75
110	Queen comforter	5	49.25

- **Ohio Sales Tax:** 5%
- **Amount Received:** $821.63

Practice Transaction. Enter Sales Receipts Window — 10Cash Cust

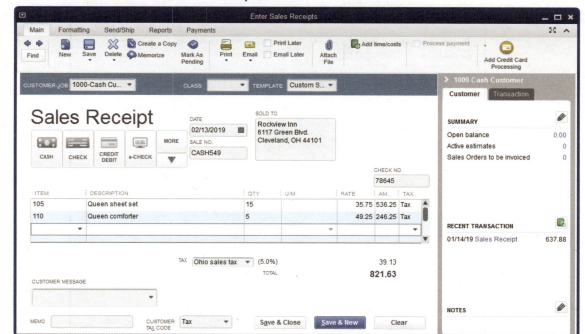

The preceding diagram shows the Enter Sales Receipts window with the practice transaction entered.

▶ *If your window is consistent with the diagram, save the transaction. If there are errors, correct them before saving. Click No if asked if you want to change all future cash sales billing addresses to Rockview's.*

Transaction Review After the Transaction Is Saved

▶ *Click Customers in the Home Center → Transactions tab → Sales Receipts → double-click Customer Payment for Sale No. CASH549. Review the information for CASH549 on the same Receive Payments window that you just prepared and correct the information if it is wrong.*

Make a Bank Deposit

▶ *Process the bank deposit of $821.63 on 02/13/2019 in the General Account following the instructions on pages 30 through 32 of the Reference book, but do not save it yet.*

Practice Transaction. Payments to Deposit Window

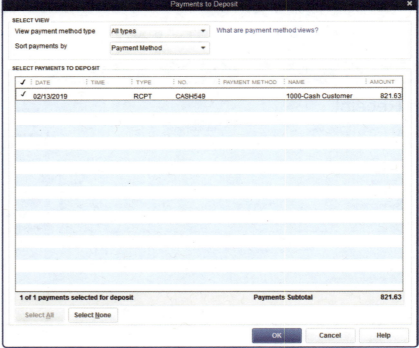

Practice Transaction. Make Deposits Window

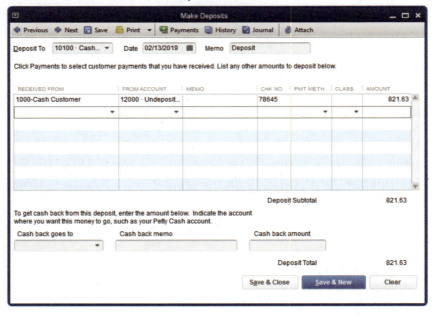

The preceding diagram shows the Make Deposits window with the deposit completed.

▶ *If your window is consistent with the diagram, save the deposit. If there are errors, correct them before saving.*

Process a Sales Return or Allowance (Credit Memo)

Reference Material

A sales return or allowance is processed through the Create Credit Memos/Refunds–Accounts Receivable window, an example of which is shown on page 37 of the Reference book. Read and understand the Process a Sales Return or Allowance (Credit Memo) overview on page 36 of the Reference book before processing the transaction. Then follow the instructions on pages 36 and 37 of the Reference book as you complete the practice section.

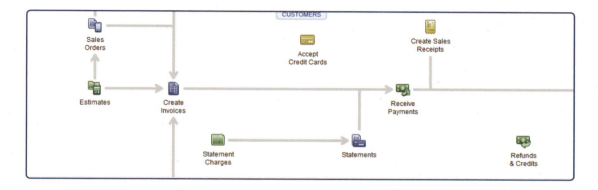

Practice Transaction #1. Create Credit Memo/Refund

▶ *Process a sales return, but do not save it yet.* Use the following information to record the sales return.

This sales return transaction is related to the February 1, 2019, sale to Columbus Inn. You processed the original sale in the Make a Credit Sale practice section on pages 7-3 and 7-4 (Practice Transaction #1). On February 13, 2019, the customer returned some of the items purchased on Invoice No. 5128. Details of the sales return are shown below and on the following page.

- Customer: Columbus Inn
- Date: February 13, 2019
- Credit No.:   1502
- P.O. No.:   5128 (original invoice #)

(continued on the following page)

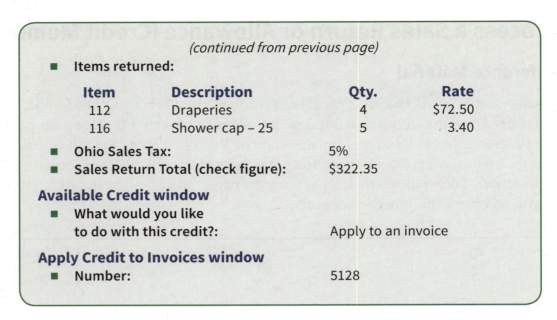

(continued from previous page)

- Items returned:

Item	Description	Qty.	Rate
112	Draperies	4	$72.50
116	Shower cap – 25	5	3.40

- Ohio Sales Tax: 5%
- Sales Return Total (check figure): $322.35

Available Credit window

- What would you like
  to do with this credit?:   Apply to an invoice

Apply Credit to Invoices window

- Number: 5128

The following diagrams show the Create Credit Memos/Refunds–Accounts Receivable, Available Credit, and Apply Credit to Invoices windows with the preceding information included.

Practice Transaction #1. Create Credit Memos/Refunds–Accounts Receivable

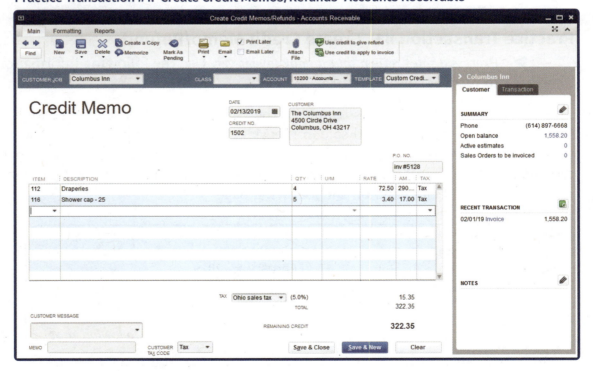

Practice Transaction #1. Available Credit Window

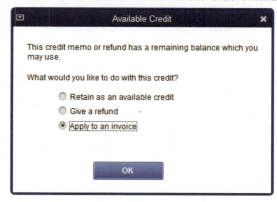

Practice Transaction #1. Apply Credit to Invoices Window

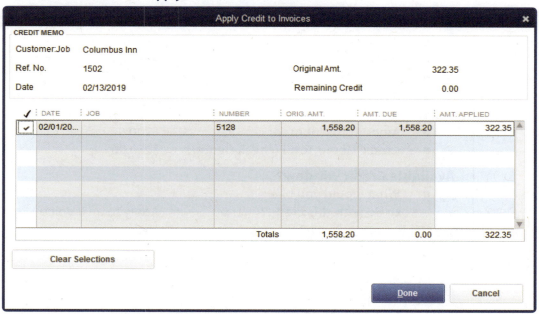

▶ *If your windows are consistent with the diagrams, save the transaction.*
If there are errors, correct them before saving.

Transaction Review After the Transaction Is Saved

▶ *Click Customers in the Home Center → Transactions tab → Credit Memos →*
double-click Credit Memo No. 1502. Review the information for Credit Memo
No. 1502 on the same Create Credit Memos/Refunds → Accounts Receivable
window that you just prepared and correct the information if it is wrong.

Practice Transaction #2. Grant a Sales Allowance

This sales allowance transaction is also related to the February 1, 2019, sale to Columbus Inn. You processed the original sale in the Make a Credit Sale practice section on pages 7-3 and 7-4. On February 13, 2019, you processed a credit memo for the customer's return of some of the items purchased on Invoice No. 5128. After considerable discussion, Jackson's management has agreed to grant an allowance of $200 for the remaining draperies, which Columbus Inn management contends are lower in quality than was expected. Credit Memo No. 1503 was approved on February 15, 2019, for the allowance.

▶ *Process a credit memo for the sales allowance to Columbus Inn using the preceding information and other information from Practice Transaction #1, but do not save it yet.*

▪ **Customer:**	Columbus Inn
▪ **Date:**	February 15, 2019
▪ **Credit No.:**	1503
▪ **P.O. No.:**	5128 (original invoice #)
▪ **Item:**	112 - Draperies
▪ **Amount Discounted:**	$200.00
▪ **Credit Memo Total (check figure):**	$210.00
Available Credit window	
▪ **What would you like to do with this credit?:**	Apply to an invoice
Apply Credit to Invoices window	
▪ **Number:**	5128

The following diagrams show the Create Credit Memos/Refunds–Accounts Receivable, Available Credit, and Apply Credit to Invoices windows with the preceding information included.

Practice Transaction #2. Create Credit Memos/Refunds–Accounts Receivable Window

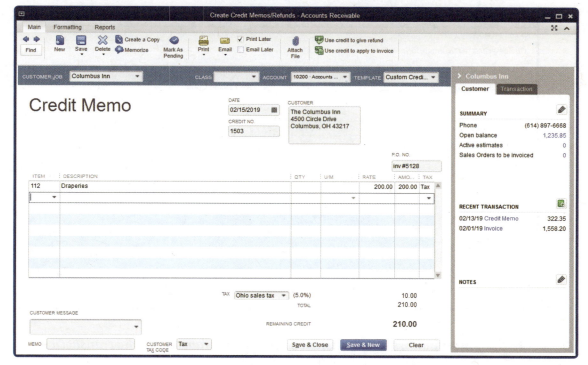

Practice Transaction #2. Available Credit Window

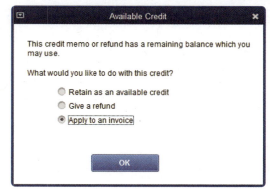

Practice Transaction #2. Apply Credit to Invoices Window

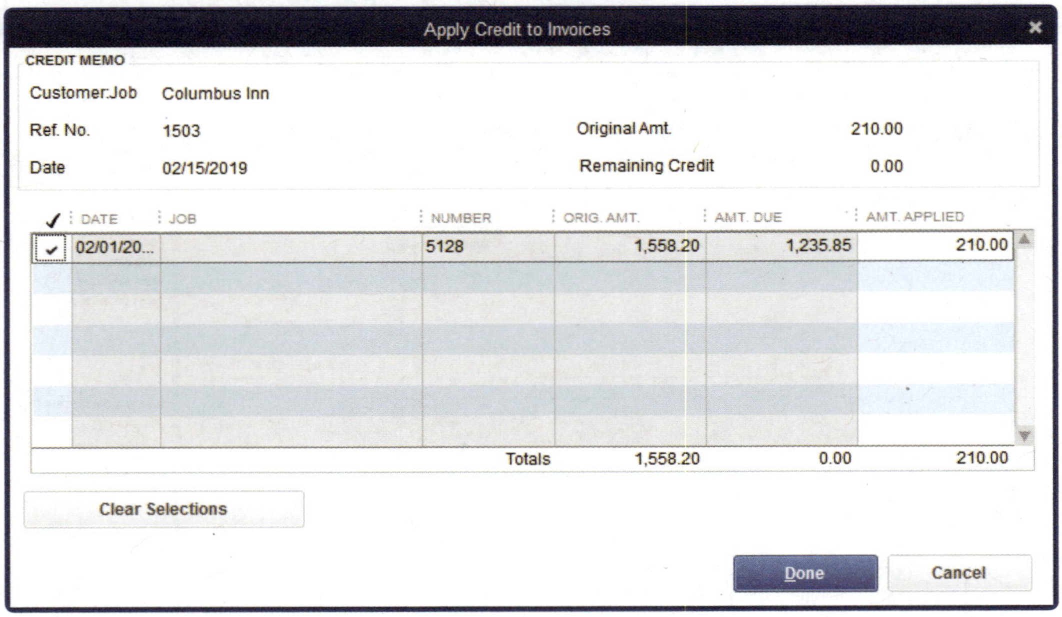

Apply Credit to Invoices	✕

CREDIT MEMO

Customer:Job	Columbus Inn			
Ref. No.	1503		Original Amt.	210.00
Date	02/15/2019		Remaining Credit	0.00

✓	DATE	JOB	NUMBER	ORIG. AMT.	AMT. DUE	AMT. APPLIED
✓	02/01/20...		5128	1,558.20	1,235.85	210.00
		Totals		1,558.20	0.00	210.00

Clear Selections

Done   Cancel

▶ *If your windows are consistent with the diagrams, save the transaction. If there are errors, correct them before saving.*

Transaction Review After the Transaction Is Saved

▶ *Click Customers in the Home Center → Transactions tab → Credit Memos → double-click Credit Memo No. 1503. Review the information for Credit Memo No. 1503 on the same Create Credit Memos/Refunds → Accounts Receivable window that you just prepared and correct the information if it is wrong.*

Write Off an Uncollectible Account Receivable

Direct and Allowance Methods of Accounting for Bad Debts and Write-off of an Uncollectible Accounts Receivable

There are two ways to write off an uncollectible accounts receivable:

- **Direct method.** In this method, the write-off of an uncollectible account receivable is a debit to bad debt expense and a credit to account receivable. Although this method is simple, it is acceptable for accrual accounting only when the potential uncollectible accounts receivable balance is immaterial at year-end. This method is used for both sample companies in *QuickBooks*. (Rock Castle Construction and Larry's Landscaping & Garden Supply).

- **Allowance method.** In this method, an estimate is made of the potential uncollectible receivables at year-end. The accounting entry for this estimate is a debit to bad debt expense and a credit to allowance to uncollectible accounts, which is a contra-account to accounts receivable. When an account becomes uncollectible, one option for the accounting entry is a debit to allowance for uncollectible accounts and a credit to accounts receivable. Another option that is common in practice is to follow the direct method for the write-off of the uncollectible accounts and then adjust the allowance to the appropriate amount at year-end. The allowance method is used in this chapter for Jackson Supply Company and in Chapter 9 for Waren Sports Supply. In both cases you will follow the direct method to write off uncollectible accounts and adjust the allowance at year-end.

When the allowance method is used in *QuickBooks*, the Account Type in Chart of Account maintenance must be Accounts Receivable for the allowance for uncollectible accounts to be included as a contra-account to accounts receivable on the balance sheet. When an account is labeled Accounts Receivable, *QuickBooks* requires that there is also a subsidiary account. In both Jackson Supply Company and Waren Sports Supply, there is a customer in the Customer & Jobs subsidiary record called Allowance account with a credit balance even though it is not an account receivable. This is illustrated below.

▶ *For Jackson Supply Company, click Home Page → Chart of Accounts. Right-click Account No. 10300, Allowance for Uncoll Accts. → Edit Account to access the chart of accounts maintenance account for the allowance. Observe that the Account Type is Accounts Receivable and the Subaccount box is checked, which makes it a contra-account to accounts receivable.*

▶ *Save and close all open windows and return to the Home Page. Click the Customers icon → Customers and Jobs tab. Observe that there is a customer called Allowance account with a credit balance of $500, even though it is not a customer.*

▶ *Click Report Center → Company & Financial → double-click on Balance Sheet Standard under Balance Sheet & Net Worth. Change the "As of" date to 01/31/2019. Observe that both accounts receivable and the allowance are shown, with net accounts receivable also included.*

▶ *Click Report Center → Customers & Receivables → double-click on A/R Aging Summary. Change the date to 01/31/2019. Observe that the allowance is again shown as a negative amount.*

Reference Material

A write-off of an uncollectible account receivable is processed through the Receive Payments window, an example of which is shown on page 39 of the Reference book. Read and understand the Write-off an Uncollectible Account Receivable overview on page 38 of the Reference book before processing the transaction. Then follow the instructions on pages 38 through 40 of the Reference book as you complete the practice section.

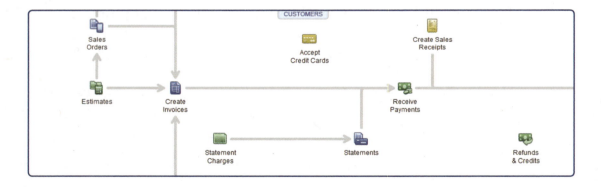

Practice Transaction. Receipts — Write-off

▶ *Process an account receivable write-off transaction following the allowance method using the following information, but do not save the transaction yet.*

McCarthy's Bed & Breakfast filed for bankruptcy protection and is unable to pay its outstanding receivable balance from Invoice No. 5126. Recall from the previous practice section that the customer remitted $3,000.00 of the invoice balance on February 8, 2019 (Practice Transaction #2 on pages 7-15 through 7-17). The remaining balance is uncollectible and is to be written off. Other details of the write-off transaction follow:

- **Customer Name:** McCarthy's Bed & Breakfast
- **Reference No.:** Write-off
- **Date:** February 25, 2019
- **Invoice No. written off:** 5126
- **Amount written off:** $138.75 (balance outstanding)
- **General ledger account information:**
   Dr. A/C No. 40900 Bad Debt Expense
   Cr. A/C No. 10200 Accounts Receivable

Practice Transaction. Discounts and Credits Window

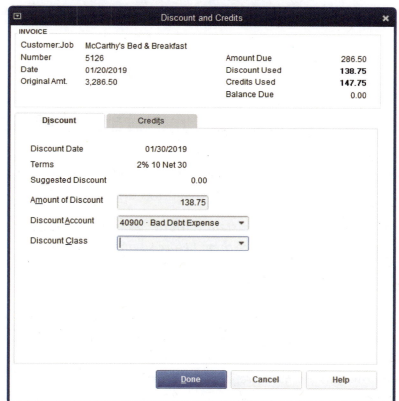

Practice Transaction. Receive Payments Window

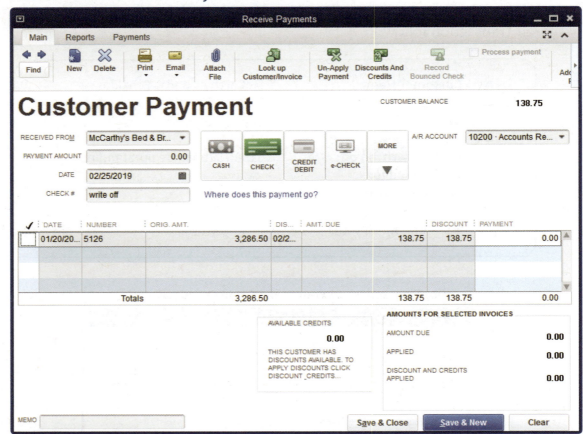

The preceding diagrams show the Discount and Credits and Receive Payments windows with the write-off practice transaction entered.

 ▶ *If your windows are consistent with the diagrams, save the transaction. If there are errors, correct them before saving.*

Transaction Review After the Transaction Is Saved

 ▶ *Click Customers in the Home Center → Transactions tab → Received Payments → double-click customer McCarthy's Bed & Breakfast for Amount - $138.75. Review the information for Amount - $138.75 on the same Receive Payments window that you just prepared and correct the information if it is wrong.*

Receive a Miscellaneous Cash Receipt

Reference Material

Miscellaneous cash receipts, such as loan proceeds or sales of fixed assets or marketable securities, are processed through the Make Deposits window. An example of the window is shown on page 43 of the Reference book. Read and understand the Receive a Miscellaneous Cash Receipt overview on page 42. Then follow the instructions on pages 42 and 43 of the Reference book as you complete the practice section.

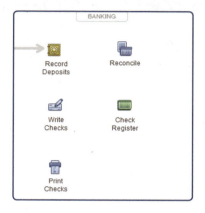

Practice Transaction. Receive a Miscellaneous Cash Receipt

▶ *Process a miscellaneous cash receipt transaction using the following information, but do not save the transaction yet.*

On February 15, 2019, the company received Check No. 83206 for $5,000 from Sun Bank for a note payable that is due February 15, 2021. The general ledger account for the debt is 21100, Long-Term Debt. *Note:* Sun Bank is not included in *QuickBooks* for Jackson. You will therefore need to add Sun Bank with the Make Deposits window open.

▶ *In Received From, type Sun Bank and press Enter. That will open the Name Not Found window. Click Quick Add. In the Select Name Type window click Other and then click Ok.*

Practice Transaction. Customer: Job Not Found Window

Practice Transaction. Make Deposits Window

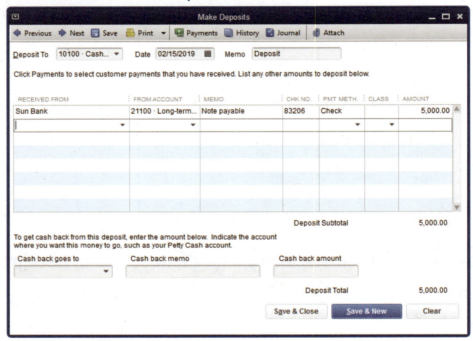

The preceding diagram shows the Make Deposits window with the bank loan transaction entered.

▶ *If your window is consistent with the diagram, save the transaction.*
If there are errors, correct them before saving.

Transaction Review After the Transaction Is Saved

▶ *Click Reports Center → Accountant & Taxes → click on Journal. Select 02/15/19 in both the From and To date boxes. Scroll to the last transaction on the list to make sure it is recorded as a debit to Cash and a credit to Long-term Debt for $5,000. If any of the information is wrong, drill down on the transaction to access the Make Deposits window and correct the information.*

Prepare a Monthly Statement for Accounts Receivable

Reference Material

Accounts receivable statements are prepared using the Create Statements window, an example of which is shown on page 45 of the Reference book. Read and understand the Prepare a Statement for Accounts Receivable overview on page 44 of the Reference book before preparing the accounts receivable statements. Then follow the instructions on pages 44 through 46 of the Reference book as you complete the practice section.

Practice Transaction. Create a Statement

▶ *Prepare and preview an accounts receivable statement at February 28, 2019, for the period January 1, 2019, to February 28, 2019, for McCarthy's Bed & Breakfast. If the statement is the same as the one shown on page 7-34, print it. If it is not the same, make corrections in the Create Statements window. If it is still not the same, you have made errors in recording the transactions for McCarthy's Bed & Breakfast. You can make those corrections by first comparing the transactions in your window to those shown below and then correcting the transactions you have recorded.*

Practice Transaction. Create Statements Window

	Create Statements	– □ ✕

SELECT STATEMENT OPTIONS

A/R Account   10200 · Accounts Receivable ▼

Statement Date   02/28/2019 📅

◉ Statement Period From   01/01/2019 📅   To   02/28/2019 📅

○ All open transactions as of Statement Date

    ☐ Include only transactions over   30   days past due date

SELECT CUSTOMERS

○ All Customers

○ Multiple Customers

◉ One Customer   McCarthy's Bed & Breakfast ▼

○ Customers of Type

○ Preferred Send Method

    [View Selected Customers...]

[Preview]   [Print]   [E-mail]

SELECT ADDITIONAL OPTIONS

Template

Intuit Standard Statement ▼   [Customize]

Create One Statement   Per Customer ▼

☐ Show invoice item details on statements

☐ Print statements by billing address zip code

☑ Print due date on transactions

Do not create statements:

☐ with a zero balance

☐ with a balance less than   0.00

☐ with no account activity

☑ for inactive customers

    [Assess Finance Charges...]

[Close]   [Help]

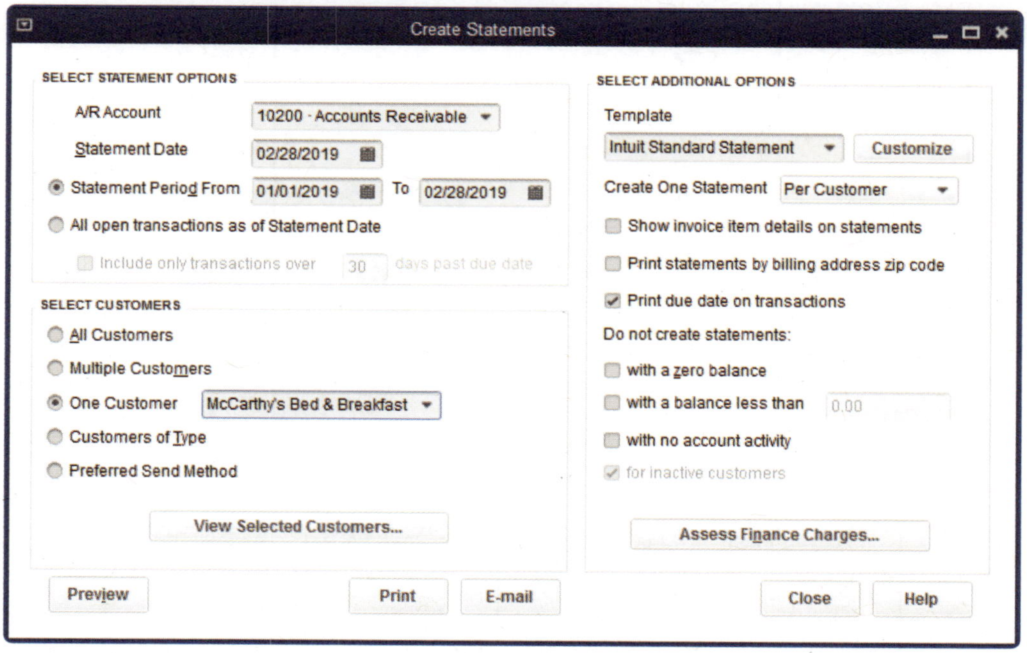

Statement

Jackson Supply Company

6211 Washburn Ave.
Columbus, OH 43216
(555) 342-4500

Date
2/28/2019

To:
McCarthy's B&B 511 Mansion Columbus, OH 43216

Amount Due	Amount Enc.
$0.00	

U/M	Date	Transaction	Amount	Balance
	12/31/2018	Balance forward		0.00
	01/20/2019	INV #5126. Due 02/19/2019.	3,286.50	3,286.50
	01/20/2019	CREDMEM #1501.	-147.75	3,138.75
	02/08/2019	PMT #7563.	-3,000.00	138.75
	02/25/2019	Discount #write off.	-138.75	0.00

CURRENT	1-30 DAYS PAST DUE	31-60 DAYS PAST DUE	61-90 DAYS PAST DUE	OVER 90 DAYS PAST DUE	Amount Due
0.00	0.00	0.00	0.00	0.00	$0.00

Chapter Summary

After completing Chapter 7, back up your data files for Jackson Supply Company using the instructions in the E-materials you downloaded from the Armond Dalton Resources website. Be sure to use a descriptive file name, such as "Jackson after Ch 07."

You have now practiced processing transactions for the seven sales and cash receipts activities. If you are satisfied with your understanding of the sales and cash receipts transactions in this chapter, you should now proceed to Chapter 7 homework assigned by your instructor either online or in the Student Problems & Cases book (consult your instructor).

Because of practice exercises and chapter problems in previous chapters, each of you may have different transactions and balances in the company datasets used in the homework. To ensure consistent answers across everyone in the class, please restore both the Rock Castle Construction and Larry's Landscaping & Garden Supply datasets using the initial backups you downloaded from the Armond Dalton Resources website. **Restore both companies before proceeding to the Chapter 7 homework assigned by your instructor.**

If you do not feel comfortable with your ability to complete each of these sales and cash receipts activities, you can practice further in one of two ways.

1. You can do the entire chapter again by following the instructions in the E-materials to restore the Jackson Supply Company dataset from the initial backup you downloaded from the Armond Dalton Resources website. Doing so will return all data to its original form and permit you to practice all procedures again.
2. You can do additional practice by completing any or most of the sections in this chapter again. You can continue to use Jackson Supply Company.

Document numbers will, of course, be different if you redo sections without restoring the company data. In addition, there are certain activities that you cannot do without first recording other information. For example, you cannot complete the Receive Goods on a Sales Return practice section without first recording the original sale in the Make a Credit Sale section. All practice sections that are dependent upon the completion of an earlier practice section are clearly identified in this chapter.

If you cannot recall how to make a periodic backup or restore a backup, refer to the E-materials located at www.armonddaltonresources.com.

This page is intentionally blank.

8 Chapter

PRACTICE — PAYROLL CYCLE AND OTHER ACTIVITIES

Introduction

In this chapter you will first learn about paying employees. Following that, you will learn about three other activities. The same approach is followed for the payroll cycle (pay employees) and other activities that was followed for the purchases and cash disbursements and sales and cash receipts cycles. After the completion of the pay employees section, the following three activities are included:

- Prepare a general journal entry
- Adjust perpetual inventory records
- Prepare a bank reconciliation

The same considerations included in Chapter 6, pages 6-3 and 6-4, are equally relevant for Chapter 8. You should carefully read these considerations again before you begin processing transactions in this chapter to avoid making mistakes.

Pay Employees

Reference Material

A payroll transaction is processed starting with the Employee Center: Payroll Center window. Although payroll is easy to process, there are several windows involved:

- Employee Center: Payroll Center
- Enter Payroll Information
- Review and Create Paychecks
- Preview Paycheck
- Confirmation and Next Steps

Read and understand the Pay Employees overview on page 48 of the Reference book before processing the transaction. Then follow the instructions on pages 48 through 51 as you complete the practice section.

Practice Transaction—Pay Employees

Recall that in Chapter 5 you increased Mark Phelps's regular pay rate to $15 and his overtime pay rate to $22.50. If you have restored the Jackson Supply Company dataset since processing this maintenance task in Chapter 5 (maintenance task #7), you will need to update his pay rates again now using the Edit Employee window.

▶ *Process the February 15, 2019, semi-monthly payroll for Jackson Supply Company using the information below by completing all steps in the Quick Reference Table on pages 48 and 49 of the Reference book.*

▶ *As you complete each window, compare the information to the relevant window that follows before continuing to the next window.* **Note:** The windows shown here assume you completed the maintenance task in Chapter 5 that updated Mark Phelps's regular pay rate to $15.00 and his overtime pay rate to $22.50. If you did not do this maintenance task in Chapter 5, either complete it now or realize that your windows will have slightly different information.

Enter Employee Information window

- **Pay Period Ends:**   02/15/2019
- **Check Date:**   02/15/2019
- **First Check No.:**   517
- **Bank Account No.:**   10100 Cash–General Account
- **Employees Selected To Pay:** 3 (all)
- **Employee Information:**

Employee	Regular Pay	Overtime Pay	Salary	Total Hours
Jennifer Brownell			*	
Kenneth Jorgensen	88 (88.00)	7.50 hrs.		95.50 hrs.
Mark Phelps	88 (88.00)	6.00 hrs.		94.00 hrs.

Totals

- **Gross Pay:** $ 6,738.50
- **Taxes:** $ 1,342.38 withheld**
- **Net Pay:** $ 5,396.12**
- **Employer Taxes:** $ 715.97**

* Leave salary blank in this window. It will appear in the final version after you preview the paychecks.

** Because the *QuickBooks* program periodically updates for changes in federal and state taxes, you may have slightly different numbers for taxes or the resulting net pay. There may be slight differences, depending on when you load *QuickBooks* and when the periodic updates happen after initial installation.

▶ *If the contents of your windows are the same as the ones shown on pages 8-5 through 8-8, go ahead and click the Create Paychecks button in the Review and Create Paychecks window per step M in the Quick Reference Table. If there are errors, go back and correct them before clicking the Create Paychecks button. See the illustrations that follow for the multiple payroll windows you will be processing. Do not be concerned if any of the payroll taxes and/ or the bank account balance shown on your screen are different from the window illustrations.*

Practice Transaction. Employee Center: Payroll Center Window

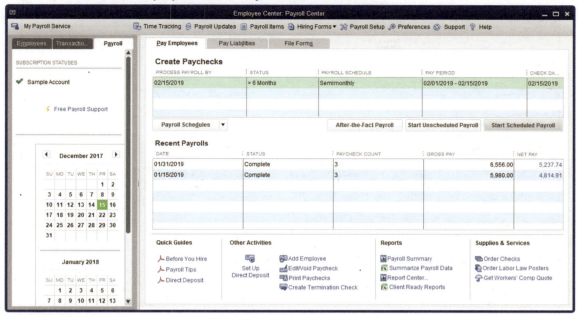

Practice Transaction. Enter Payroll Information Window

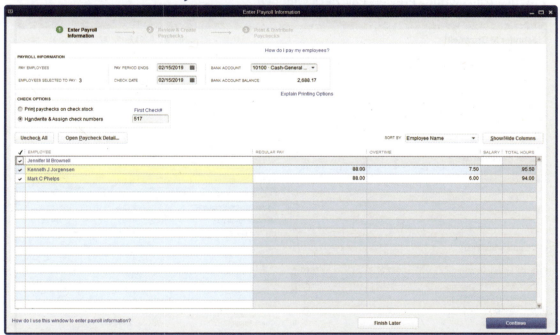

Practice Transaction. Review and Create Paychecks Window

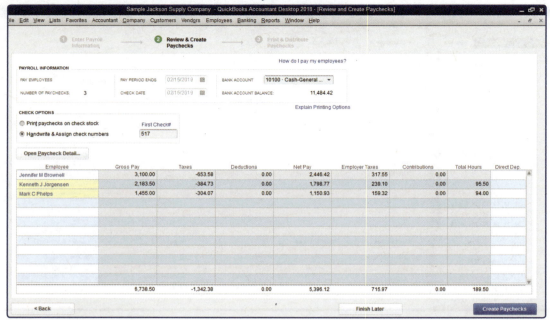

Practice Transaction. Preview Paycheck Window

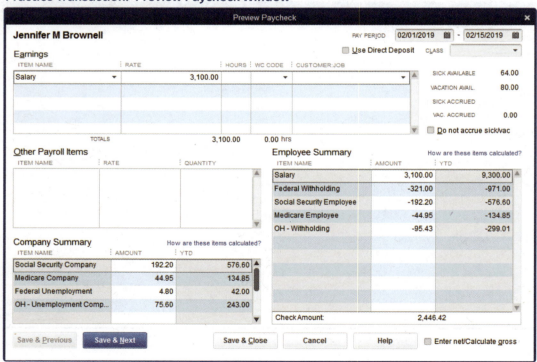

Chapter 8: Practice — Payroll Cycle And Other Activities

Practice Transaction. Preview Paycheck Window

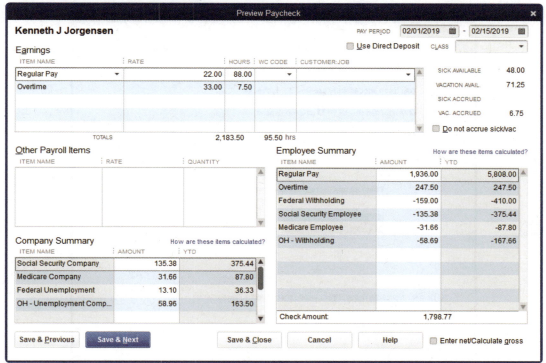

Practice Transaction. Preview Paycheck Window

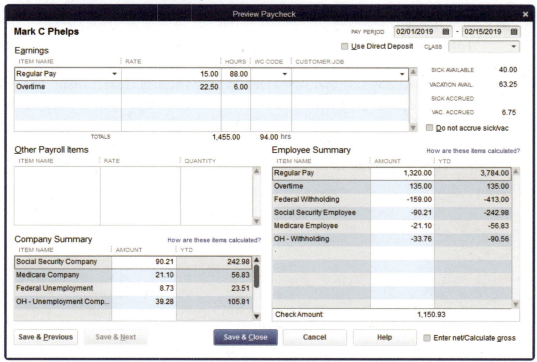

Practice Transaction. Confirmation and Next Steps Window

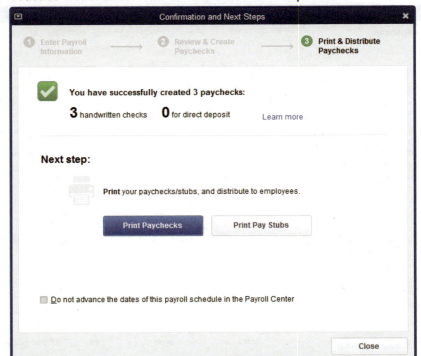

Transaction Review After the Transactions Are Saved

▶ *Click Report Center → Employees & Payroll → Payroll Item Detail. Select "This Calendar Year" in the Dates column. Click the check number for each of the three employees and compare the amounts to the amounts in the Preview Paycheck windows. If there are errors it is possible to correct them, but it typically requires preparing a new check and retrieving it from the employee if it has already been issued. If state or federal tax forms have already been sent, the change may require an amendment of these forms. It is therefore essential that processing payroll be done with extreme care.*

Prepare a General Journal Entry

General journal entries are transactions that can affect any of the three cycles. In *QuickBooks*, a general journal entry is used only when a transaction cannot be recorded with any of the activities already discussed in the previous chapters. For example, the write-off of an uncollectible accounts receivable is processed by completing a discount to bad debt expense in the Receive Payments window that was shown in the revenue cycle section of this book. The monthly or annual provision for bad debt expense is processed with a Make General Journal Entries activity because it is not included as a separate activity.

Reference Material

A general journal entry is posted through the Make General Journal Entries window, an example of which is shown on page 53 of the Reference book. Read and understand the Prepare a General Journal Entry overview on page 52 of the Reference book before processing the transaction. Then follow the instructions on pages 52 and 53 of the Reference book as you complete the practice section.

Practice Transaction #1. Accrue Interest Expense on Long-Term Debt

On February 28, 2019, Jackson Supply Company records interest expense on its long-term debt.

▶ *Record a general journal entry at 2/28/2019 for two months interest on long-term debt of $800 (6% annually for two months on an $80,000 note). The general journal entry is number 1.*

> - **Date:**   02/28/2019
> - **General Journal Entry No.:**   909
> - **Account:**
>
> | | **Amount** | **Memo** |
> |---|---|---|
> | Debit: 40800 Interest Expense | $800.00 | Interest Pmt LT Debt |
> | Credit: 21000 Interest Payable | 800.00 | Interest Pmt LT Debt |
>
> - **Name:**   leave blank

The diagram at the top of the following page shows the Make General Journal Entries window with the interest expense general journal entry displayed. Note that a descriptive entry was made in the Memo box to clearly identify the general journal entry.

Practice Transaction #1. Make General Journal Entries Window—Quarterly Interest Expense

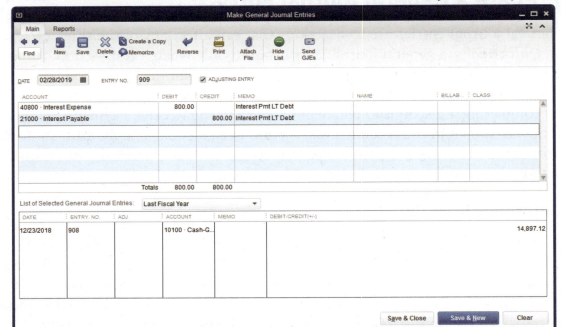

▶ *If your window is consistent with the diagram, save the general journal entry. If there are errors, correct them before saving. Don't be concerned if your window shows a different number in the Entry No. box.*

General Journal Entry Review After the Journal Entry Is Saved

▶ *Click Report Center → Accountant & Taxes in the left column → Adjusting Journal Entries. Change the From & To dates to 2/28/2019, and then locate the journal entry you just processed. Determine that the information is correct in the journal listing. If any of the information is wrong, drill down on the entry to access the Make General Journal Entries window and correct the information.*

Practice Transaction #2. Adjust Bad Debt Expense

Jackson Supply Company management wants to adjust bad debt expense and the allowance for uncollectible accounts at 2/28/2019. Bad debt expense on the first two months sales is estimated to be 1% of sales (Account No. 30100).

▶ *Record the general journal entry for bad debts at 2/28/2019. You must also select the subsidiary account named "Allowance account" in the name column on the same line as the credit for the allowance for uncollectible accounts. (See pages 7-27 through 7-30 in Chapter 7 for an explanation of the treatment of bad debt expense and the allowance for uncollectible accounts.)*

Note: Regardless of the balance in your general ledger account for sales and bad debt expense, assume that sales has a balance of $52,041.00 and bad debt expense of $138.75. One percent of sale of $52,041.00 is $520.41. Since the existing balance is $138.75, the adjustment is for the difference, which is $381.66 ($520.41 - $138.75). The general journal entry is therefore a debit to bad debt expense of $381.66 and a credit to the allowance for uncollectible accounts of the same amount.

■ Date:		02/28/2019	
■ General Journal Entry No.:		910	
■ Account:			
		Amount	**Memo**
Debit: 40900 Bad Debt Expense		$381.66	Bad Debt Expense Adj.
Credit: 10300 Allowance Uncoll Accts		$381.66	Bad Debt Expense Adj.
■ Name:		Leave blank for debit but select Allowance account using the Name drop-down list for the credit.	

The diagram that follows shows the Make General Journal Entries window with the bad debt expense general journal entry displayed. Note that a descriptive entry was made in the Memo box to clearly identify the general journal entry. If you cannot see account # 10300, you might have to expand the Make General Journal Entries window to see it.

Practice Transaction #2. Make General Journal Entries Window

▶ *If your window is consistent with the diagram, save the general journal entry. If there are errors, correct them before saving. Don't be concerned if your window shows a different number in the Entry No. box.*

General Journal Entry Review After the Journal Entry Is Saved

▶ *Follow the same procedure that you used for the first general journal entry. If any of the information is wrong, drill down on the entry to access the Make General Journal Entries window and correct the information.*

Adjust Perpetual Inventory Records

Reference Material

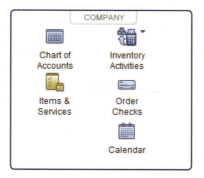

An adjustment of the perpetual inventory records is processed through the Adjust Quantity/Value on Hand window, an example of which is shown on page 55 of the Reference book. Inventory adjustments are usually made to make the perpetual records equal a physical count, but they can also be made for such things as inventory transfers, returned goods, and discarding obsolete inventory. Read and understand the Adjust Perpetual Inventory Records overview on page 54 of the Reference book before processing the transaction. Then follow the instructions on pages 54 and 55 of the Reference book as you complete the practice section.

Practice Transaction

▶ *Record an adjustment to the perpetual inventory records in QuickBooks using the following information, but do not save it yet.*

On February 28, 2019, a physical inventory count was taken by company personnel. There were differences between the physical count and the perpetual quantities recorded in **QuickBooks** for two inventory items. The physical count is the correct quantity for each of the items. The adjustment account is Cost of Goods Sold.

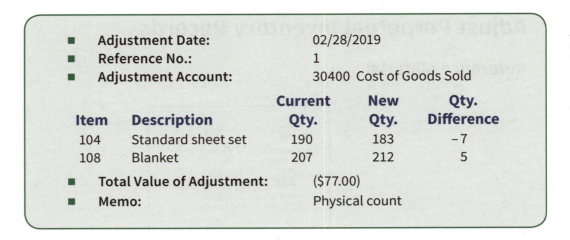

- **Adjustment Date:** 02/28/2019
- **Reference No.:** 1
- **Adjustment Account:** 30400 Cost of Goods Sold

Item	Description	Current Qty.	New Qty.	Qty. Difference
104	Standard sheet set	190	183	– 7
108	Blanket	207	212	5

- **Total Value of Adjustment:** ($77.00)
- **Memo:** Physical count

The diagram below shows the Adjust Quantity/Value on Hand window for the two inventory adjustments. Don't be concerned if your window contains different current quantities as long as the new quantities are correct.

Practice Transaction. Adjust Quantity/Value on Hand Window

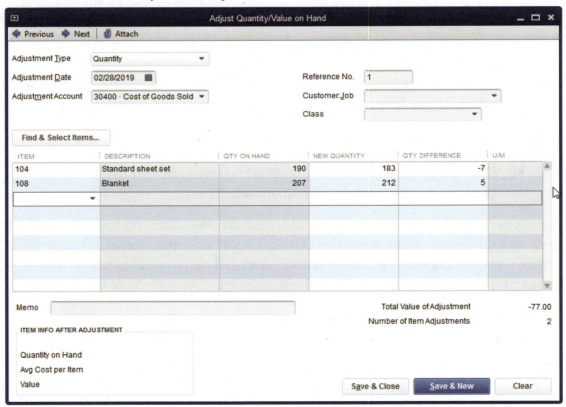

▶ *If your window is consistent with the diagram for each item, Save & Close the transaction. If there are errors, correct them before saving.*

Transaction Review After the Inventory Adjustment Is Saved

▶ *Click Lists on the Menu Bar → Item List to open the Item List window. Examine the two inventory items you adjusted to determine if the correct quantities are now included.* If any of the information is wrong, return to the Adjust Quantity/Value on Hand window and make corrections.

Prepare a Bank Reconciliation

Reference Material

Bank reconciliations are prepared through the Begin Reconciliation and Reconcile windows, examples of which are shown on page 57 of the Reference book. Read and understand the Prepare a Bank Reconciliation overview on page 56 of the Reference book before you perform the activity. Then follow the instructions on pages 56 and 57 of the Reference book as you complete the practice section.

There is no way for *QuickBooks* to know which deposits, checks, and other bank transactions have cleared the bank. Therefore, the person preparing the bank reconciliation must determine whether recorded *QuickBooks* transactions have or have not cleared the bank and whether any transactions that have cleared the bank were not recorded in *QuickBooks*. This is done by the person preparing the bank reconciliation examining all bank statement transactions and comparing them to the amounts included on the *QuickBooks* account reconciliation window. In some cases this may result in additions or changes to recorded amounts in the *QuickBooks* records.

You are to prepare the bank reconciliation for the month of January 2019. The transactions processed in this chapter do not affect the bank reconciliation because they are all in February, not in the period you are doing the bank reconciliation. Keep in mind that you can make most corrections prior to reconciling the bank statement by clicking the appropriate box and correcting the error.

Practice Activity

► *Prepare the January bank reconciliation for the 10100 Cash–General Account in QuickBooks using the information on the following page.* **Note:** Select the check box next to "Hide transactions after the statement end date" in the upper-right corner of the Reconcile–Cash–General Account window to hide transactions after 1/31/19.

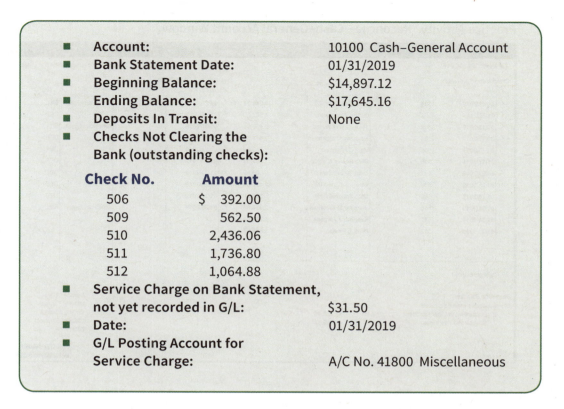

Account:	10100 Cash–General Account
Bank Statement Date:	01/31/2019
Beginning Balance:	$14,897.12
Ending Balance:	$17,645.16
Deposits In Transit:	None

- Checks Not Clearing the Bank (outstanding checks):

Check No.	Amount
506	$ 392.00
509	562.50
510	2,436.06
511	1,736.80
512	1,064.88

Service Charge on Bank Statement, not yet recorded in G/L:	$31.50
Date:	01/31/2019
G/L Posting Account for Service Charge:	A/C No. 41800 Miscellaneous

Practice Activity. Bank Reconciliation—Begin Reconciliation Window

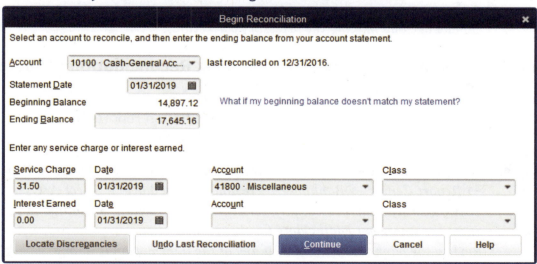

The diagram on the following page shows the Account Reconciliation window before the "Reconcile Now" button is clicked.

Practice Activity. Reconcile—Cash–General Account Window

Reconcile - Cash-General Account							_ □ x

For period: 01/31/2019   ☑ Hide transactions after the statement's end date

Checks and Payments

✓	DATE ▲	CHK #	PAYEE	AMOUNT
✓	01/03/2019	501	American Linen Su...	14,048.30
✓	01/07/2019	502	National Insurance ...	1,502.00
✓	01/15/2019	503	Jennifer M Brownell	2,436.06
✓	01/15/2019	504	Kenneth J Jorgensen	1,479.03
✓	01/15/2019	505	Mark C Phelps	899.82
	01/17/2019	506	Ohio Power & Light	392.00
✓	01/21/2019	507	Omni Incorporated	1,609.65
✓	01/25/2019	508	American Linen Su...	22,475.00
	01/26/2019	509	Ohio National Bank	562.50
	01/31/2019	510	Jennifer M Brownell	2,436.06
	01/31/2019	511	Kenneth J Jorgens...	1,736.80
	01/31/2019	512	Mark C Phelps	1,064.88

Deposits and Other Credits

✓	DATE ▲	CHK #	PAYEE	TYPE	AMOUNT
✓	12/27/2018		American Linen ...	BILL	
✓	01/07/2019			DEP	637.88
	01/07/2019		American Linen...	BILL	
✓	01/10/2019			DEP	1,557.65
✓	01/13/2019			DEP	10,140.79
✓	01/19/2019			DEP	1,089.38
✓	01/20/2019			DEP	11,602.50
✓	01/21/2019			DEP	12,142.20
✓	01/25/2019			DEP	10,059.00
✓	01/28/2019		American Linen ...	BILL	

☑ Highlight Marked   [Mark All]   [Unmark All]   [Go To]   [Columns to Display...]

Beginning Balance	14,897.12	
Items you have marked cleared		
9 Deposits and Other Credits	47,229.40	
7 Checks and Payments	44,449.86	

[Modify]

Service Charge	-31.50
Interest Earned	0.00
Ending Balance	17,645.16
Cleared Balance	17,645.16
Difference	0.00

[Reconcile Now]   [Leave]

▶ *If your window is consistent with the diagram for each item, click the Reconcile Now button. If there is a problem, click the Leave button. You can then go back to correct the reconciliation in process by selecting the Reconcile button from the Home Page and making the necessary corrections.*

Bank Reconciliation Review After the Bank Reconciliation Is Saved

After clicking the Reconcile Now button when the reconciliation is complete, *QuickBooks* allows you to view and print the reconciliation reports immediately. If you want to view them later, complete the following step.

▶ *You can review your bank reconciliation at any time by navigating to Reports → Report Center → Banking → Previous Reconciliation. Select A/C No. 10100 (Cash–General Account), select Summary, Detail, or Both, and then select the statement ending date for the reconciliation you want to review. Review the bank reconciliation. If any of the information is wrong, go back and correct it.*

Chapter Summary

After completing Chapter 8, you should again back up your data files for Jackson Supply Company using the instructions in the E-materials you downloaded from the Armond Dalton Resources website. Be sure to give the file a descriptive name, such as "Jackson after Ch 08."

You have now practiced all of the primary transactions and other activities included in Chapters 6 through 8. You will use the knowledge you have gained in these chapters as you do Chapters 9 and 10. You are not, however, expected to be able to process transactions and do other activities without the use of the Reference book. You should plan to use the Reference book for all remaining parts of the project.

One way for you to summarize the information you have learned is to reread the brief overview in the Reference book for each section. This overview emphasizes what is being accomplished using *QuickBooks*, not how to do the activity.

You should now proceed to Chapter 8 homework assigned by your instructor either online or in the Student Problems & Cases book (consult your instructor). Because of practice exercises and chapter problems in previous chapters, each of you may have different transactions and balances in the company datasets used in the homework. To ensure consistent answers across everyone in the class, please restore both the Rock Castle Construction and Larry's Landscaping & Garden Supply datasets using the initial backups that you downloaded from the Armond Dalton Resources website. **Restore both companies before proceeding to the Chapter 8 homework assigned by your instructor.**

If you are satisfied with your understanding of the payroll transactions and other activities in this chapter and have completed all of this chapter's problems assigned by your instructor, continue to Chapter 9, which applies the knowledge about *QuickBooks* that you have learned in the first eight chapters.

If you do not feel comfortable with your ability to complete each of the transactions and other activities in this chapter, you can practice further in one of two ways.

1. You can do the entire chapter again by following the instructions in the E-materials to restore the Jackson Supply Company data. Doing so will return all data to its original form and permit you to practice all procedures again.

2. You can do additional practice by completing any or most of the sections in this chapter again. You can continue to use Jackson Supply Company.

Document numbers will, of course, be different if you redo sections without restoring the company data.

If you cannot recall how to make a periodic backup or restore a backup, refer to the E-materials located at www.armonddaltonresources.com.

This page is intentionally blank.

RECORDING TRANSACTIONS, PERFORMING MONTH-END PROCEDURES, RECORDING YEAR-END ADJUSTING ENTRIES, AND PRINTING REPORTS

Overview of Options A, B, and C

This chapter has three options depending upon your instructor's preference and your previous experience with a manual version systems project called the *Systems Understanding Aid*, written by Arens & Ward, two of the authors of this computerized project. You will only do one of the three options. Your instructor will inform you which option to do.

Following is the option that is typically, but not always, assigned.

- **Option A** — normally assigned to students who have already completed the *Systems Understanding Aid* using the *blue* version of Transactions List A (*Blue* Document No. 1).
- **Option B** — normally assigned to students who have already completed the *Systems Understanding Aid* using the *green* version of Transactions List B (*Green* Document No. 1).
- **Option C** — normally assigned to students who have not used the *Systems Understanding Aid*.

Option A — Introduction

Waren Sports Supply Dataset

As described in the E-materials, you should have downloaded an initial backup file for Waren Sports Supply to either your hard drive or to a USB flash drive after registering on the Armond Dalton Resources website (www.armonddaltonresources. com). If you correctly followed the instructions in the E-materials, then Waren Sports Supply has been successfully restored into *QuickBooks*.

If you haven't yet completed these tasks from the E-materials, complete the following step. **If you have already completed these tasks from the E-materials, skip this step.**

> ▶ *Follow the instructions in the E-materials for downloading and restoring the Waren Sports Supply initial company backup file (Step 5 in the E-materials for personal computer users and Step 3 in the E-materials for lab users).*

Chapter Overview

In this chapter, you will record the same December 16–31, 2017, transactions for Waren Sports Supply that you did in the *Systems Understanding Aid* (*SUA*). You will also complete other activities commonly done with accounting software and print several items to submit to your instructor.

An important difference in this assignment compared to previous ones is the lack of detailed instructions for using *QuickBooks* to record transactions and perform other activities. You are expected to use the Reference book for guidance if you need any.

If, at any time, you decide that you want to start the chapter over, you may do so by restoring the Waren Sports Supply dataset using the initial backup you down-loaded from the Armond Dalton Resources website (armonddaltonresources.com). If you have not downloaded this initial backup yet, take the time to do that now, following the instructions in the E-materials. You may restore this backup at any time during your work on this chapter if you have made errors that are too difficult to correct or if you believe that you do not understand the material in this chapter.

In recording the transactions and performing the other activities for Waren Sports Supply, you will need several things:

1. **Optional Items from the *SUA*** (not required; all necessary information from the *SUA* has been incorporated into this chapter):

   ■ Instructions, Flowcharts, and Ledgers book
   ■ Journals book
   ■ All year-end financial statements and schedules you prepared for Waren Sports Supply in the *SUA*.

2. **Information in this chapter.** The material instructs you what to record or do.

3. **Reference Book Guide (see inside front cover of Reference book).** Use this to locate the appropriate pages in the Reference book for recording transactions or doing other activities.

4. **Reference book.** Open the Reference book to the appropriate pages for the transaction you are recording or other activity you are doing and follow the instructions. In some cases, you will not have practiced using a Reference book section. You should not be concerned about this lack of practice.

5. **Student Problems & Cases book.** After you have completed the transactions and other requirements for Waren Sports Supply, you will answer the questions for Chapter 9 either online at the Armond Dalton Resources website or in hard copy for submission to your instructor along with the other requirements of this chapter. *Consult your instructor about which method to use.*

6. ***QuickBooks* software and company data sets.**

7. **A printer connected to your computer and/or the ability to print reports to PDF format.**

When you have completed this chapter, you will submit several reports to your instructor, along with the questions for Chapter 9 in the Student Problems & Cases book. Consult your instructor as to whether you should submit these items using the online grading page of the Armond Dalton Resources website or in hard copy.

Jim Adams has recorded all transactions for the year ended 12/31/17 through December 15, 2017, using *QuickBooks*. This is consistent with the *SUA*. For this assignment, you will do the following using *QuickBooks*:

- Perform maintenance for inventory sales prices.
- Record December 16–31 transactions and perform related maintenance.
- Perform December 2017 month-end procedures.
- Record 2017 year-end adjusting entries.
- Print financial statements and other reports.

When you are finished, the financial statements and other results will be comparable to the correct solution for the *Systems Understanding Aid*.

Perform Maintenance for Inventory Sales Prices

The price list on page 9-6, taken from the *SUA*, reflects the current selling prices for Waren's thirty products. In this section, you will compare the selling price of each product to the amount included in *QuickBooks* and update the software for any differences.

▶ *If you are not already working in QuickBooks, open the program and open Waren Sports Supply.*

▶ *Perform maintenance for inventory sales prices (Sales Information section in the Edit Item Window) for each inventory item, following the guidance on page 69 of the Reference book. Be sure to click the OK button after each inventory item is changed before proceeding to the next inventory item.*

Alternative Method

▶ *Perform maintenance for inventory sales prices for multiple inventory items by first highlighting any of the Inventory Part items in the Item List. Then, select the Add/Edit Multiple List Entries function from the Item drop-down list in the lower-left corner of the Item List. Edit the entries in the Sales Price column that need to be changed, using the directional buttons to move from item to item for editing. When you have finished editing your sales prices, make sure you select the Save Changes button in the lower-right corner of the Item List.*

PRICE LIST
As of December 15, 2017

Item No.	Description	Selling Price
BA-054	Premium aluminum bat	$209.00
BA-158	Baseballs–12 game balls	63.00
BA-199	Fielding glove	66.00
BA-281	60 lb. dry line marker	103.00
BA-445	Catcher's mask	69.00
BA-507	Baseball equipment bag	39.00
BA-667	Ball bucket with seat–set of 3	33.00
BA-694	Batting gloves–1 pair	34.00
BA-807	Pitching machine	245.00
BA-859	Set of bases	172.00
BB-008	Basketball	35.00
BB-019	Basketball pole pad	135.00
BB-113	Scoreboard and timer	400.00
BB-267	Goal and rim set	142.00
BB-358	Backboard	117.00
BB-399	Basketball net	16.00
BB-431	Whistle and lanyard–set of 6	38.00
BB-538	Basketball bag	40.00
BB-688	Portable inflation pump	107.00
BB-926	Trainer's first aid kit	42.00
FB-027	Shoulder pad set	127.00
FB-091	Hip, tail, arm pad set	58.00
FB-225	Football helmet	87.00
FB-344	Football	29.00
FB-513	Portable storage locker	210.00
FB-573	Kicking tees–set of 6	24.00
FB-650	Football post pad	147.00
FB-812	Collapsible cones–set of 8	39.00
FB-874	Sideline repair kit	124.00
FB-952	Portable hand warmer	37.00

Be sure that you also adjust the 'RET' (Inventory Return Items) sales prices to the correct amount as you just did for the regular inventory items. *QuickBooks* handles returns as separate inventory items for Waren, so if you have to change the sales price of an item, you will need to do it twice: once for the item and once for the 'RET' of that same item.

After you complete inventory sales price maintenance, the *QuickBooks* files will contain the correct default sales price for each inventory item. These sales prices are used for all sales of inventory between December 16 and December 31, 2017. You should recheck the amounts to make sure they are all correct before you proceed.

Record December 16-31 Transactions and Perform Related Maintenance

The transactions on pages 9-8 through 9-14 for December 16–31 are the same as the transactions you recorded in the *SUA*, except where noted. Events and information that are not necessary to process the transactions in *QuickBooks* have been removed. In addition, supplemental information from the *SUA* documents has been incorporated into the *QuickBooks* transactions list so that you do not need the *SUA* transactions list or the *SUA* documents to complete this section. Deal with each transaction in the order listed. Some transactions must be recorded whereas others require only maintenance.

The information you have already learned about *QuickBooks* in Chapters 2 through 8 is used throughout the rest of this chapter. You should use the *QuickBooks* Reference book and information from prior chapters to the extent you need it. Also, if you want to check whether information was recorded and correctly included, use the Home Page buttons (for Centers), the Icon Bar, or the Company, Customer, Vendor, or Employee lists menu to view transactions, as you did in earlier chapters. You can also compare each transaction you record in *QuickBooks* to the result you obtained in the *SUA* journals book when you did the *SUA*, although this is not required.

All default general ledger distribution accounts are correct for Waren's December 16–31 transactions unless otherwise noted on pages 9-8 through 9-14.

WAREN SPORTS SUPPLY TRANSACTIONS FOR DECEMBER 16–31, 2017

▶ *Record each of the following transactions (No. 1 through 16) using QuickBooks.*

> ■ Hints are provided in boxed areas like this.

Use care in recording each transaction. **You should follow each step in the Quick Reference Table for each transaction.** It may take slightly more time, but it will almost certainly help you avoid serious errors. Find the appropriate Quick Reference Table for each transaction or other activity by using the Reference Book Guide located on the inside front cover of the Reference book.

Trans. No.	Dec.	
1	18	**Prepare a purchase order:** Ordered the following inventory on account from Velocity Sporting Goods (Vendor 252) using Purchase Order No. 328. The goods will be received at the warehouse at a later date. The purchase order total is $17,680.

Units	Item No.	Description
120	BB-019	Basketball pole pad
80	BB-538	Basketball bag
30	BB-688	Portable inflation pump
75	BB-926	Trainer's first aid kit

Trans. No.	Dec.	
2	18	**Receive a miscellaneous cash receipt:** Borrowed and deposited $60,000 from First American Bank & Trust (Received from 264) by issuing a two-year note payable. Check No. 545 for $60,000 was received and deposited.

> ■ The credit portion of the transaction should be posted to A/C #21000 (Notes Payable).
> ■ Miscellaneous cash receipts are processed through the Banking Center Make Deposits window (see pages 42 and 43 of the Reference book).

Trans. No.	Dec.	
3	19	**Make a credit sale:** Received customer Purchase Order No. 37225 in the mail from University of Southern Iowa (Customer 409), approved their credit, prepared Invoice No. 731 totaling $18,647 and shipped the goods from the warehouse. The following goods were shipped (only 56 shoulder pad sets were available for shipment out of the 65 that were ordered):

Units	Item No.	Description
30	BB-267	Goal and rim set
25	BB-358	Backboard
56	FB-027	Shoulder pad set
50	FB-225	Football helmet

Trans. No.	Dec.	
4	20	**Change an employee record (employee maintenance):** Increased employee salary and wage rates, effective December 16. Recall that for hourly employees, overtime is paid at 1.5 times the regular hourly rate. There were no changes in filing status or withholding allowances.

Employee	New Salary / Wage Rate
Ray Kramer	$3,700, semi-monthly (enter $88,800 Annual Rate)
Jim Adams	$22.50 per hour regular $33.75 per hour overtime
Nancy Ford	$18.20 per hour regular $27.30 per hour overtime

Trans. No.	Dec.	
5	22	**Receive goods on a purchase order:** Received merchandise from Velocity Sporting Goods (Vendor 252) as listed on Purchase Order No. 328, along with Invoice No. 34719. The payment terms on the invoice are 2/10, Net 30. All merchandise listed on the purchase order was delivered in good condition and in the quantities ordered, except that only 76 basketball pole pads (Item No. BB-019) were received. The total of the invoice is $13,544. The goods were placed immediately in the inventory warehouse.

Trans. No.	Dec.	
6	22	**Collect an account receivable and make a deposit:** Received Check No. 28564 for $1,622.88 from Branch College (Received From 408) for payment in full of sales Invoice No. 730, and deposited the check. The early payment discount taken by Branch College was $33.12.

- After recording the customer payment, record the deposit into the bank on the same day using the Make Deposits window (see pages 30 through 32 of the Reference book).

7	22	**Write off an uncollectible account receivable:** Received legal notification from Benson, Rosenbrook, and Martinson, P.C., attorneys at law, that Stevenson College (Received From 411) is unable to pay any of its outstanding debts to its suppliers. The $2,900 balance remaining on Invoice No. 719 should therefore be written off as uncollectible.

- Recall that Waren uses the allowance method for recording bad debt expense at year-end, but uses the direct write-off method during the year.
- Follow the instructions on pages 38 through 40 of the Reference book carefully so that you use the direct write-off method correctly (debit to bad debt expense, credit to accounts receivable). The allowance for uncollectible accounts and bad debt expense will be adjusted during the year-end procedures later in the chapter.

8	26	**Process a sales return or allowance (credit memo):** Eastern Wisconsin University (Customer 410) returned 5 basketball pole pads (RET BB-019) and 7 scoreboard and timer sets (RET BB-113) that were originally purchased on Invoice No. 729. Waren previously authorized EWU by phone to return the goods for credit against their account balance. EWU's Return Request No. R8034 was received with the goods. Sales return document CM 42 was issued for $3,475 and applied to Invoice No. 729.

- When you select Save in the Create Credit Memos/Refunds window, the Available Credit window will appear. Make sure you select Apply to Invoice, and select the correct invoice for the credit memo.

Trans. No.	Dec.	
9	26	**Collect an account receivable and make a deposit:** Received and deposited Check No. 49326 for $10,000 from Eastern Wisconsin University (Received from 410) in partial payment of the remaining amount (after sales return) on Invoice No. 729.

- Be sure that the entry in the A/R Account box says "10200 Accounts Receivable."
- After recording the customer payment, record the deposit into the bank on the same day using the Make Deposits window (see pages 30 through 32 of the Reference book).

10	26	**Purchase non-inventory items or services without a purchase order—payment made at time of purchase:** Received Freight Bill No. 26425 for $346.75 from Interstate Motor Freight (Pay to the Order of 255) and immediately issued Check No. 1152 for payment in full. The freight bill is for the merchandise received from Purchase Order No. 328. The payment terms on the freight bill are Net 30.
11	26	**Make a cash sale and make a deposit:** Received and deposited Check No. 65891 for $7,855 from Andrews College (CASHCUSTOMER) for a cash sale. The goods were shipped from the warehouse and the cash sale was processed and recorded (Invoice No. C-30 in the *SUA*). All goods ordered were shipped as follows:

Units	Item No.	Description
125	BB-008	Basketball
20	FB-091	Hip, tail, arm pad set
80	FB-344	Football

- After recording the cash receipt, record the deposit into the bank on the same day using the Make Deposits window (see pages 30 through 32 of the Reference book).

Trans. No.	Dec.	
12	26	**Receive goods on a purchase order:** Received office supplies from Chicago Office Supply (Vendor 253) as listed on Purchase Order No. 327, which is shown as an open purchase order in *QuickBooks*. Chicago Office Supply's vendor Invoice No. 2378 was received with the goods, totaling $602.61 including sales tax of $34.11. All supplies ordered on Purchase Order No. 327 were received in good condition and taken to the warehouse.

- Because Waren purchases similar items from Chicago Office Supply on a regular basis, each office supply item is kept track of in the inventory module as a non-inventory part. Office supplies expense is debited instead of inventory for these non-inventory parts.
- Remember to enter the sales tax portion of the invoice in the Expenses tab as a debit to the appropriate expense account (see step O in the Quick Reference Table).

Trans. No.	Dec.	
13	28	**Pay a vendor's outstanding invoice:** Issued Check No. 1153 for $13,273.12 to Velocity Sporting Goods (Pay to the Order of 252) for payment in full of Invoice No. 34719 for goods received December 23. The early payment discount taken by Waren was $270.88.

- Be sure to select account #30700–Purchases Discounts—in the Discount Account box of the Discount and Credits window.
- If account #10100–Cash—is not already selected in the Account box, select it before clicking the Pay Selected Bills button.

Trans. No.	Dec.	
14	29	**Pay employees:** Finished the payroll for the semi-monthly pay period December 16–31, 2017, and issued Check Nos. 1154 through 1156. Regular and overtime hours for hourly employees were as follows:

Employee	Regular Hours	Overtime Hours
Jim Adams	80	7.3
Nancy Ford	80	4.7
Ray Kramer	N/A	N/A

- You do not need to enter anything for Ray Kramer's hours because he is a salaried employee.
- If you receive a message about payroll liabilities, click OK.

15	29	**Receive goods on a purchase order:** Received but did not pay for ten 11-inch tablets from Chicago Office Supply (Vendor 253) ordered on Waren's Purchase Order No. 325, which is shown as an open purchase order in *QuickBooks*. Also received vendor's Invoice No. 2423 from Chicago Office Supply, totaling $3,699.40 including sales tax of $209.40. The tablets were received in new and undamaged condition in the warehouse. After they were unpacked and tested, they were taken directly to the office.

- You must enter A/C #10800 (Fixed Assets) in the Account column when entering sales tax in the Expenses tab because the default will be office supplies expense and the purchase should be debited to fixed assets.

Trans. No.	Dec.	
16	29	**Purchase non-inventory items or services without a purchase order—payment made at time of purchase:** Received vendor Invoice No. 72654 for $1,470 from the University Athletic News (Pay to the Order of 254) for advertisements Waren ran during the Christmas season and immediately issued Check No. 1157 for payment in full.

You should, but are not required to, perform backup procedures for Waren Sports Supply before proceeding, to reduce the potential for having to reenter the transactions. See E-materials for backup procedures.

Perform December 2017 Month-end Procedures

Because many of Waren's month-end procedures are done automatically by *QuickBooks*, the only month-end procedures you will need to perform are:

- Prepare the December bank reconciliation.
- Print a customer monthly statement.

Check Figure for Your Cash Balance

Before starting the December bank reconciliation, be sure that your cash balance is correct by completing the following steps:

▶ *Click the Chart of Accounts icon and then double-click on the cash account (account #10100).*

Examine the ending balance in the cash account at 12/31/17. The balance should be $93,316.79. If the cash balance in your window differs significantly from this amount, return to the December 16–31 transactions to locate and correct any errors before starting the bank reconciliation. Because *QuickBooks* performs periodic automatic updates online, your cash balance may differ slightly (less than $15.00) due to changes in federal tax tables downloaded by the program. When your cash balance is correct, continue with the requirements that follow.

Bank Reconciliation Information, Process, and Printing

The following information is taken from the December bank statement and the November bank reconciliation, neither of which is included in these materials:

- The December 31, 2017, bank statement balance is $103,372.26.
- The following checks have not cleared the bank as of December 31: Check Nos. 1118, 1143, 1152, 1153, 1154, 1155, 1156, and 1157.
- The December 26 deposits from Eastern Wisconsin University and Andrews College have not cleared the bank as of December 31.
- A service charge of $25.50 is included on the December bank statement. *Note:* The bank service charge should be posted to A/C #41000 (Other Operating Expense).

▶ *Prepare the December bank reconciliation. The cutoff date for the bank reconciliation is December 31, 2017.*

▶ *When the reconciliation is correct, click the Print button to print the bank reconciliation either in hard copy or in PDF format if you are submitting your work online. The Select Type of Reconciliation Report window will appear with the option to select Summary, Detail, or print Both reports; select Both.*

▶ *Review your printed reconciliation for accuracy and acceptability.* You will submit these reports to your instructor along with year-end reports.

Print a Customer Monthly Statement

At the end of each month, Waren sends monthly statements to all customers with an outstanding balance. For this section, you are to print the December monthly statement for Eastern Wisconsin University.

▶ *Follow the instructions on pages 44 through 46 of the Reference book to print a December 2017 customer statement for Eastern Wisconsin University either in hard copy or in PDF format if you are submitting your work online.*

▶ *In the Statement Date field select December 31, 2017, and check the Statement Period From and To field and enter the month of December date range.*

▶ *Click the Preview button to preview the statement. Click Print to print a copy to submit to your instructor with other chapter requirements.*

Print a General Ledger Trial Balance for Check Figures Prior to Year-end Adjusting Entries

The trial balance on page 9-22 shows the correct balances in all general ledger accounts after the December month-end procedures are completed. You will use the Memorized tab of the Report Center to access some reports that have been specifically set up for this project.

▶ *Click Report Center → Memorized tab (if not already opened) → Trial Balance to print a 12/31/17 trial balance to the screen.*

Compare the amounts on your printed trial balance with those on page 9-22. If any amounts are different, return to the December 16–31 transactions and the month-end procedures you processed in *QuickBooks* and make the necessary corrections using the procedures you learned in earlier chapters. When all errors are corrected, print a corrected trial balance either in hard copy or in PDF format if you are submitting your work online.

You can also compare the amounts on your printed trial balance with those included in the *SUA* year-end unadjusted trial balance (part of the year-end worksheet). All account balances should agree if your solution to the *SUA* was correct, except inventory-related accounts, bad debt expense, and the allowance for doubtful accounts. ***Note:*** Ignore any minor rounding differences in the cash and payroll-related accounts. Inventory-related account balances do not agree because of the use of different inventory methods. The balances in inventory-related accounts, bad debt expense, and the allowance account will agree after adjusting entries are completed. See the following page for a discussion of the inventory methods used in the *SUA* and in the *QuickBooks* project.

When your balances agree with those on page 9-22, go to the following section where you will record year-end adjusting entries.

Record 2017 Year-end Adjusting Entries

The next step at the end of an accounting year before printing output is to record year-end adjusting entries. The following are the types of year-end adjustments required for Waren:

- Inventory adjustment to the physical count
- Depreciation expense
- Accrued interest payable
- Bad debt expense and allowance
- Cost of goods sold for freight and sales discounts taken
- Federal income taxes

Each of the year-end adjustments is explained in a section that follows. Perform the procedures in the order listed.

Adjust Perpetual Inventory Records

Recall that in the *SUA* you were provided with the ending dollar balance in inventory and you adjusted to that total. That system was a periodic inventory system. *QuickBooks* uses a perpetual system, which provides a current inventory balance after each transaction. At year-end, the perpetual records are adjusted to a physical count to adjust for obsolescence, theft, or accounting errors.

The physical count was taken on December 31. A comparison of the physical count and the perpetual records showed a difference for certain items. Management is concerned about these inventory differences, but knows that the physical count is accurate. Thus, the perpetual records must be adjusted as follows to agree with the physical count:

Item No.	Quantity on Description	Quantity per Perpetual Records	Physical Count
BB-019	Basketball pole pad	92 (77 regular, 15 returns)	95 (80 regular, 15 returns)
FB-027	Shoulder pad set	78 (28 regular, 50 returns)	66 (16 regular, 50 returns)
BA-158	Baseballs–12 balls	156 (all regular)	162 (all regular)

▶ *Record the inventory adjustments in QuickBooks following the guidance in the Reference book. Use Cost of Goods Sold as the Adjustment Account.*

After the inventory adjustments have been processed, record the remaining five year-end adjusting entries through the General Journal Entry window.

▶ *Use the information in the following five sections to record each of the remaining year-end adjusting entries by preparing a general journal entry in QuickBooks following the guidance in the Reference book pages 52 and 53.*

Depreciation Expense

Recall from the *SUA* that depreciation expense is calculated once at the end of each year. Depreciation is calculated using the straight-line method over the estimated useful lives of the assets (five or ten years for Waren's existing fixed assets). Waren's depreciation expense for Waren for 2017 totaled $35,023.64.

Accrued Interest Payable

Recall from Transaction No. 2 on page 9-8 that Waren has a $60,000 two-year note payable to First American Bank & Trust, dated December 18, 2017. The stated annual interest rate on the note is 5%. The terms of the note payable call for the following payments:

- $3,000 interest payments on 12/18/18 and 12/18/19
- $60,000 principal payment on 12/18/19

Recall from the *SUA* that interest accruals are calculated using a 365-day year with the day after the note was made counting as the first day. General ledger account numbers for the journal entry are: A/C #40800 (Interest Expense) and A/C #20900 (Interest Payable). Either enter the correct amount on the online grading page of the Armond Dalton Resources website or show your calculation on the Chapter 9 pages of the Student Problems & Cases book (consult your instructor).

Bad Debt Expense and Allowance

Bad debt expense is estimated once annually at the end of each year as 1/5 of one percent (0.002) of net sales and is recorded in the general journal as of December 31. As explained in Chapter 7, Waren uses the direct write-off method during the year and then the allowance method at year-end. General ledger account numbers for the journal entry are: A/C #40900 (Bad Debt Expense) and A/C #10300 (Allowance for Doubtful Account). In order to balance out the Allowance for Doubtful Account and Bad Debt Expense account due to the direct write-off to Bad Debt Expense, you must readjust the Bad Debt Expense account to equal the 1/5 of one percent (0.002) calculation of net sales. Either enter the correct amount on the online grading page of the Armond Dalton Resources website or show your calculation on the Chapter 9 pages of the Student Problems & Cases book (consult your instructor).

- Determine the amount of net sales by examining the 2017 income statement on the screen. For your convenience, the 2017 income statement has been included in the Memorized tab of the Report Center.
- *QuickBooks* requires you to add a customer to the Name box. Type "write off" in the Name box and press Enter. If "write off" doesn't exist as a customer, select Quick Add and select the Customer category.

Cost of Goods Sold

QuickBooks automatically debits cost of goods sold and credits inventory for the product cost for each sale. The inventory account is also automatically updated for inventory purchases and purchases returns. Therefore, the *QuickBooks* data does not include the following accounts from the *SUA*: A/C #30500 (Purchases) and A/C #30600 (Purchase Returns and Allowance). Waren treats purchase discounts taken and freight-in as a part of cost of goods sold, but records them in separate accounts during the accounting period. Therefore, these two accounts must be closed to A/C #30400 (Cost of Goods Sold): A/C #30700 (Purchases Discounts Taken) and A/C #30800 (Freight-In).

> ■ Before preparing the general journal entry, determine the balance in each account being closed to cost of goods sold. Determine the balance in the accounts to be closed by examining the income statement.

Federal Income Taxes

Recall from the *SUA* that corporate income tax rate brackets applicable to Waren for 2017 are: 15% of the first $50,000 of pre-tax income, plus 25% of the next $25,000, plus 34% of the next $25,000, plus 39% of the next $235,000. The remaining tax brackets are not listed here because they aren't applicable to Waren's level of pre-tax income. General ledger account numbers for the journal entry are: A/C #40700 (Federal Income Tax Expense) and A/C #20700 (Federal Income Taxes Payable). Either enter the correct amount on the online grading page of the Armond Dalton Resources website or show your calculation on the Chapter 9 pages of the Student Problems & Cases book (consult your instructor).

> ■ After all other adjusting entries are recorded, determine 2017 pre-tax income by examining the 2017 income statement from the Memorized tab in the Report Center.
> ■ Do not be concerned if your amount is slightly different from the *SUA* amount for federal income taxes. The minor payroll differences will affect the calculation slightly.

Print a General Ledger Trial Balance for Check Figures After Year-end Adjusting Entries

The trial balance on page 9-23 shows the correct balances in all general ledger accounts after the year-end adjusting entries are recorded.

▶ *Open and use the general ledger trial balance for Waren to compare the amounts in your window to the correct balances. If there are differences, return to the year-end adjusting entries and make the necessary corrections. After you determine that your trial balance is correct, print the 12/31/17 trial balance either in hard copy or in PDF format if you are submitting your work online.* You will submit this report to your instructor along with year-end reports.

You can also compare the amounts on your printed trial balance with those included in the *SUA* year-end adjusted trial balance (part of the year-end worksheet). All account balances should agree if your *SUA* solution was correct, except for possible minor differences for cash, payroll-related accounts, and federal income tax expense because of the difference in the way payroll taxes are calculated in the *SUA* and *QuickBooks*.

When your balances agree with those on page 9-23, go to the following section where you will print financial statements and other reports. All entries have now been recorded.

Print Financial Statements and Other Reports

(all of these are to be submitted to your instructor either in hard copy or uploaded to the online grading page of Armond Dalton Resources)

▶ *Print the following reports.* Each of these reports has already been set up in the Memorized tab of the Report Center.

1. 12/31/17 balance sheet
2. 2017 income statement
3. General journal for December 2017
4. Accounts receivable aged trial balance as of 12/31/17
5. Accounts payable aged trial balance as of 12/31/17
6. Inventory valuation summary as of 12/31/17
7. Employee earnings register for December 2017
8. Sales journal for December 2017
9. Cash receipts journal for December 2017
10. Purchases journal for December 2017
11. Cash disbursements journal for December 2017
12. Payroll journal for December 2017

You can compare the reports printed using *QuickBooks* to the manual reports you prepared in the *SUA*.

Submit Reports and Answers to Assigned Questions

Submit the following to your course instructor either in hard copy or on the online grading page of the Armond Dalton Resources site *(consult your instructor)*:

- All twelve reports just listed
- December 2017 bank reconciliation that you already printed
- Customer monthly statement for Eastern Wisconsin University that you already printed
- Trial balance after year-end adjustments that you already printed
- All questions and print requirements as listed on the Chapter 9 pages of the Student Problems & Cases book.

All procedures are now complete for this chapter. Now that you have completed Chapter 9, you should back up your data files for Waren Sports Supply following the instructions in the E-Materials.

Check Figures

Waren Sports Supply
Trial Balance
As of December 31, 2017

	Dec 31, 17	
	Debit	Credit
10100 · Cash	93,291.29	
10200 · Accounts Receivable	48,149.00	
10300 · Allowance for Doubtful Accts.		3,250.81
10400 · Inventory	192,501.00	
10600 · Marketable Securities	24,000.00	
12000 · Undeposited Funds	0.00	
10800 · Fixed Assets	331,731.40	
10900 · Accumulated Depreciation		81,559.50
20100 · Accounts Payable		7,952.01
*Sales Tax Payable	0.00	
Payroll Liabilities	0.00	
20300 · Federal Income Tax Withheld		1,437.00
20400 · State Unemployment Taxes Pay.		110.43
20500 · Fed. Unemployment Taxes Pay.		18.66
20600 · FICA Taxes Payable		2,229.32
20700 · Federal Income Taxes Payable	0.00	
21000 · Notes Payable		60,000.00
Opening Bal Equity	0.00	
26000 · Common Stock		225,000.00
29000 · Retained Earnings		90,264.99
30100 · Sales		1,590,883.00
30200 · Sales Returns and Allowances	61,106.00	
30300 · Sales Discounts Taken	15,405.82	
31200 · Miscellaneous Revenue		825.00
30400 · Cost of Goods Sold	1,017,894.00	
30700 · Purchases Discounts Taken		16,554.48
30800 · Freight-in	24,506.44	
40100 · Rent Expense	57,600.00	
40200 · Advertising Expense	22,275.00	
40300 · Office Supplies Expense	5,664.91	
40500 · Wages and Salaries Expense	140,663.35	
40600 · Payroll Tax Expense	11,611.24	
40900 · Bad Debt Expense	4,400.00	
41000 · Other Operating Expense	29,285.75	
TOTAL	**2,080,085.20**	**2,080,085.20**

Waren Sports Supply
Trial Balance
As of December 31, 2017

	Dec 31, 17	
	Debit	**Credit**
10100 · Cash	93,291.29	
10200 · Accounts Receivable	48,149.00	
10300 · Allowance for Doubtful Accts.		1,879.55
10400 · Inventory	191,967.00	
10600 · Marketable Securities	24,000.00	
12000 · Undeposited Funds	0.00	
10800 · Fixed Assets	331,731.40	
10900 · Accumulated Depreciation		116,583.14
20100 · Accounts Payable		7,952.01
*Sales Tax Payable	0.00	
Payroll Liabilities	0.00	
20300 · Federal Income Tax Withheld		1,437.00
20400 · State Unemployment Taxes		110.43
Pay. 20500 · Fed. Unemployment Taxes		18.66
Pay. 20600 · FICA Taxes Payable		2,229.32
20700 · Federal Income Taxes Payable		54,837.13
20900 · Interest Payable		106.85
21000 · Notes Payable		60,000.00
Opening Bal Equity	0.00	
26000 · Common Stock		225,000.00
29000 · Retained Earnings		90,264.99
30100 · Sales		1,590,883.00
30200 · Sales Returns and Allowances	61,106.00	
30300 · Sales Discounts Taken	15,405.82	
31200 · Miscellaneous Revenue		825.00
30400 · Cost of Goods Sold	1,026,379.96	
30700 · Purchases Discounts Taken	0.00	
30800 · Freight-in	0.00	
40100 · Rent Expense	57,600.00	
40200 · Advertising Expense	22,275.00	
40300 · Office Supplies Expense	5,664.91	
40400 · Depreciation Expense	35,023.64	
40500 · Wages and Salaries Expense	140,663.35	
40600 · Payroll Tax Expense	11,611.24	
40700 · Federal Income Tax Expense	54,837.13	
40800 · Interest Expense	106.85	
40900 · Bad Debt Expense	3,028.74	
41000 · Other Operating Expense	29,285.75	
TOTAL	**2,152,127.08**	**2,152,127.08**

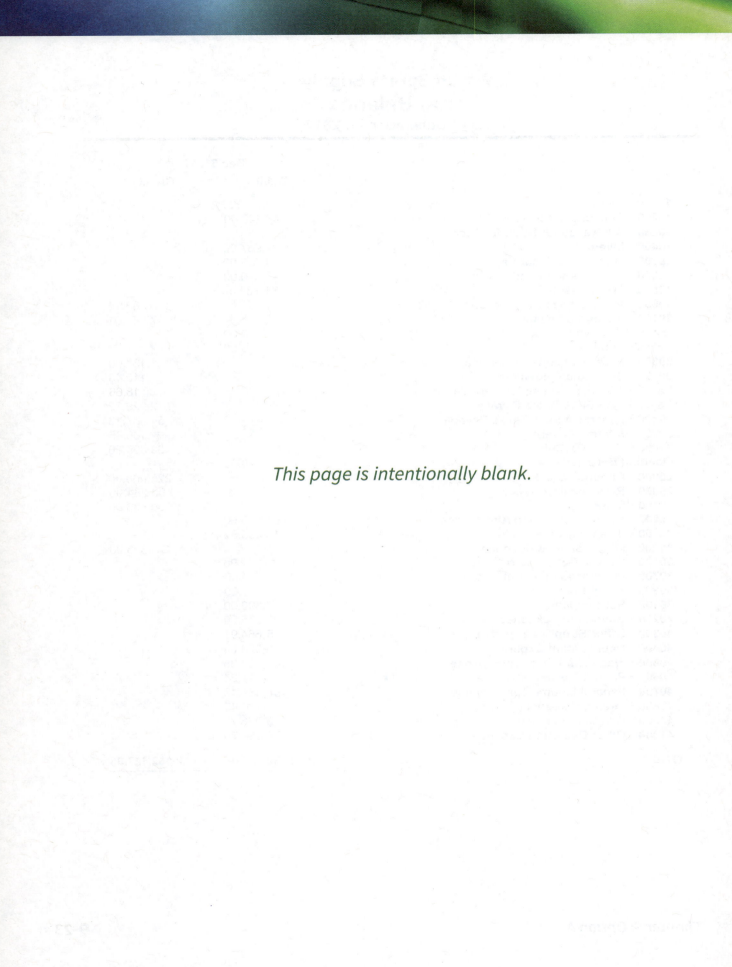

This page is intentionally blank.

Option B — Introduction

Waren Sports Supply Dataset

As described in the E-materials, you should have downloaded an initial backup file for Waren Sports Supply to either your hard drive or to a USB flash drive after registering on the Armond Dalton Resources website (www.armonddaltonresources. com). If you correctly followed the instructions in the E-materials, then Waren Sports Supply has been successfully restored into *QuickBooks*.

If you haven't yet completed these tasks from the E-materials, complete the following step. ***If you have already completed these tasks from the E-materials, skip this step.***

> ▶ *Follow the instructions in the E-materials for downloading and restoring the Waren Sports Supply initial company backup file (Step 5 in the E-materials for personal computer users and Step 3 in the E-materials for lab users).*

Chapter Overview

In this chapter, you will record the same December 16–31, 2017, transactions for Waren Sports Supply that you did in the *Systems Understanding Aid* (*SUA*). You will also complete other activities commonly done with accounting software and print several items to submit to your instructor.

An important difference in this assignment compared to previous ones is the lack of detailed instructions for using *QuickBooks* to record transactions and perform other activities. You are expected to use the Reference book for guidance if you need any.

If, at any time, you decide that you want to start the chapter over, you may do so by restoring the Waren Sports Supply dataset using the initial backup you downloaded from the Armond Dalton Resources website (armonddaltonresources.com). If you have not downloaded this initial backup yet, take the time to do that now, following the instructions in the E-materials. You may restore this backup at any time during your work on this chapter if you have made errors that are too difficult to correct or if you believe that you do not understand the material in this chapter.

In recording the transactions and performing the other activities for Waren Sports Supply, you will need several things:

1. **Optional Items from the *SUA*** (not required; all necessary information from the *SUA* has been incorporated into this chapter):

   - Instructions, Flowcharts, and Ledgers book
   - Journals book
   - All year-end financial statements and schedules you prepared for Waren Sports Supply in the *SUA*.

2. **Information in this chapter.** The material instructs you what to record or do.

3. **Reference Book Guide (see inside front cover of Reference book).** Use this to locate the appropriate pages in the Reference book for recording transactions or doing other activities.

4. **Reference book.** Open the Reference book to the appropriate pages for the transaction you are recording or other activity you are doing and follow the instructions. In some cases, you will not have practiced using a Reference book section. You should not be concerned about this lack of practice.

5. **Student Problems & Cases book.** After you have completed the transactions and other requirements for Waren Sports Supply, you will answer the questions for Chapter 9 either online at the Armond Dalton Resources website or in hard copy for submission to your instructor along with the other requirements of this chapter. *Consult your instructor about which method to use.*

6. ***QuickBooks*** software and company data sets.

7. **A printer connected to your computer and/or the ability to print reports to PDF format.**

When you have completed this chapter, you will submit several reports to your instructor, along with the questions for Chapter 9 in the Student Problems & Cases book. Consult your instructor as to whether you should submit these items using the online grading page of the Armond Dalton Resources website or in hard copy.

Jim Adams has recorded all transactions for the year ended 12/31/17 through December 15, 2017, using *QuickBooks*. This is consistent with the *SUA*. For this assignment, you will do the following using *QuickBooks*:

- Perform maintenance for inventory sales prices.
- Record December 16–31 transactions and perform related maintenance.
- Perform December 2017 month-end procedures.
- Record 2017 year-end adjusting entries.
- Print financial statements and other reports.

When you are finished, the financial statements and other results will be comparable to the correct solution for the *Systems Understanding Aid*.

Perform Maintenance for Inventory Sales Prices

The price list on page 9-28, taken from the *SUA*, reflects the current selling prices for Waren's thirty products. In this section, you will compare the selling price of each product to the amount included in *QuickBooks* and update the software for any differences.

▶ *If you are not already working in QuickBooks, open the program and open Waren Sports Supply.*

▶ *Perform maintenance for inventory sales prices (Sales Information section in the Edit Item Window) for each inventory item, following the guidance on page 69 of the Reference book. Be sure to click the OK button after each inventory item is changed before proceeding to the next inventory item.*

Alternative Method

▶ *Perform maintenance for inventory sales prices for multiple inventory items by first highlighting any of the Inventory Part items in the Item List. Then, select the Add/Edit Multiple List Entries function from the Item drop-down list in the lower-left corner of the Item List.*

Edit the entries in the Sales Price column that need to be changed, using the directional buttons to move from item to item for editing. When you have finished editing your sales prices, make sure you select the Save Changes button in the lower-right corner of the Item List.

PRICE LIST
As of December 15, 2017

Item No.	Description	Selling Price
BA-054	Premium aluminum bat	$209.00
BA-158	Baseballs–12 game balls	63.00
BA-199	Fielding glove	66.00
BA-281	60 lb. dry line marker	103.00
BA-445	Catcher's mask	69.00
BA-507	Baseball equipment bag	39.00
BA-667	Ball bucket with seat–set of 3	33.00
BA-694	Batting gloves–1 pair	34.00
BA-807	Pitching machine	245.00
BA-859	Set of bases	172.00
BB-008	Basketball	35.00
BB-019	Basketball pole pad	135.00
BB-113	Scoreboard and timer	400.00
BB-267	Goal and rim set	142.00
BB-358	Backboard	117.00
BB-399	Basketball net	16.00
BB-431	Whistle and lanyard–set of 6	38.00
BB-538	Basketball bag	40.00
BB-688	Portable inflation pump	107.00
BB-926	Trainer's first aid kit	42.00
FB-027	Shoulder pad set	127.00
FB-091	Hip, tail, arm pad set	58.00
FB-225	Football helmet	87.00
FB-344	Football	29.00
FB-513	Portable storage locker	210.00
FB-573	Kicking tees–set of 6	24.00
FB-650	Football post pad	147.00
FB-812	Collapsible cones–set of 8	39.00
FB-874	Sideline repair kit	124.00
FB-952	Portable hand warmer	37.00

Be sure that you also adjust the 'RET' (Inventory Return Items) sales prices to the correct amount as you just did for the regular inventory items. *QuickBooks* handles returns as separate inventory items for Waren, so if you have to change the sales price of an item, you will need to do it twice: once for the item and once for the 'RET' of that same item.

After you complete inventory sales price maintenance, the *QuickBooks* files will contain the correct default sales price for each inventory item. These sales prices are used for all sales of inventory between December 16 and December 31, 2017. You should recheck the amounts to make sure they are all correct before you proceed.

Record December 16-31 Transactions and Perform Related Maintenance

The transactions on pages 9-30 through 9-41 for December 16–31 are the same as the transactions you recorded in the *SUA*, except where noted. Events and information that are not necessary to process the transactions in *QuickBooks* have been removed. In addition, supplemental information from the *SUA* documents has been incorporated into the *QuickBooks* transactions list so that you do not need the *SUA* transactions list or the *SUA* documents to complete this section. Deal with each transaction in the order listed. Some transactions must be recorded whereas others require only maintenance.

The information you have already learned about *QuickBooks* in Chapters 2 through 8 is used throughout the rest of this chapter. You should use the *QuickBooks* Reference book and information from prior chapters to the extent you need it. Also, if you want to check whether information was recorded and correctly included, use the Home Page buttons (for Centers), the Icon Bar, or the Company, Customer, Vendor, or Employee lists menu to view transactions, as you did in earlier chapters. You can also compare each transaction you record in *QuickBooks* to the result you obtained in the *SUA* journals book when you did the *SUA*, although this is not required.

All default general ledger distribution accounts are correct for Waren's December 16–31 transactions unless otherwise noted on pages 9-30 through 9-41.

WAREN SPORTS SUPPLY TRANSACTIONS FOR DECEMBER 16–31, 2017

▶ *Record each of the following transactions (No. 1 through 27) using QuickBooks.*

> ■ Hints are provided in boxed areas like this.

Use care in recording each transaction. **You should follow each step in the Quick Reference Table for each transaction.** It may take slightly more time, but it will almost certainly help you avoid serious errors. Find the appropriate Quick Reference Table for each transaction or other activity by using the Reference Book Guide located on the inside front cover of the Reference book.

Trans. No.	Dec.	
1	18	**Prepare a purchase order:** Ordered the following inventory on account from Velocity Sporting Goods (Vendor 252) using Purchase Order No. 328. The goods will be received at the warehouse at a later date. The purchase order total is $17,680.

Units	Item No.	Description
120	BB-019	Basketball pole pad
80	BB-538	Basketball bag
30	BB-688	Portable inflation pump
75	BB-926	Trainer's first aid kit

2	18	**Receive a miscellaneous cash receipt:** Borrowed and deposited $60,000 from First American Bank & Trust (Received from 264) by issuing a two-year note payable. Check No. 545 for $60,000 was received and deposited.

> ■ The credit portion of the transaction should be posted to A/C #21000 (Notes Payable).
> ■ Miscellaneous cash receipts are processed through the Banking Center Make Deposits window (see pages 42 and 43 of the Reference book).

Trans. No.	Dec.	
3	19	**Make a credit sale:** Received customer Purchase Order No. 37225 in the mail from University of Southern Iowa (Customer 409), approved their credit, prepared Invoice No. 731 totaling $18,647, and shipped the goods from the warehouse. The following goods were shipped (only 56 shoulder pad sets were available for shipment out of the 65 that were ordered):

Units	Item No.	Description
30	BB-267	Goal and rim set
25	BB-358	Backboard
56	FB-027	Shoulder pad set
50	FB-225	Football helmet

Trans. No.	Dec.	
4	20	**Change an employee record (employee maintenance):** Increased employee salary and wage rates, effective December 16. Recall that for hourly employees, overtime is paid at 1.5 times the regular hourly rate. There were no changes in filing status or withholding allowances.

Employee	New Salary / Wage Rate
Ray Kramer	$3,700, semi-monthly (enter $88,800 Annual Rate)
Jim Adams	$22.50 per hour regular $33.75 per hour overtime
Nancy Ford	$18.20 per hour regular $27.30 per hour overtime

Trans. No.	Dec.	
5	22	**Receive goods on a purchase order:** Received merchandise from Velocity Sporting Goods (Vendor 252) as listed on Purchase Order No. 328, along with Invoice No. 34719. The payment terms on the invoice are 2/10, Net 30. All merchandise listed on the purchase order was delivered in good condition and in the quantities ordered, except that only 76 basketball pole pads (Item No. BB-019) were received. The total of the invoice is $13,544. The goods were placed immediately in the inventory warehouse.

Trans. No.	Dec.	
6	22	**Collect an account receivable and make a deposit:** Received Check No. 28564 for $1,622.88 from Branch College (Received from 408) for payment in full of sales Invoice No. 730, and deposited the check. The early payment discount taken by Branch College was $33.12.

- After recording the customer payment, record the deposit into the bank on the same day using the Make Deposits window (see pages 30 through 32 of the Reference book).

Trans. No.	Dec.	
7	22	**Write off an uncollectible account receivable:** Received legal notification from Benson, Rosenbrook, and Martinson, P.C., attorneys at law, that Stevenson College (Received from 411) is unable to pay any of its outstanding debts to its suppliers. The $2,900 balance remaining on Invoice No. 719 should therefore be written off as uncollectible.

- Recall that Waren uses the allowance method for recording bad debt expense at year-end, but uses the direct write-off method during the year.
- Follow the instructions on pages 38 through 40 of the Reference book carefully so that you use the direct write-off method correctly (debit to bad debt expense, credit to accounts receivable). The allowance for uncollectible accounts and bad debt expense will be adjusted during the year-end procedures later in the chapter.

Trans. No.	Dec.	
8	26	**Process a sales return or allowance (credit memo):** Eastern Wisconsin University (Customer 410) returned 5 basketball pole pads (RET BB-019) and 7 scoreboard and timer sets (RET BB-113) that were originally purchased on Invoice No. 729. Waren previously authorized EWU by phone to return the goods for credit against their account balance. EWU's Return Request No. R8034 was received with the goods. Sales return document CM 42 was issued for $3,475 and applied to Invoice No. 729.

- When you select Save in the Create Credit Memos/Refunds window, the Available Credit window will appear. Make sure you select Apply to Invoice, and select the correct invoice for the credit memo.

Trans. No.	Dec.	
9	26	**Collect an account receivable and make a deposit:** Received and deposited Check No. 49326 for $10,000 from Eastern Wisconsin University (Received from 410) in partial payment of the remaining amount (after sales return) on Invoice No. 729.

- Be sure that the entry in the A/R Account box says "10200 Accounts Receivable."
- After recording the customer payment, record the deposit into the bank on the same day using the Make Deposits window (see pages 30 through 32 of the Reference book).

Trans. No.	Dec.	
10	26	**Purchase non-inventory items or services without a purchase order—payment made at time of purchase:** Received Freight Bill No. 26425 for $346.75 from Interstate Motor Freight (Pay to the Order of 255) and immediately issued Check No. 1152 for payment in full. The freight bill is for the merchandise received from Purchase Order No. 328. The payment terms on the freight bill are Net 30.

11	26	**Make a cash sale and make a deposit:** Received and deposited Check No. 65891 for $7,855 from Andrews College (CASHCUSTOMER) for a cash sale. The goods were shipped from the warehouse and the cash sale was processed and recorded (Invoice No. C-30 in the *SUA*). All goods ordered were shipped as follows:

Units	Item No.	Description
125	BB-008	Basketball
20	FB-091	Hip, tail, arm pad set
80	FB-344	Football

- After recording the cash receipt, record the deposit into the bank on the same day using the Make Deposits window (see pages 30 through 32 of the Reference book).

12	26	**Make a credit sale:** Made a special promotional sale on account to Rosemont University (Customer 406), using Invoice No. 732, totaling $9,876. For the promotion, Waren agreed to a 20% reduction in the selling prices of the items sold. Rosemont University did not submit a purchase order for the sale. The following goods were shipped from the warehouse for this sale:

Units	Item No.	Description
45	BA-054	Premium aluminum bat
12	BA-807	Pitching machine

Trans. No.	Dec.	
13	26	**Receive goods on a purchase order:** Received office supplies from Chicago Office Supply (Vendor 253) as listed on Purchase Order No. 327, which is shown as an open purchase order in *QuickBooks*. Chicago Office Supply's vendor Invoice No. 2378 was received with the goods, totaling $602.61 including sales tax of $34.11. All supplies ordered on Purchase Order No. 327 were received in good condition and taken to the warehouse.

- Because Waren purchases similar items from Chicago Office Supply on a regular basis, each office supply item is kept track of in the inventory module as a non-inventory part. Office supplies expense is debited instead of inventory for these non-inventory parts.
- Remember to enter the sales tax portion of the invoice in the Expenses tab as a debit to the appropriate expense account (see step O in the Quick Reference Table).

14	26	**Prepare a general journal entry to sell a fixed asset:** Sold a desk to an employee, Jim Adams, for $250. Jim will pay this amount to Waren in January of 2018. The desk was purchased for a cost of $1,225 on September 30, 2011. The desk was fully depreciated at the end of 2017.

- The receivable from Jim Adams should be posted to A/C #10210 (Accounts Receivable from Employees) and the gain/loss on the sale should be posted to A/C #30900 (Gain/Loss on Sale of Fixed Assets).
- Type Adams in the Name box on the line for account 10210 and press Enter. Use Quick Add to add Adams in the Customer category when prompted. *QuickBooks* will not allow posting to an account receivable account without a corresponding customer record.

Trans. No.	Dec.	
15	26	**Receive a miscellaneous cash receipt:** Received Check No. 4014 from Central Brokerage (Received from 262) for $1,100 of dividend income. The dividends were earned on various common stocks in the marketable securities general ledger account.

- The credit portion of the transaction should be posted to A/C #31100 (Interest Dividend Income).
- Miscellaneous cash receipts are processed through the Banking Center Make Deposits window (see pages 42 and 43 of the Reference book).

16	28	**Pay a vendor's outstanding invoice:** Issued Check No. 1153 for $13,273.12 to Velocity Sporting Goods (Pay to the Order of 252) for payment in full of Invoice No. 34719 for goods received December 23. The early payment discount taken by Waren was $270.88.

- Be sure to select account #30700–Purchases Discounts–in the Discount Account box of the Discount and Credits window.
- If account 10100–Cash–is not already selected in the Account box, select it before clicking the Pay Selected Bills button.

17	28	**Collect an account receivable and make a deposit:** Received Check No. 14002 for $9,678.48 from Rosemont University (Received From 406) in payment of Invoice No. 732 (see Transaction No. 12 on page 9-34 for the original sale transaction). The early payment discount taken by Rosemont University was $197.52.

- Be sure that the entry in the A/R Account box says "10200 Accounts Receivable."
- After recording the customer payment, record the deposit into the bank on the same day using the Make Deposits window (see pages 30 through 32 of the Reference book).

Trans. No.	Dec.	
18	29	**Pay employees:** Finished the payroll for the semi-monthly pay period December 16–31, 2017, and issued Check Nos. 1154 through 1156. Regular and overtime hours for hourly employees were as follows:

Employee	Regular Hours	Overtime Hours
Jim Adams	80	7.3
Nancy Ford	80	4.7
Ray Kramer	N/A	N/A

- You do not need to enter anything for Ray Kramer's hours because he is a salaried employee.
- If you receive a message about payroll liabilities, click OK.

Trans. No.	Dec.	
19	29	**Receive goods on a purchase order:** Received but did not pay for ten 11-inch tablets from Chicago Office Supply (Vendor 253) ordered on Waren's Purchase Order No. 325, which is shown as an open Purchase Order in *QuickBooks*. Also received vendor's Invoice No. 2423 from Chicago Office Supply, totaling $3,699.40 including sales tax of $209.40. The tablets were received in new and undamaged condition in the warehouse. After they were unpacked and tested, they were taken directly to the office.

- You must enter A/C #10800 (Fixed Assets) in the Account column when entering sales tax in the Expenses tab because the default will be office supplies expense and the purchase should be debited to fixed assets.

Trans. No.	Dec.	
20	29	**Purchase non-inventory items or services without a purchase order–payment made at time of purchase:** Received vendor Invoice No. 72654 for $1,470 from the University Athletic News (Pay to the Order of 254) for advertisements Waren ran during the Christmas season and immediately issued Check No. 1157 for payment in full.
21	29	**Purchase non-inventory items or services without a purchase order–payment made at time of purchase:** Purchased 175 shares of Naretta Corporation (Pay to the Order of 262) common stock for $21.00 per share plus commission of $110. Immediately issued Check No. 1158 payable to Central Brokerage. Either recompute the amounts for this transaction now, or obtain the correct information from the *SUA* general journal.

- The debit portion of the transaction should be posted to A/C #10600 (Marketable Securities).

Trans. No.	Dec.	
22	29	**Prepare a general journal entry:** The Board of Directors declared a $2.50 per share dividend on the 3,000 shares of $75 par value common stock outstanding. The dividends will be payable on January 31, 2018, to all stockholders of record as of January 25, 2018. Either recompute the amounts for this transaction now, or obtain the correct information from the *SUA* general journal.

> - Use the following accounts for the general journal entry: A/C #29010 (Dividends Declared) and A/C #20800 (Dividends Payable).

Trans. No.	Dec.	
23	29	**Purchase non-inventory items or services without a purchase order—payment made at time of purchase:** Issued Check No. 1159 to First American Bank & Trust (Pay to the Order of 264) for $13,000 for partial payment on the bank note, which included no payment for interest. The terms on the back of the bank note stipulate that prepayments such as this one can be made without early payment penalty. For purposes of your year-end adjusting journal entry for interest, note that this payment will not reach the bank until January 2, 2018.

> - The debit portion of the transaction should be posted to A/C #21000 (Notes Payable).

Trans. No.	Dec.	
24	29	**Purchase non-inventory items or services without a purchase order—payment made at time of purchase:** Issued Check No. 1160 for $4,200 to First Security Insurance (Pay to the Order of 260) for the premium on Waren's six-month liability insurance policy. The policy period runs from December 1, 2017, to June 1, 2018. First Security's invoice number was 6822.

> - The expense portion of the transaction should be posted to A/C #41000 (Other Operating Expenses), while the prepaid portion of the transaction should be posted to A/C #10500 (Prepaid Expenses).

Trans. No.	Dec.	
25	29	**Receive a miscellaneous cash receipt:** Sold 150 shares of Green Corporation common stock for $26.00 per share. The shares were originally purchased on July 11, 2017, for $22.00 per share plus a commission of $85. Central Brokerage retained a commission of $75 on the sale and forwarded Check No. 4289 for the net sale proceeds to Waren of $3,825. Either recompute the amounts for this transaction now or obtain the correct amounts from the *SUA* cash receipts journal.

- Use two separate lines in the Make Deposits window to record the stock sale: one for the amount to be credited to marketable securities and one to be credited to gain/loss on the sale of marketable securities.
- The following accounts should be used for this transaction: A/C #10100 (Cash), A/C #10600 (Marketable Securities), and A/C #31000 (Gain/Loss on Sale of Marketable Securities).

26	29	**Purchase non-inventory items or services without a purchase order—payment made at time of purchase:** Loaned $7,000 to Maple Valley Electric (Pay to the Order of 263) by issuing a four-year note receivable with a stated annual interest rate of 6%. The funds were loaned by issuing Check No. 1161. Interest payments of $420 are due on December 31 of each year, beginning in 2018, and no interest is receivable on 12/31/17. The entire principal is due four years from December 31, 2017.

- The debit portion of the transaction should be posted to A/C #11000 (Notes Receivable).

Trans. No.	Dec.	
27	29	**Receive a miscellaneous cash receipt:** The company signed an agreement to sublease an office in its building to Campos & Associates at a monthly rental rate of $1,100. In connection with the sublease, Waren received a $3,300 check (Check No. 16644) from Campos & Associates covering rent for the first quarter of 2018.

- Type Campos & Associates in the Received From box of the Make Deposits window and press Enter. Then use Quick Add to add Campos & Associates as an "Other" name type.
- The credit portion of the transaction should be posted to A/C #21100 (Unearned Revenue).

You should, but are not required to, perform backup procedures for Waren Sports Supply before proceeding, to reduce the potential for having to reenter the transactions. See E-materials for backup procedures.

Perform December 2017 Month-end Procedures

Because many of Waren's month-end procedures are done automatically by *QuickBooks*, the only month-end procedures you will need to perform are:

- Prepare the December bank reconciliation.
- Print a customer monthly statement.

Check Figure for Your Cash Balance

Before starting the December bank reconciliation, be sure that your cash balance is correct by completing the following steps:

▶ *Click the Chart of Accounts icon and then double-click on the cash account (account #10100).*

Examine the ending balance in the cash account at 12/31/17. The balance should be $83,235.27. If the cash balance in your window differs significantly from this amount, return to the December 16–31 transactions to locate and correct any errors before starting the bank reconciliation. Because *QuickBooks* performs periodic automatic updates online, your cash balance may differ slightly (less than $15.00) due to changes in federal tax tables downloaded by the program. When your cash balance is correct, continue with the requirements that follow.

Bank Reconciliation Information, Process, and Printing

The following information is taken from the December bank statement and the November bank reconciliation, neither of which is included in these materials:

- The December 31, 2017, bank statement balance is $103,372.26.
- The following checks have not cleared the bank as of December 31: Check Nos. 1118, 1143, 1152, 1153, 1154, 1155, 1156, 1157, 1158, 1159, 1160, and 1161.
- The following deposits have not cleared the bank as of December 31:
  - (1) The December 26 account receivable collection from Eastern Wisconsin University.
  - (2) The December 26 cash sale proceeds from Andrews College.
  - (3) The December 26 deposit of dividend income.
  - (4) The December 28 account receivable collection from Rosemont University.
  - (5) The December 29 deposit from the Green Corporation stock sale.
  - (6) The December 29 deposit of rent from Campos & Associates.
- A service charge of $25.50 is included on the December bank statement. *Note:* The bank service charge should be posted to A/C #41000 (Other Operating Expense).

▶ *Prepare the December bank reconciliation. The cutoff date for the bank reconciliation is December 31, 2017.*

▶ *When the reconciliation is correct, click the Print button to print the bank reconciliation either in hard copy or in PDF format if you are submitting your work online. The Select Type of Reconciliation Report window will appear with the option to select Summary, Detail, or print Both reports; select Both.*

▶ *Review your printed reconciliation for accuracy and acceptability. You will submit these reports to your instructor along with year-end reports.*

Print a Customer Monthly Statement

At the end of each month, Waren sends monthly statements to all customers with an outstanding balance. For this section, you are to print the December monthly statement for Eastern Wisconsin University.

▶ *Follow the instructions on pages 44 through 46 of the Reference book to print a December 2017 customer statement for Eastern Wisconsin University either in hard copy or in PDF format if you are submitting your work online.*

▶ *In the Statement Date field select December 31, 2017, and check the Statement Period From and To field and enter the month of December date range.*

▶ *Click the Preview button to preview the statement. Click Print to print a copy to submit to your instructor with other chapter requirements.*

Print a General Ledger Trial Balance for Check Figures Prior to Year-end Adjusting Entries

The trial balance on page 9-49 shows the correct balances in all general ledger accounts after the December month-end procedures are completed. You will use the Memorized tab of the Report Center to access some reports that have been specifically set up for this project.

▶ *Click Report Center → Memorized tab (if not already opened) → Trial Balance to print a 12/31/17 trial balance to the screen.*

Compare the amounts on your printed trial balance with those on page 9-49. If any amounts are different, return to the December 16–31 transactions and the month-end procedures you processed in *QuickBooks* and make the necessary corrections using the procedures you learned in earlier chapters. When all errors are corrected, print a corrected trial balance either in hard copy or in PDF format if you are submitting your work online.

You can also compare the amounts on your printed trial balance with those included in the *SUA* year-end unadjusted trial balance (part of the year-end worksheet). All account balances should agree if your solution to the *SUA* was correct, except inventory-related accounts, bad debt expense, and the allowance for doubtful accounts. *Note:* Ignore any minor rounding differences in the cash and payroll-related accounts. Inventory-related account balances do not agree because of the use of different inventory methods. The balances in inventory-related accounts, bad debt expense, and the allowance account will agree after adjusting entries are completed. See the bottom of this page for a discussion of the inventory methods used in the *SUA* and in the *QuickBooks* project.

When your balances agree with those on page 9-49, go to the following section where you will record year-end adjusting entries.

Record 2017 Year-end Adjusting Entries

The next step at the end of an accounting year before printing output is to record year-end adjusting entries. The following are the types of year-end adjustments required for Waren:

- Inventory adjustment to the physical count
- Depreciation expense
- Accrued interest payable
- Bad debt expense and allowance
- Cost of goods sold for freight and sales discounts taken
- Federal income taxes

Each of the year-end adjustments is explained in a section that follows. Perform the procedures in the order listed.

Adjust Perpetual Inventory Records

Recall that in the *SUA* you were provided with the ending dollar balance in inventory and you adjusted to that total. That system was a periodic inventory system. *QuickBooks* uses a perpetual system, which provides a current inventory balance after each transaction. At year-end, the perpetual records are adjusted to a physical count to adjust for obsolescence, theft, or accounting errors.

The physical count was taken on December 31. A comparison of the physical count and the perpetual records showed a difference for certain items. Management is concerned about these inventory differences, but knows that the physical count is accurate. Thus, the perpetual records must be adjusted as follows to agree with the physical count:

Item No.	Description	Quantity on Perpetual Records	Quantity per Physical Count
BB-019	Basketball pole pad	92 (77 regular, 15 returns)	95 (80 regular, 15 returns)
FB-027	Shoulder pad set	78 (28 regular, 50 returns)	66 (16 regular, 50 returns)
BA-158	Baseballs–12 balls	156 (all regular)	162 (all regular)

▶ *Record the inventory adjustments in QuickBooks following the guidance in the Reference book. Use Cost of Goods Sold as the Adjustment Account.*

After the inventory adjustments have been processed, record the remaining five year-end adjusting entries through the General Journal Entry window.

▶ *Use the information in the following five sections to record each of the remaining year-end adjusting entries by preparing a general journal entry in QuickBooks following the guidance in the Reference book pages 52 and 53.*

Depreciation Expense

Recall from the *SUA* that depreciation expense is calculated once at the end of each year. Depreciation is calculated using the straight-line method over the estimated useful lives of the assets (five or ten years for Waren's existing fixed assets). Waren's depreciation expense for Waren for 2017 totaled $35,023.64.

Accrued Interest Payable

Recall from Transaction No. 2 on page 9-30 that Waren has a $60,000 two-year note payable to First American Bank & Trust, dated December 18, 2017. The stated annual interest rate on the note is 5%. The terms of the note payable call for the following payments:

- $3,000 interest payments on 12/18/18 and 12/18/19
- $60,000 principal payment on 12/18/19

Recall from the *SUA* that interest accruals are calculated using a 365-day year with the day after the note was made counting as the first day. General ledger account numbers for the journal entry are: A/C #40800 (Interest Expense) and A/C #20900 (Interest Payable). Either enter the correct amount on the online grading page of the Armond Dalton Resources website or show your calculation on the Chapter 9 pages of the Student Problems & Cases book (consult your instructor).

Bad Debt Expense and Allowance

Bad debt expense is estimated once annually at the end of each year as 1/5 of one percent (0.002) of net sales and is recorded in the general journal as of December 31. As explained in Chapter 7, Waren uses the direct write-off method during the year and then the allowance method at year-end. General ledger account numbers for the journal entry are: A/C #40900 (Bad Debt Expense) and A/C #10300 (Allowance for Doubtful Account). In order to balance out the Allowance for Doubtful Account and Bad Debt Expense account due to the direct write-off to Bad Debt Expense, you must readjust the Bad Debt Expense account to equal the 1/5 of one percent (0.002) calculation of net sales. Either enter the correct amount on the online grading page of the Armond Dalton Resources website or show your calculation on the Chapter 9 pages of the Student Problems & Cases book (consult your instructor).

- Determine the amount of net sales by examining the 2017 income statement on the screen. For your convenience, the 2017 income statement has been included in the Memorized tab of the Report Center.
- *QuickBooks* requires you to add a customer to the Name box. Type "write off" in the Name box and press Enter. If "write off" doesn't exist as a customer, select Quick Add and select the Customer category.

Cost of Goods Sold

QuickBooks automatically debits cost of goods sold and credits inventory for the product cost for each sale. The inventory account is also automatically updated for inventory purchases and purchases returns. Therefore, the *QuickBooks* data does not include the following accounts from the *SUA*: A/C #30500 (Purchases) and A/C #30600 (Purchase Returns and Allowance). Waren treats purchase discounts taken and freight-in as a part of cost of goods sold, but records them in separate accounts during the accounting period. Therefore, these two accounts must be closed to A/C #30400 (Cost of Goods Sold): A/C #30700 (Purchases Discounts Taken) and A/C #30800 (Freight-In).

- Before preparing the general journal entry, determine the balance in each account being closed to cost of goods sold. Determine the balance in the accounts to be closed by examining the income statement.

Federal Income Taxes

Recall from the *SUA* that corporate income tax rate brackets applicable to Waren for 2017 are: 15% of the first $50,000 of pre-tax income, plus 25% of the next $25,000, plus 34% of the next $25,000, plus 39% of the next $235,000. The remaining tax brackets are not listed here because they aren't applicable to Waren's level of pre-tax income. General ledger account numbers for the journal entry are: A/C #40700 (Federal Income Tax Expense) and A/C #20700 (Federal Income Taxes Payable). Either enter the correct amount on the online grading page of the Armond Dalton Resources website or show your calculation on the Chapter 9 pages of the Student Problems & Cases book (consult your instructor).

- After all other adjusting entries are recorded, determine 2017 pre-tax income by examining the 2017 income statement from the Memorized tab in the Report Center.
- Do not be concerned if your amount is slightly different from the *SUA* amount for federal income taxes. The minor payroll differences will affect the calculation slightly.

Print a General Ledger Trial Balance for Check Figures After Year-end Adjusting Entries

The trial balance on page 9-50 shows the correct balances in all general ledger accounts after the year-end adjusting entries are recorded.

▶ *Open and use the general ledger trial balance for Waren to compare the amounts in your window to the correct balances. If there are differences, return to the year-end adjusting entries and make the necessary corrections. After you determine that your trial balance is correct, print the 12/31/17 trial balance either in hard copy or in PDF format if you are submitting your work online. You will submit in this report to your instructor along with year-end reports.*

You can also compare the amounts on your printed trial balance with those included in the *SUA* year-end adjusted trial balance (part of the year-end worksheet). All account balances should agree if your *SUA* solution was correct, except for possible minor differences for cash, payroll-related accounts, and federal income tax expense because of the difference in the way payroll taxes are calculated in the *SUA* and *QuickBooks*.

When your balances agree with those on page 9-50, go to the following section where you will print financial statements and other reports. All entries have now been recorded.

Print Financial Statements and Other Reports

(all of these are to be submitted to your instructor either in hard copy or uploaded to the online grading page of Armond Dalton Resources)

▶ *Print the following reports.* Each of these reports has already been set up in the Memorized tab of the Report Center.

1. 12/31/17 balance sheet
2. 2017 income statement
3. General journal for December 2017
4. Accounts receivable aged trial balance as of 12/31/17
5. Accounts payable aged trial balance as of 12/31/17
6. Inventory valuation summary as of 12/31/17
7. Employee earnings register for December 2017
8. Sales journal for December 2017
9. Cash receipts journal for December 2017
10. Purchases journal for December 2017
11. Cash disbursements journal for December 2017
12. Payroll journal for December 2017

You can compare the reports printed using *QuickBooks* to the manual reports you prepared in the *SUA*.

Submit Reports and Answers to Assigned Questions

Submit the following to your course instructor either in hard copy or on the online grading page of the Armond Dalton Resources site (*consult your instructor*):

- All twelve reports just listed
- December 2017 bank reconciliation that you already printed
- Customer monthly statement for Eastern Wisconsin University that you already printed
- Trial balance after year-end adjustments that you already printed
- All questions and print requirements as listed on the Chapter 9 pages of the Student Problems & Cases book.

All procedures are now complete for this chapter. Now that you have completed Chapter 9, you should back up your data files for Waren Sports Supply following the instructions in the E-materials.

Check Figures

Waren Sports Supply
Trial Balance
As of December 31, 2017

	Dec 31, 17	
	Debit	**Credit**
10100 · Cash	83,209.77	
10200 · Accounts Receivable	48,149.00	
10210 · Accts. Rec. from Employees	250.00	
10300 · Allowance for Doubtful Accts.		3,250.81
10400 · Inventory	184,278.00	
10500 · Prepaid Expenses	3,500.00	
10600 · Marketable Securities	24,400.00	
12000 · Undeposited Funds	0.00	
10800 · Fixed Assets	330,506.40	
10900 · Accumulated Depreciation		80,334.50
11000 · Notes Receivable	7,000.00	
20100 · Accounts Payable		7,952.01
*Sales Tax Payable	0.00	
Payroll Liabilities	0.00	
20300 · Federal Income Tax Withheld		1,437.00
20400 · State Unemployment Taxes Pay.		110.43
20500 · Fed. Unemployment Taxes Pay.		18.66
20600 · FICA Taxes Payable		2,229.32
20700 · Federal Income Taxes Payable	0.00	
20800 · Dividends Payable		7,500.00
21100 · Unearned Revenue		3,300.00
21000 · Notes Payable		47,000.00
Opening Bal Equity	0.00	
26000 · Common Stock		225,000.00
29000 · Retained Earnings		90,264.99
29010 · Dividends Declared	7,500.00	
30100 · Sales		1,600,759.00
30200 · Sales Returns and Allowances	61,106.00	
30300 · Sales Discounts Taken	15,603.34	
30900 · G/L on Sale of Fixed Assets		250.00
31000 · G/L on Sale of Mkt. Securities		440.00
31100 · Interest/Dividend Income		1,100.00
31200 · Miscellaneous Revenue		825.00
30400 · Cost of Goods Sold	1,026,117.00	
30700 · Purchases Discounts Taken		16,554.48
30800 · Freight-in	24,506.44	
40100 · Rent Expense	57,600.00	
40200 · Advertising Expense	22,275.00	
40300 · Office Supplies Expense	5,664.91	
40500 · Wages and Salaries Expense	140,663.35	
40600 · Payroll Tax Expense	11,611.24	
40800 · Interest Expense	0.00	
40900 · Bad Debt Expense	4,400.00	
41000 · Other Operating Expense	29,985.75	
TOTAL	**2,088,326.20**	**2,088,326.20**

Waren Sports Supply
Trial Balance
As of December 31, 2017

	Dec 31, 17	
	Debit	Credit
10100 · Cash	83,209.77	
10200 · Accounts Receivable	48,149.00	
10210 · Accts. Rec. from Employees	250.00	
10300 · Allowance for Doubtful Accts.		1,898.91
10400 · Inventory	183,744.00	
10500 · Prepaid Expenses	3,500.00	
10600 · Marketable Securities	24,400.00	
12000 · Undeposited Funds	0.00	
10800 · Fixed Assets	330,506.40	
10900 · Accumulated Depreciation		115,358.14
11000 · Notes Receivable	7,000.00	
20100 · Accounts Payable		7,952.01
*Sales Tax Payable	0.00	
Payroll Liabilities	0.00	
20300 · Federal Income Tax Withheld		1,437.00
20400 · State Unemployment Taxes Pay.		110.43
20500 · Fed. Unemployment Taxes Pay.		18.66
20600 · FICA Taxes Payable		2,229.32
20700 · Federal Income Taxes Payable		55,822.32
20800 · Dividends Payable		7,500.00
20900 · Interest Payable		106.85
21100 · Unearned Revenue		3,300.00
21000 · Notes Payable		47,000.00
Opening Bal Equity	0.00	
26000 · Common Stock		225,000.00
29000 · Retained Earnings		90,264.99
29010 · Dividends Declared	7,500.00	
30100 · Sales		1,600,759.00
30200 · Sales Returns and Allowances	61,106.00	
30300 · Sales Discounts Taken	15,603.34	
30900 · G/L on Sale of Fixed Assets		250.00
31000 · G/L on Sale of Mkt. Securities		440.00
31100 · Interest/Dividend Income		1,100.00
31200 · Miscellaneous Revenue		825.00
30400 · Cost of Goods Sold	1,034,602.96	
30700 · Purchases Discounts Taken	0.00	
30800 · Freight-in	0.00	
40100 · Rent Expense	57,600.00	
40200 · Advertising Expense	22,275.00	
40300 · Office Supplies Expense	5,664.91	
40400 · Depreciation Expense	35,023.64	
40500 · Wages and Salaries Expense	140,663.35	
40600 · Payroll Tax Expense	11,611.24	
40700 · Federal Income Tax Expense	55,822.32	
40800 · Interest Expense	106.85	
40900 · Bad Debt Expense	3,048.10	
41000 · Other Operating Expense	29,985.75	
TOTAL	**2,161,372.63**	**2,161,372.63**

Option C — Introduction

Waren Sports Supply Dataset

As described in the E-materials, you should have downloaded an initial backup file for Waren Sports Supply to either your hard drive or to a USB flash drive after registering on the Armond Dalton Resources website (www.armonddaltonresources. com). If you correctly followed the instructions in the E-materials, then Waren Sports Supply has been successfully restored into *QuickBooks*.

If you haven't yet completed these tasks from the E-materials, complete the following step. ***If you have already completed these tasks from the E-materials, skip this step.***

> ▶ *Follow the instructions in the E-materials for downloading and restoring the Waren Sports Supply initial company backup file (Step 5 in the E-materials for personal computer users and Step 3 in the E-materials for lab users).*

Chapter Overview

In this chapter, you will record transactions for an existing company, Waren Sports Supply, for December 16–31, 2017. You will also complete other activities commonly done with accounting software and print several items to submit to your instructor.

An important difference in this assignment compared to previous ones is the lack of detailed instructions for using *QuickBooks* to record transactions and perform other activities. You are expected to use the Reference book for guidance if you need any.

If, at any time, you decide that you want to start the chapter over, you may do so by restoring the Waren Sports Supply dataset using the initial backup you downloaded from the Armond Dalton Resources website (armonddaltonresources.com). If you have not downloaded this initial backup yet, take the time to do that now, following the instructions in the E-materials. You may restore this backup at any time during your work on this chapter if you have made errors that are too difficult to correct or if you believe that you do not understand the material in this chapter.

In recording the transactions and performing the other activities for Waren Sports Supply, you will need several things:

1. **Information in this chapter.** The material instructs you what to record or do.
2. **Reference Book Guide (see inside front cover of Reference book).** Use this to locate the appropriate pages in the Reference book for recording transactions or doing other activities.
3. **Reference book.** Open the Reference book to the appropriate pages for the transaction you are recording or other activity you are doing and follow the instructions. In some cases, you will not have practiced using a Reference book section. You should not be concerned about this lack of practice.
4. **Student Problems & Cases book.** After you have completed the transactions and other requirements for Waren Sports Supply, you will answer the questions for Chapter 9 either online at the Armond Dalton Resources website or in hard copy for submission to your instructor along with the other requirements of this chapter. *Consult your instructor about which method to use.*
5. *QuickBooks* **software and company data sets.**
6. **A printer connected to your computer and/or the ability to print reports to PDF format.**

When you have completed this chapter, you will submit several reports to your instructor, along with the questions for Chapter 9 in the Student Problems & Cases book. Consult your instructor as to whether you should submit these items using the online grading page of the Armond Dalton Resources website or in hard copy.

Waren Sports Supply is a distributor of sporting goods to colleges and universities in the Midwest. Waren's accountant has recorded all transactions for the year ended 12/31/17 through December 15, 2017, using *QuickBooks*. For this assignment, you will do the following using *QuickBooks*:

- Perform maintenance for inventory sales prices.
- Record December 16–31 transactions and perform related maintenance.
- Perform December 2017 month-end procedures.
- Record 2017 year-end adjusting entries.
- Print financial statements and other reports.

Perform Maintenance for Inventory Sales Prices

The price list on page 9-54 reflects the current selling prices for Waren's thirty products. Waren purchases all products for resale from one vendor, Velocity Sporting Goods. Waren sells each inventory item at the same price to all customers. A new price list is prepared each time there is a change in an item's cost or selling price.

In this section, you will compare the selling price of each product to the amount included in *QuickBooks* and update the software for any differences.

► *If you are not already working in QuickBooks, open the program and open Waren Sports Supply.*

► *Perform maintenance for inventory sales prices (Sales Information section in the Edit Item Window) for each inventory item, following the guidance on page 69 of the Reference book. Be sure to click the OK button after each inventory item is changed before proceeding to the next inventory item.*

Alternative Method

► *Perform maintenance for inventory sales prices for multiple inventory items by first highlighting any of the Inventory Part items in the Item List. Then select the Add/Edit Multiple List Entries function from the Item drop-down list in the lower-left corner of the Item List. Edit the entries in the Sales Price column that need to be changed, using the directional buttons to move from item to item for editing. When you have finished editing your sales prices, make sure you select the Save Changes button in the lower-right corner of the Item List.*

PRICE LIST
As of December 15, 2017

Item No.	Description	Selling Price
BA-054	Premium aluminum bat	$209.00
BA-158	Baseballs–12 game balls	63.00
BA-199	Fielding glove	66.00
BA-281	60 lb. dry line marker	103.00
BA-445	Catcher's mask	69.00
BA-507	Baseball equipment bag	39.00
BA-667	Ball bucket with seat–set of 3	33.00
BA-694	Batting gloves–1 pair	34.00
BA-807	Pitching machine	245.00
BA-859	Set of bases	172.00
BB-008	Basketball	35.00
BB-019	Basketball pole pad	135.00
BB-113	Scoreboard and timer	400.00
BB-267	Goal and rim set	142.00
BB-358	Backboard	117.00
BB-399	Basketball net	16.00
BB-431	Whistle and lanyard–set of 6	38.00
BB-538	Basketball bag	40.00
BB-688	Portable inflation pump	107.00
BB-926	Trainer's first aid kit	42.00
FB-027	Shoulder pad set	127.00
FB-091	Hip, tail, arm pad set	58.00
FB-225	Football helmet	87.00
FB-344	Football	29.00
FB-513	Portable storage locker	210.00
FB-573	Kicking tees–set of 6	24.00
FB-650	Football post pad	147.00
FB-812	Collapsible cones–set of 8	39.00
FB-874	Sideline repair kit	124.00
FB-952	Portable hand warmer	37.00

Be sure that you also adjust the 'RET' (Inventory Return Items) sales prices to the correct amount as you just did for the regular inventory items. *QuickBooks* handles returns as separate inventory items for Waren, so if you have to change the sales price of an item, you will need to do it twice: once for the item and once for the 'RET' of that same item.

After you complete inventory sales price maintenance, the *QuickBooks* files will contain the correct default sales price for each inventory item. These sales prices are used for all sales of inventory between December 16 and December 31, 2017. You should recheck the amounts to make sure they are all correct before you proceed.

Record December 16-31 Transactions and Perform Related Maintenance

The transactions on pages 9-57 through 9-63 for December 16–31 should be dealt with in the order listed. Some transactions must be recorded whereas others require only maintenance.

The information you have already learned about *QuickBooks* in Chapters 2 through 8 is used throughout the rest of this chapter. You should use the *QuickBooks* Reference book and information from prior chapters to the extent you need it. Also, if you want to check whether information was recorded and correctly included, use the Home Page buttons (for Centers), the Icon Bar, or the Company, Customer, Vendor, or Employee lists menu to view transactions, as you did in earlier chapters.

All default general ledger distribution accounts are correct for Waren's December 16–31 transactions unless otherwise noted on pages 9-57 through 9-63.

The following is background information that you will need to record Waren's December 16–31 transactions.

Bank

Waren uses only one bank, First American Bank & Trust, for all deposits and checks, including payroll.

Credit Terms for Waren Sports Supply

Waren requires most of its customers to prepay for goods ordered. For these cash sales, the customer sends a check with its purchase order and Waren ships the merchandise. All trade discounts are already factored into the price list. Only a few favored customers with long-standing relationships with Waren and those who buy larger quantities are granted credit. These favored customers receive the following cash discount for early payment: 2/10, Net 30.

Waren receives a similar cash discount from its main inventory supplier, Velocity Sporting Goods (2/10, Net 30). No cash discount is offered by Chicago Office Supply, whose invoices are payable upon receipt. All discount terms have already been included as default information in *QuickBooks*.

Sales Tax for Waren Sports Supply

Waren Sports Supply makes only wholesale sales, which are exempt from state sales tax. Because Waren purchases all of its inventory items for resale, there is also no sales tax on its inventory purchases. Sales tax of 6% applies to office supplies and fixed asset purchases.

Inventory Method

Waren uses the perpetual inventory method. All purchases of inventory are debited directly to the inventory account. Cost of goods sold for each sale is calculated automatically by *QuickBooks*. Waren conducts a year-end physical inventory count and adjusts the perpetual inventory records as necessary. You will make those adjustments later.

WAREN SPORTS SUPPLY TRANSACTIONS FOR DECEMBER 16–31, 2017

▶ *Record each of the following transactions (No. 1 through 16) using QuickBooks.*

> ■ Hints are provided in boxed areas like this.

Use care in recording each transaction. **You should follow each step in the Quick Reference Table for each transaction.** It may take slightly more time, but it will almost certainly help you avoid serious errors. Find the appropriate Quick Reference Table for each transaction or other activity by using the Reference Book Guide located on the inside front cover of the Reference book.

Trans. No.	Dec.	
1	18	**Prepare a purchase order:** Ordered the following inventory on account from Velocity Sporting Goods (Vendor 252) using Purchase Order No. 328. The goods will be received at the warehouse at a later date. The purchase order total is $17,680.

Units	Item No.	Description
120	BB-019	Basketball pole pad
80	BB-538	Basketball bag
30	BB-688	Portable inflation pump
75	BB-926	Trainer's first aid kit

2	18	**Receive a miscellaneous cash receipt:** Borrowed and deposited $60,000 from First American Bank & Trust (Received from 264) by issuing a two-year note payable. Check No. 545 for $60,000 was received and deposited.

> ■ The credit portion of the transaction should be posted to A/C #21000 (Notes Payable).
> ■ Miscellaneous cash receipts are processed through the Banking Center Make Deposits window (see pages 42 and 43 of the Reference book).

Trans. No.	Dec.	
3	19	**Make a credit sale:** Received customer Purchase Order No. 37225 in the mail from University of Southern Iowa (Customer 409), approved their credit, prepared Invoice No. 731 totaling $18,647, and shipped the goods from the warehouse. All goods ordered were shipped as follows:

Units	Item No.	Description
30	BB-267	Goal and rim set
25	BB-358	Backboard
56	FB-027	Shoulder pad set
50	FB-225	Football helmet

Trans. No.	Dec.	
4	20	**Change an employee record (employee maintenance):** Increased employee salary and wage rates, effective December 16. For hourly employees, overtime is paid at 1.5 times the regular hourly rate. There were no changes in filing status or withholding allowances.

Employee	New Salary / Wage Rate
Ray Kramer	$3,700, semi-monthly (enter $88,800 Annual Rate)
Jim Adams	$22.50 per hour regular $33.75 per hour overtime
Nancy Ford	$18.20 per hour regular $27.30 per hour overtime

Trans. No.	Dec.	
5	22	**Receive goods on a purchase order:** Received merchandise from Velocity Sporting Goods (Vendor 252) as listed on Purchase Order No. 328, along with Invoice No. 34719. The payment terms on the invoice are 2/10, Net 30. All merchandise listed on the purchase order was delivered in good condition and in the quantities ordered, except that only 76 basketball pole pads (Item No. BB-019) were received. The total of the invoice is $13,544. The goods were placed immediately in the inventory warehouse.

Trans. No.	Dec.	
6	22	**Collect an account receivable and make a deposit:** Received Check No. 28564 for $1,622.88 from Branch College (Received from 408) for payment in full of sales Invoice No. 730, and deposited the check. The early payment discount taken by Branch College was $33.12.

- After recording the customer payment, record the deposit into the bank on the same day using the Make Deposits window (see pages 30 through 32 of the Reference book).

Trans. No.	Dec.	
7	22	**Write off an uncollectible account receivable:** Received legal notification from Benson, Rosenbrook, and Martinson, P.C., attorneys at law, that Stevenson College (Received from 411) is unable to pay any of its outstanding debts to its suppliers. The $2,900 balance remaining on Invoice No. 719 should therefore be written off as uncollectible.

- Recall that Waren uses the allowance method for recording bad debt expense at year-end, but uses the direct write-off method during the year.
- Follow the instructions on pages 38 through 40 of the Reference book carefully so that you use the direct write-off method correctly (debit to bad debt expense, credit to accounts receivable). The allowance for uncollectible accounts and bad debt expense will be adjusted during the year-end procedures later in the chapter.

Trans. No.	Dec.	
8	26	**Process a sales return or allowance (credit memo):** Eastern Wisconsin University (Customer 410) returned 5 basketball pole pads (RET BB-019) and 7 scoreboard and timer sets (RET BB-113) that were originally purchased on Invoice No. 729. Waren previously authorized EWU by phone to return the goods for credit against their account balance. EWU's Return Request No. R8034 was received with the goods. Sales return document CM 42 was issued for $3,475 and applied to Invoice No. 729.

- When you select Save in the Create Credit Memos/Refunds window, the Available Credit window will appear. Make sure you select Apply to Invoice, and select the correct invoice for the credit memo.

Trans. No.	Dec.	
9	26	**Collect an account receivable and make a deposit:** Received and deposited Check No. 49326 for $10,000 from Eastern Wisconsin University (Received from 410) in partial payment of the remaining amount (after sales return) on Invoice No. 729.

- Be sure that the entry in the A/R Account box says "10200 Accounts Receivable."
- After recording the customer payment, record the deposit into the bank on the same day using the Make Deposits window (see pages 30 through 32 of the Reference book).

10	26	**Purchase non-inventory items or services without a purchase order—payment made at time of purchase:** Received Freight Bill No. 26425 for $346.75 from Interstate Motor Freight (Pay to the Order of 255) and immediately issued Check No. 1152 for payment in full. The freight bill is for the merchandise received from Purchase Order No. 328. The payment terms on the freight bill are Net 30.
11	26	**Make a cash sale and make a deposit:** Received and deposited Check No. 65891 for $7,855 from Andrews College (CASHCUSTOMER) for a cash sale (Invoice No. C-30). The goods were shipped from the warehouse and the cash sale was processed and recorded. All goods ordered were shipped as follows:

Units	Item No.	Description
125	BB-008	Basketball
20	FB-091	Hip, tail, arm pad set
80	FB-344	Football

- After recording the cash receipt, record the deposit into the bank on the same day using the Make Deposits window (see pages 30 through 32 of the Reference book).

Trans. No.	Dec.	
12	26	**Receive goods on a purchase order:** Received office supplies from Chicago Office Supply (Vendor 253) as listed on Purchase Order No. 327, which is shown as an open purchase order in *QuickBooks*. Chicago Office Supply's vendor Invoice No. 2378 was received with the goods, totaling $602.61 including sales tax of $34.11. All supplies ordered on Purchase Order No. 327 were received in good condition and taken to the warehouse.

- Because Waren purchases similar items from Chicago Office Supply on a regular basis, each office supply item is kept track of in the inventory module as a non-inventory part. Office supplies expense is debited instead of inventory for these non-inventory parts.
- Remember to enter the sales tax portion of the invoice in the Expenses tab as a debit to the appropriate expense account (see step O in the Quick Reference Table).

Trans. No.	Dec.	
13	28	**Pay a vendor's outstanding invoice:** Issued Check No. 1153 for $13,273.12 to Velocity Sporting Goods (Pay to the Order of 252) for payment in full of Invoice No. 34719 for goods received December 23. The early payment discount taken by Waren was $270.88.

- Be sure to select account #30700–Purchases Discounts—in the Discount Account box of the Discount and Credits window.
- If account #10100–Cash—is not already selected in the Account box, select it before clicking the Pay Selected Bills button.

Trans. No.	Dec.	
14	29	**Pay employees:** Finished the payroll for the semi-monthly pay period December 16–31, 2017, and issued Check Nos. 1154 through 1156. Regular and overtime hours for hourly employees were as follows:

Employee	Regular Hours	Overtime Hours
Jim Adams	80	7.3
Nancy Ford	80	4.7
Ray Kramer	N/A	N/A

- You do not need to enter anything for Ray Kramer's hours because he is a salaried employee.
- If you receive a message about payroll liabilities, click OK.

Trans. No.	Dec.	
15	29	**Receive goods on a purchase order:** Received but did not pay for ten 11-inch tablets from Chicago Office Supply (Vendor 253) ordered on Waren's Purchase Order No. 325, which is shown as an open purchase order in *QuickBooks*. Also received vendor's Invoice No. 2423 from Chicago Office Supply, totaling $3,699.40 including sales tax of $209.40. The tablets were received in new and undamaged condition in the warehouse. After they were unpacked and tested, they were taken directly to the office.

- You must enter A/C #10800 (Fixed Assets) in the Account column when entering sales tax in the Expenses tab because the default will be office supplies expense and the purchase should be debited to fixed assets.

Trans. No.	Dec.	
16	29	**Purchase non-inventory items or services without a purchase order—payment made at time of purchase:** Received vendor Invoice No. 72654 for $1,470 from the University Athletic News (Pay to the Order of 254) for advertisements Waren ran during the Christmas season and immediately issued Check No. 1157 for payment in full.

You should, but are not required to, perform backup procedures for Waren Sports Supply before proceeding, to reduce the potential for having to reenter the transactions. See E-materials for backup procedures.

Perform December 2017 Month-end Procedures

Because many of Waren's month-end procedures are done automatically by *QuickBooks*, the only month-end procedures you will need to perform are:

- Prepare the December bank reconciliation.
- Print a customer monthly statement.

Check Figure for Your Cash Balance

Before starting the December bank reconciliation, be sure that your cash balance is correct by completing the following steps:

▶ *Click the Chart of Accounts icon and then double-click on the cash account (account #10100).*

Examine the ending balance in the cash account at 12/31/17. The balance should be $93,316.79. If the cash balance in your window differs significantly from this amount, return to the December 16–31 transactions to locate and correct any errors before starting the bank reconciliation. Because *QuickBooks* performs periodic automatic updates online, your cash balance may differ slightly (less than $15.00) due to changes in federal tax tables downloaded by the program. When your cash balance is correct, continue with the requirements that follow.

Bank Reconciliation Information, Process, and Printing

The following information is taken from the December bank statement and the November bank reconciliation, neither of which is included in these materials:

- The December 31, 2017, bank statement balance is $103,372.26.
- The following checks have not cleared the bank as of December 31: Check Nos. 1118, 1143, 1152, 1153, 1154, 1155, 1156, and 1157.
- The December 26 deposits from Eastern Wisconsin University and Andrews College have not cleared the bank as of December 31.
- A service charge of $25.50 is included on the December bank statement. *Note:* The bank service charge should be posted to A/C #41000 (Other Operating Expense).

▶ *Prepare the December bank reconciliation. The cutoff date for the bank reconciliation is December 31, 2017.*

▶ *When the reconciliation is correct, click the Print button to print the bank reconciliation either in hard copy or in PDF format if you are submitting your work online. The Select Type of Reconciliation Report window will appear with the option to select Summary, Detail, or print Both reports; select Both.*

> *Review your printed reconciliation for accuracy and acceptability.* You will submit these reports to your instructor along with year-end reports.

Print a Customer Monthly Statement

At the end of each month, Waren sends monthly statements to all customers with an outstanding balance. For this section, you are to print the December monthly statement for Eastern Wisconsin University.

> ▶ *Follow the instructions on pages 44 through 46 of the Reference book to print a December 2017 customer statement for Eastern Wisconsin University either in hard copy or in PDF format if you are submitting your work online.*
>
> ▶ *In the Statement Date field select December 31, 2017, and check the Statement Period From and To field and enter the month of December date range.*
>
> ▶ *Click the Preview button to preview the statement. Click Print to print a copy to submit to your instructor with other chapter requirements.*

Print a General Ledger Trial Balance for Check Figures Prior to Year-end Adjusting Entries

The trial balance on page 9-70 shows the correct balances in all general ledger accounts after the December month-end procedures are completed. You will use the Memorized tab of the Report Center to access some reports that have been specifically set up for this project.

> ▶ *Click Report Center → Memorized tab (if not already opened) → Trial Balance to print a 12/31/17 trial balance to the screen.*

Compare the amounts on your printed trial balance with those on page 9-70. If any amounts are different, return to the December 16–31 transactions and the month-end procedures you processed in *QuickBooks* and make the necessary corrections using the procedures you learned in earlier chapters. When all errors are corrected, print a corrected trial balance either in hard copy or in PDF format if you are submitting your work online.

When your balances agree with those on page 9-70, go to the following section where you will record year-end adjusting entries.

Record 2017 Year-end Adjusting Entries

The next step at the end of an accounting year before printing output is to record year-end adjusting entries. The following are the types of year-end adjustments required for Waren:

- Inventory adjustment to the physical count
- Depreciation expense
- Accrued interest payable
- Bad debt expense and allowance
- Cost of goods sold for freight and sales discounts taken
- Federal income taxes

Each of the year-end adjustments is explained in a section that follows. Perform the procedures in the order listed.

Adjust Perpetual Inventory Records

The physical count was taken on December 31. A comparison of the physical count and the perpetual records showed a difference for certain items. Management is concerned about these inventory differences, but knows that the physical count is accurate. Thus, the perpetual records must be adjusted as follows to agree with the physical count:

Item No.	Description	Quantity on Perpetual Records	Quantity per Physical Count
BB-019	Basketball pole pad	92 (77 regular, 15 returns)	95 (80 regular, 15 returns)
FB-027	Shoulder pad set	78 (28 regular, 50 returns)	66 (16 regular, 50 returns)
BA-158	Baseballs–12 balls	156 (all regular)	162 (all regular)

▶ *Record the inventory adjustments in QuickBooks following the guidance in the Reference book. Use Cost of Goods Sold as the Adjustment Account.*

After the inventory adjustments have been processed, record the remaining five year-end adjusting entries through the General Journal Entry window.

▶ *Use the information in the following five sections to record each of the remaining year-end adjusting entries by preparing a general journal entry in QuickBooks following the guidance in the Reference book pages 52 and 53.*

Depreciation Expense

Depreciation expense is calculated once at the end of each year. Depreciation is calculated using the straight-line method over the estimated useful lives of the assets (five or ten years for Waren's existing fixed assets). Waren's depreciation expense for Waren for 2017 totaled $35,023.64.

Accrued Interest Payable

Recall from Transaction No. 2 on page 9-57 that Waren has a $60,000 two-year note payable to First American Bank & Trust, dated December 18, 2017. The stated annual interest rate on the note is 5%. The terms of the note payable call for the following payments:

- $3,000 interest payments on 12/18/18 and 12/18/19
- $60,000 principal payment on 12/18/19

Interest accruals are calculated using a 365-day year with the day after the note was made counting as the first day. General ledger account numbers for the journal entry are: A/C #40800 (Interest Expense) and A/C #20900 (Interest Payable). Either enter the correct amount on the online grading page of the Armond Dalton Resources website or show your calculation on the Chapter 9 pages of the Student Problems & Cases book (consult your instructor).

Bad Debt Expense and Allowance

Bad debt expense is estimated once annually at the end of each year as 1/5 of one percent (0.002) of net sales and is recorded in the general journal as of December 31. As explained in Chapter 7, Waren uses the direct write-off method during the year and then the allowance method at year-end. General ledger account numbers for the journal entry are: A/C #40900 (Bad Debt Expense) and A/C #10300 (Allowance for Doubtful Account). In order to balance out the Allowance for Doubtful Account and Bad Debt Expense account due to the direct write-off to Bad Debt Expense, you must readjust the Bad Debt Expense account to equal the 1/5 of one percent (0.002) calculation of net sales. Either enter the correct amount on the online grading page of the Armond Dalton Resources website or show your calculation on the Chapter 9 pages of the Student Problems & Cases book (consult your instructor).

- Determine the amount of net sales by examining the 2017 income statement on the screen. For your convenience, the 2017 income statement has been included in the Memorized tab of the Report Center.
- *QuickBooks* requires you to add a customer to the Name box. Type "write off" in the Name box and press Enter. If "write off" doesn't exist as a customer, select Quick Add and select the Customer category.

Cost of Goods Sold

QuickBooks automatically debits cost of goods sold and credits inventory for the product cost for each sale. The inventory account is also automatically updated for inventory purchases and purchases returns. Waren treats purchase discounts taken and freight-in as a part of cost of goods sold, but records them in separate accounts during the accounting period. Therefore, these two accounts must be closed to A/C #30400 (Cost of Goods Sold): A/C #30700 (Purchases Discounts Taken) and A/C #30800 (Freight-In).

> ■ Before preparing the general journal entry, determine the balance in each account being closed to cost of goods sold. Determine the balance in the accounts to be closed by examining the income statement.

Federal Income Taxes

Assume that corporate income tax rate brackets for 2017 are: 15% of the first $50,000 of pre-tax income, plus 25% of the next $25,000, plus 34% of the next $25,000, plus 39% of the next $235,000. The remaining tax brackets are not listed here because they aren't applicable to Waren's level of pre-tax income. General ledger account numbers for the journal entry are: A/C #40700 (Federal Income Tax Expense) and A/C #20700 (Federal Income Taxes Payable). Either enter the correct amount on the online grading page of the Armond Dalton Resources website or show your calculation on the Chapter 9 pages of the Student Problems & Cases book (consult your instructor).

> ■ After all other adjusting entries are recorded, determine 2017 pre-tax income by examining the 2017 income statement from the Memorized tab in the Report Center.

Print a General Ledger Trial Balance for Check Figures After Year-end Adjusting Entries

The trial balance on page 9-71 shows the correct balances in all general ledger accounts after the year-end adjusting entries are recorded.

> ▶ *Open and use the general ledger trial balance for Waren to compare the amounts in your window to the correct balances. If there are differences, return to the year-end adjusting entries and make the necessary corrections. After you determine that your trial balance is correct, print the 12/31/17 trial balance. You will submit this report to your instructor along with year-end reports.*

When your balances agree with those on page 9-71, go to the following section where you will print financial statements and other reports. All entries have now been recorded.

Print Financial Statements and Other Reports

(all of these are to be submitted to your instructor either in hard copy or uploaded to the online grading page of Armond Dalton Resources)

▶ *Print the following reports.* Each of these reports has already been set up in the Memorized tab of the Report Center.

1. 12/31/17 balance sheet
2. 2017 income statement
3. General journal for December 2017
4. Accounts receivable aged trial balance as of 12/31/17
5. Accounts payable aged trial balance as of 12/31/17
6. Inventory valuation summary as of 12/31/17
7. Employee earnings register for December 2017
8. Sales journal for December 2017
9. Cash receipts journal for December 2017
10. Purchases journal for December 2017
11. Cash disbursements journal for December 2017
12. Payroll journal for December 2017

Submit Reports and Answers to Assigned Questions

Submit the following to your course instructor either in hard copy or on the online grading page of the Armond Dalton Resources site *(consult your instructor)*:

- All twelve reports just listed
- December 2017 bank reconciliation that you already printed
- Customer monthly statement for Eastern Wisconsin University that you already printed
- Trial balance after year-end adjustments that you already printed
- All questions and print requirements as listed on the Chapter 9 pages of the Student Problems & Cases book.

All procedures are now complete for this chapter. Now that you have completed Chapter 9, you should back up your data files for Waren Sports Supply following the instructions in the E-materials.

Check Figures

Waren Sports Supply
Trial Balance
As of December 31, 2017

	Dec 31, 17	
	Debit	Credit
10100 · Cash	93,291.29	
10200 · Accounts Receivable	48,149.00	
10300 · Allowance for Doubtful Accts.		3,250.81
10400 · Inventory	192,501.00	
10600 · Marketable Securities	24,000.00	
12000 · Undeposited Funds	0.00	
10800 · Fixed Assets	331,731.40	
10900 · Accumulated Depreciation		81,559.50
20100 · Accounts Payable		7,952.01
*Sales Tax Payable	0.00	
Payroll Liabilities	0.00	
20300 · Federal Income Tax Withheld		1,437.00
20400 · State Unemployment Taxes Pay.		110.43
20500 · Fed. Unemployment Taxes Pay.		18.66
20600 · FICA Taxes Payable		2,229.32
20700 · Federal Income Taxes Payable	0.00	
21000 · Notes Payable		60,000.00
Opening Bal Equity	0.00	
26000 · Common Stock		225,000.00
29000 · Retained Earnings		90,264.99
30100 · Sales		1,590,883.00
30200 · Sales Returns and Allowances	61,106.00	
30300 · Sales Discounts Taken	15,405.82	
31200 · Miscellaneous Revenue		825.00
30400 · Cost of Goods Sold	1,017,894.00	
30700 · Purchases Discounts Taken		16,554.48
30800 · Freight-in	24,506.44	
40100 · Rent Expense	57,600.00	
40200 · Advertising Expense	22,275.00	
40300 · Office Supplies Expense	5,664.91	
40500 · Wages and Salaries Expense	140,663.35	
40600 · Payroll Tax Expense	11,611.24	
40900 · Bad Debt Expense	4,400.00	
41000 · Other Operating Expense	29,285.75	
TOTAL	2,080,085.20	2,080,085.20

Waren Sports Supply
Trial Balance
As of December 31, 2017

	Dec 31, 17	
	Debit	Credit
10100 · Cash	93,291.29	
10200 · Accounts Receivable	48,149.00	
10300 · Allowance for Doubtful Accts.		1,879.55
10400 · Inventory	191,967.00	
10600 · Marketable Securities	24,000.00	
12000 · Undeposited Funds	0.00	
10800 · Fixed Assets	331,731.40	
10900 · Accumulated Depreciation		116,583.14
20100 · Accounts Payable		7,952.01
*Sales Tax Payable	0.00	
Payroll Liabilities	0.00	
20300 · Federal Income Tax Withheld		1,437.00
20400 · State Unemployment Taxes Pay.		110.43
20500 · Fed. Unemployment Taxes Pay.		18.66
20600 · FICA Taxes Payable		2,229.32
20700 · Federal Income Taxes Payable		54,837.13
20900 · Interest Payable		106.85
21000 · Notes Payable		60,000.00
Opening Bal Equity	0.00	
26000 · Common Stock		225,000.00
29000 · Retained Earnings		90,264.99
30100 · Sales		1,590,883.00
30200 · Sales Returns and Allowances	61,106.00	
30300 · Sales Discounts Taken	15,405.82	
31200 · Miscellaneous Revenue		825.00
30400 · Cost of Goods Sold	1,026,379.96	
30700 · Purchases Discounts Taken	0.00	
30800 · Freight-in	0.00	
40100 · Rent Expense	57,600.00	
40200 · Advertising Expense	22,275.00	
40300 · Office Supplies Expense	5,664.91	
40400 · Depreciation Expense	35,023.64	
40500 · Wages and Salaries Expense	140,663.35	
40600 · Payroll Tax Expense	11,611.24	
40700 · Federal Income Tax Expense	54,837.13	
40800 · Interest Expense	106.85	
40900 · Bad Debt Expense	3,028.74	
41000 · Other Operating Expense	29,285.75	
TOTAL	2,152,127.08	2,152,127.08

This page is intentionally blank.

10
Chapter

NEW COMPANY SETUP

Introduction

When a company uses *QuickBooks* for the first time, considerable maintenance is sometimes required before transactions can be entered. When the company has used an earlier version of *QuickBooks* the software imports the data easily, including maintenance information and transactions that were previously recorded. Similarly, *QuickBooks* imports maintenance and transactions from other accounting software such as *Sage 50 Accounting*. However, when a company is converting from a manual system to *QuickBooks*, such importing cannot be done. The initial setup in this case usually requires considerable time, especially for larger companies.

In this chapter you will set up a new company in *QuickBooks*. Super Office Furniture Plus is a furniture retailer that has been in operation for five years. The company has maintained a manual accounting system since its inception. The company's management has decided to change to a computerized system using *QuickBooks* as of January 1, 2019. To learn setup, yet keep the amount of setup time minimized, an extremely small company is used.

If, at any time, you decide that you want to start the chapter over, you may do so by restoring the Super Office Furniture Plus dataset following the instructions in the E-materials, which you should have already downloaded from the Armond Dalton Resources website (www.armonddaltonresources.com).

Create a New Company

The first step in new company setup is to create a new company in *QuickBooks*. In this case, you are converting from a manual system.

The authors have created a shell company for you to start with in order to save time for you to perform other setup procedures for vendors, customers, inventory, etc. In the real world, there are preliminary steps that you would go through to create the shell company, including selecting the industry, the type of entity, the fiscal year, and other information. For purposes of this project, Super Office Furniture Plus has been created as a retail shop, a regular corporation, and has a January fiscal year start. You will now enter other details for the company.

▶ *Open Super Office Furniture Plus. If you receive a password prompt, enter Admin11 and click OK. Note that you may need to enter this password each time you open the company in QuickBooks.*

▶ *Click Company → My Company to open the My Company window.*

▶ *Click the Edit button [🖉] in the top-right corner to open the Company Information window.*

The following illustrations show the completed windows for the first four tabs of the Company Information window. You will not use the fifth tab. Only some of the basic information has already been entered in these tabs when you first open them.

▶ *Enter any remaining items until your windows look like the ones that follow. Click OK when you are done entering information and then close the My Company window.*

▶ *If a window opens at any time asking if you want to add this file to your payroll subscription, click Skip.*

Company Information Window—Contact Information Tab

Company Information Window—Legal Information Tab

Company Information Window—Company Identification Tab

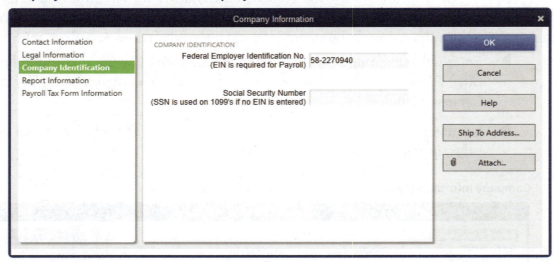

Company Information Window—Report Information Tab

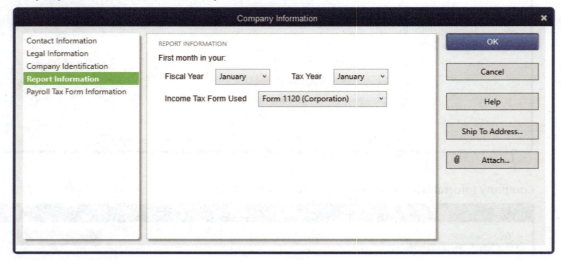

You will do most of the remainder of setup by using the maintenance windows that you have already used in previous chapters in the order listed below:

- Chart of accounts
- Vendors
- Customers
- Inventory
- Employees
- Entering beginning account balances

Observe that all of these except the last item are maintenance items you dealt with previously. You can access any of these at any time starting with the Home Page.

Change the Default Chart of Accounts

When setting up a new company in *QuickBooks,* a user can either enter the company's chart of accounts manually or choose a default chart of accounts from a list provided by *QuickBooks.*

Recall that Super Office Furniture Plus was originally set up as a retail store operating as a corporation. The default accounts have already been included as part of the general system setup. You will change the default chart of accounts to match the one used in the company's manual accounting system.

▶ *Click the Home Page icon for Super Office Furniture Plus (if it is not already open)* → *Chart of Accounts (in the Company section) to access the default chart of accounts for a retail store.* Observe that the chart of accounts includes account number, account name, and Type. The entries in the Type column are used to format financial statements. There are no balances included yet, but you will enter those later.

▶ *Use the scroll bar and the up and down scroll arrows to review the company's default chart of accounts.*

To match the chart of accounts Super Office Furniture Plus uses in its manual accounting system, you need to make some changes. The information for the chart of accounts is included in the table at the bottom of page 10-27. You can ignore the beginning balances for now because you will just be adding, editing, and deleting accounts right now. Later you will be entering beginning balances. Before you make any changes to the default chart of accounts, the following describes how to make each change.

▶ *Perform general ledger account maintenance following the instructions on pages 82 through 85 of the Reference book.* The boxes on the following page show the information to be added, deleted, and edited.

Note: Click No if you receive a message to "Set Up Bank Feed."

ACCOUNTS TO BE ADDED

Account Type	Number	Account Name and Description
Bank	10200	Savings Account
Accounts Receivable	11500	Allowance for Doubtful Accounts
Other Current Asset	12000	Inventory
Other Current Asset	14000	Prepaid Expenses
Accounts Payable	20000	Trade Accounts Payable
Other Current Liability	25000	Income Taxes Payable
Long-Term Liability	27000	Notes Payable

ACCOUNT NAMES TO BE REVISED

Account Type	Number	Existing Acct. Name and Desc.
Fixed Asset	15000	Furniture and Equipment
Income	46000	Merchandise Sales
Cost of Goods Sold	51800	Merchant Account Fees

Account Type	Number	Revised Acct. Name and Desc.
Fixed Asset	15000	Equipment
Income	46000	Sales
Cost of Goods Sold	51800	Cost of Goods Sold

ACCOUNTS TO BE DELETED

Number	Account Name
18700	Security Deposits Asset
30200	Dividends Paid
64300	Meals and Entertainment
68500	Uniforms
80000	Ask My Accountant

▶ *Click the Reports icon → Accountant & Taxes → Account Listing to access and then print the revised account listing.*

▶ *Click the Customize Report button, choose Account in the "Sort by" box, click the "Ascending order" radio button, then click OK.*

▶ *Compare the accounts, account numbers, and types of accounts to the ones in the table on the following page. If there are differences, change them now using the information you learned in this section.*

Super Office Furniture Plus
Account Listing
December 15, 2019

Account	Type	Balance Total	Description	Accnt. #
10100 · Checking Account	Bank	0.00		10100
10200 · Savings Account	Bank	0.00	Savings Account	10200
10300 · Undeposited Funds	Other Current Asset	0.00	Funds received, but not yet deposited to a bank acco...	10300
11000 · Accounts Receivable	Accounts Receivable	0.00	Unpaid or unapplied customer invoices and credits	11000
11500 · Allowance for Doubtful ...	Accounts Receivable	0.00	Allowance for Doubtful Accounts	11500
12000 · Inventory	Other Current Asset	0.00	Inventory	12000
14000 · Prepaid Expenses	Other Current Asset	0.00	Prepaid Expenses	14000
15000 · Equipment	Fixed Asset	0.00	Equipment	15000
17000 · Accumulated Depreciati...	Fixed Asset	0.00	Accumulated depreciation on equipment, buildings a...	17000
20000 · Trade Accounts Payable	Accounts Payable	0.00	Trade Accounts Payable	20000
23000 · Sales Tax Payable	Other Current Liability	0.00	Unpaid sales taxes. Amounts charged on sales, but ...	23000
24000 · Payroll Liabilities	Other Current Liability			24000
25000 · Income Taxes Payable	Other Current Liability	0.00	Income Taxes Payable	25000
27000 · Notes Payable	Long Term Liability	0.00	Notes Payable	27000
30000 · Opening Balance Equity	Equity	0.00	Opening balances during setup post to this account. ...	30000
30100 · Capital Stock	Equity	0.00	Value of corporate stock	30100
32000 · Retained Earnings	Equity		Undistributed earnings of the corporation	32000
46000 · Sales	Income		Sales	46000
48300 · Sales Discounts	Income		Discounts given to customers	48300
51800 · Cost of Goods Sold	Cost of Goods Sold		Cost of Goods Sold	51800
60000 · Advertising and Promoti...	Expense		Advertising, marketing, graphic design, and other pro...	60000
60200 · Automobile Expense	Expense		Fuel, oil, repairs, and other automobile maintenance ...	60200
60400 · Bank Service Charges	Expense		Bank account service fees, bad check charges and o...	60400
61700 · Computer and Internet ...	Expense		Computer supplies, off-the-shelf software, online fee...	61700
62400 · Depreciation Expense	Expense		Depreciation on equipment, buildings and improvem....	62400
63300 · Insurance Expense	Expense		Insurance expenses	63300
63400 · Interest Expense	Expense		Interest payments on business loans, credit card bal...	63400
63500 · Janitorial Expense	Expense		Janitorial expenses and cleaning supplies	63500
64900 · Office Supplies	Expense		Office supplies expense	64900
66000 · Payroll Expenses	Expense			66000
66700 · Professional Fees	Expense		Payments to accounting professionals and attorneys ...	66700
67100 · Rent Expense	Expense		Rent paid for company offices or other structures use...	67100
67200 · Repairs and Maintenance	Expense		Incidental repairs and maintenance of business asse...	67200
68100 · Telephone Expense	Expense		Telephone and long distance charges, faxing, and ot...	68100
68600 · Utilities	Expense		Water, electricity, garbage, and other basic utilities e...	68600
90100 · Purchase Orders	Non-Posting		Purchase orders specifying items ordered from vend...	90100
90200 · Estimates	Non-Posting			90200
90201 · Sales Orders	Non-Posting		Orders from customers to be filled, work to be perfor...	90201

Add Vendor Records and Account Balances

Entering vendor records for a new company can be a time-consuming process if there are a lot of vendors. You will set up only three vendors, including setting up balances as of 1/1/2019. Vendors are added through the Vendors Center.

▶ *Make sure you are in the Super Office Furniture Plus Home Page.*

You will enter information in the New Vendor window for each new vendor using the instructions on pages 64 through 67 of the Reference book. You have used this window several times in earlier chapters. Complete only those boxes for each vendor where information that follows is provided. You will enter the Beginning balances later when you enter historical data. Do not be concerned about making errors because it is easy to correct them at any time.

▶ *Use the information in the following box to complete the window for the first new vendor.* If no information for a box is provided, leave the box blank. If there is default information already included, do not change it unless new information is provided.

FIRST VENDOR

- **Vendor Name:**   OAK
- **Opening Balance:**   (leave blank)
- **As of:**   01/01/2019

Address Info tab
- **Company Name:**   The Oak Factory
- **Main Phone:**   810-555-1200
- **Main Email:**   oak@oakfactory.com
- **Website:**   www.oakfactory.com
- **Fax:**   810-555-5227
- **Address Details (Billed From and Shipped From):**   The Oak Factory
566 Chilson Ave.
Romeo, MI 48605

Payment Settings tab
- **Account Number:**   54678
- **Payment Terms:**   2% 10, Net 30

Account Settings tab
- **General Ledger Account:**   12000 (Inventory)

▶ *Save the new record after you have entered the information for the first vendor. Then enter the information for the second and third vendors using the following information.*

SECOND VENDOR

- Vendor Name: GAR
- Opening Balance: (leave blank)
- As of: 01/01/2019

Address Info tab
- Company Name: Garner Properties
- Main Phone: 616-555-5784
- Fax: 616-555-5910
- Address Details (Billed From and Shipped From):   Garner Properties
15700 W. Huron St.
Otsego, MI 49078

Payment Settings tab
- Account Number: 15445
- Payment Terms: 1% 10, Net 30

Account Settings tab
- General Ledger Account: 67100 (Rent Expense)

THIRD VENDOR

- Vendor Name: TAX
- Opening Balance: (leave blank)
- As of: 01/01/2019

Address Info tab
- Company Name: Michigan Department of Taxation
- Main Phone: 517-412-1672
- Website: www.mitax.org
- Fax: 517-412-1770
- Address Details (Billed From and Shipped From):   Michigan Department of Taxation
6478 Capital Ave.
Lansing, MI 48826

Payment Settings tab
- Account Number: 37211
- Payment Terms: Net 30

Account Settings tab
- General Ledger Account: 23000 (Sales Tax Payable)

► *After entering and saving the information for the last two vendors, Click OK.* You are now in the Vendors Center window, which should show all three vendors each with no balance total. Don't be concerned if you see other default-type vendors included. Those are typically included in shell companies.

► *Click GAR.* Observe that the maintenance window shows the information that you just entered.

► *Make sure the information for each of the three vendors is consistent with the information provided. If it isn't, make corrections before closing the Vendor Center and proceeding to the following section.*

Add Sales Tax Preferences

You will add customers and inventory later in the chapter. Before doing so, you need to add sales tax preferences to enable *QuickBooks* to automatically charge customers for sales tax. In this case all sales are taxable to customers at a 6% rate.

► *Click Edit → Preferences → Sales Tax in the Preferences window (fifth item from the bottom) → Company Preferences tab.*

► *Click the "Yes" radio button in the "Do you charge sales tax?" line if it is not already selected.* Your screen should now look like the following illustration.

▶ Click the "Add sales tax item" button to open the New Item window shown below.

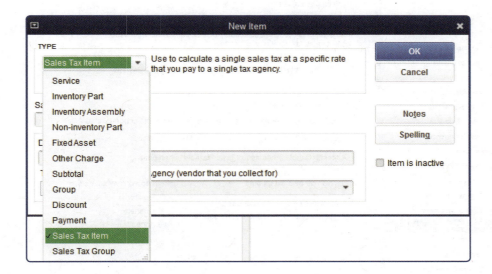

▶ Press Tab to accept Sales Tax Item, which is the default selection.

▶ Enter or select the following information and click OK to return to the Preferences window.

- **Sales Tax Name:** Michigan Sales Tax
- **Description:** Sales Tax
- **Tax Rate (%):** 6
- **Tax Agency**
  (vendor that you collect for): Select TAX (the vendor you entered earlier)

▶ Select Michigan Sales Tax in the "Your most common sales tax item" drop-down box. Click the Quarterly radio button on the bottom of the Preference window. Examine the window to make sure it includes the same information as the following window. Make changes if necessary.

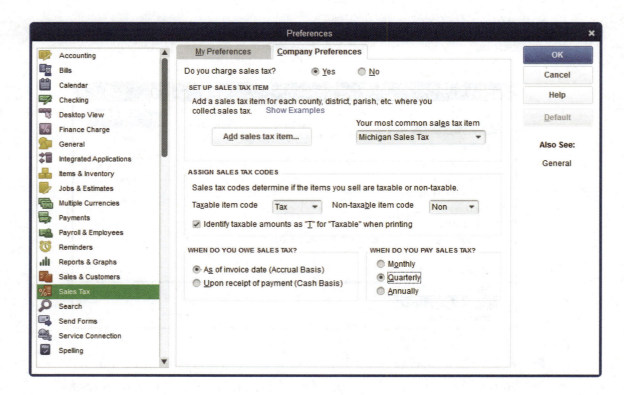

▶ *Click OK. If additional windows open relating to the sales tax you just created, click OK in each window.*

Add Customer Records and Account Balances

Recall that instructions for adding customer records are included on pages 60 through 63 of the Reference book.

▶ *Use the Reference book instructions and the information in the following boxes to add two customers for the company. Save each new record.*

> ### FIRST CUSTOMER
>
> - **Customer Name:**   LEVERETT
> - **Opening Balance:**   (leave blank)
> - **As of:**   01/01/2019
>
> #### Address Info tab
> - **Company Name:**   Leverett Enterprises
> - **Main Phone:**   616-555-7800
> - **Fax:**   616-555-8889
>
> *(continued on the following page)*

FIRST CUSTOMER
(continued from previous page)

Address Info tab *(continued)*
- **Address Details**
  (Invoice/Bill To and Ship To): Leverett Enterprises
  600 Columbia, Suite 602
  Holland, MI 49024

Payment Settings tab
- **Account No.:** 63699
- **Credit Limit:** $10,000
- **Payment Terms:** 2% 10, Net 30
- **Preferred Delivery Method:** Mail
- **Preferred Payment Method:** Check

Sales Tax Settings tab
- **Tax Code:** Tax
- **Tax Item:** Michigan Sales Tax default is correct (This is a result of the sales tax preference work you just completed.)

SECOND CUSTOMER

- **Customer Name:** THURGOOD
- **Opening Balance:** (leave blank)
- **As of:** 01/01/2019

Address Info tab
- **Company Name:** Thurgood Insurance Company
- **Main Phone:** 616-555-0002
- **Main Email:** daphne@thurgood.com
- **Fax:** 616-555-5231
- **Address Details**
  (Invoice/Bill To and Ship To): Thurgood Insurance Company
  8900 Wellston Blvd.
  Grand Haven, MI 49417

Payment Settings tab
- **Account No.:** 68319
- **Credit Limit:** $20,000
- **Payment Terms:** Net 30
- **Preferred Delivery Method:** Mail
- **Preferred Payment Method:** Check

Sales Tax Settings tab
- **Tax Code:** Tax
- **Tax Item:** Michigan Sales Tax default is correct (This is a result of the sales tax preference work you just completed.)

▶ *Close the Customer Center after adding and saving both customers.*

Add Inventory Items and Beginning Inventory Balances

The process of adding inventory items, their beginning quantities, and unit costs is similar to what you have already done for customers and vendors. Instructions for adding inventory items are included on pages 68 through 72 of the Reference book.

▶ *Add three inventory items using the Reference book instructions and the information that follows. As you did for previous maintenance items, leave a box blank and default information unchanged unless new or different information is provided. Click OK after you have entered all information for each inventory item. If you receive a message about a transaction being more than 30 days in the future, click Yes.*

FIRST INVENTORY ITEM

- **Type:** Inventory Part
- **Item Name/Number:** Basic Desk
- **Description on Purchases and Sales Transactions:** Basic Desk
- **Cost:** 236.00
- **COGS Account:** 51800 Cost of Goods Sold
  (you will need to change the default)
- **Preferred Vendor:** OAK (The Oak Factory)
- **Sales Price:** 350.00
- **Income Account:** 46000 Sales
- **Asset Account:** 12000 Inventory
  (you will need to change the default)
- **Reorder Point:** 30
- **On Hand:** 172
- **As of:** 01/01/2019

SECOND INVENTORY ITEM

- **Type:** Inventory Part
- **Item Name/Number:** Desk Chair
- **Description on Purchases and Sales Transactions:** Desk Chair
- **Cost:** 122.00

(continued on the following page)

SECOND INVENTORY ITEM

(continued from previous page)

- COGS Account: 51800 Cost of Goods Sold
                         (change default if necessary)
- Preferred Vendor: OAK (The Oak Factory)
- Sales Price: 175.00
- Income Account: 46000 Sales
- Asset Account: 12000 Inventory
                         (change default if necessary)
- Reorder Point: 50
- On Hand: 225
- As of: 01/01/2019

THIRD INVENTORY ITEM

- Type: Inventory Part
- Item Name/Number: Desk Lamp
- Description on Purchases
  and Sales Transactions: Desk Lamp
- Cost: 28.00
- COGS Account: 51800 Cost of Goods Sold
                                                 (change default if necessary)
- Preferred Vendor: OAK (The Oak Factory)
- Sales Price: 42.00
- Income Account: 46000 Sales
- Asset Account: 12000 Inventory
                                                 (change default if necessary)
- Reorder Point: 75
- On Hand: 312
- As of: 01/01/2019

▶ *Close the New Item window after entering the last item to return to the Home Page.*

Add Employees

Adding employees is similar to what you have already done for chart of accounts, vendors, customers, and inventory except there are more windows to complete. You will add two employees through the New Employee maintenance window.

Before adding the new employees, you first need to make sure that the payroll preference is set to full payroll.

▶ *Click Edit → Preferences → Payroll & Employees → Company Preferences tab. If the Full radio button is not selected, select it and click OK to close the window.*

Instructions for adding employees are included on pages 74 through 80 of the Reference book.

▶ *Use the Reference book instructions and the information in the boxes that follow to complete the New Employee window for the two employees. There are multiple steps for payroll, so extra guidance is provided. As done with previous maintenance items, leave a box blank, a tab unopened, and default information unchanged unless new or different information is provided.*

▶ *Enter the following information in the Personal, Address & Contact, Additional Info, and Employment Info tabs.*

FIRST EMPLOYEE–HOURLY

Personal tab
- First Name: Regina
- Last Name: Rexrode
- Social Security Number: 549-56-3982
- Gender: Female
- Date of Birth: 05/16/1974

Address & Contact tab
- Address: 462 Sibley Lane
- City: Holland
- State: MI
- Zip: 49024
- Main Phone: 616-555-8888

Additional Info tab
- Account No.:   REX

Employment Info tab
- Hire Date: 04/03/2013

▶ *Change tabs to the Payroll Info tab and enter the pay frequency and pay rates information listed in the box that follows.*

> **Payroll Info tab**
> - Pay Frequency: Biweekly
> - Earnings:
>   - Item Name (drop-down arrow):   Regular Pay
>   - Hourly/Annual Rate (regular):   14.00
>   - Item Name (drop-down arrow):   Overtime
>   - Hourly/Annual Rate (overtime):   21.00

▶ *Click the Taxes button to enter the "Taxes for Regina Rexrode" window. Enter the information below:*

> **Federal tab**
> - Filing Status: Married
> - Allowances: 3
>
> **State tab**
> - State Worked: MI
> - State Subject to Withholding: MI
> - Allowances: 3*
>
> * You cannot enter this last line of information until the Michigan state withholding tax is set up.

After entering the last "MI" in the window, the Payroll Item Not Found window opens. You will enter information in several windows for Michigan withholding and unemployment taxes. This is not difficult because *QuickBooks* provides guidance for both.

▶ *Click Set Up to open the Add new payroll item (MI–State Withholding Tax) window. The name in the box is satisfactory → click Next → select TAX from the drop-down arrow list in the "Enter name of agency to which liability is paid" box. Skip the next box. If the liability account is 24000-Payroll Liabilities, click Next → keep accepting information using Next, Finish, or OK. If the Schedule Payments window opens, click OK. Enter the state allowances for this employee and click OK in the State tab. This should open the Payroll Item Not Found window for Michigan Unemployment. Click the Set Up button.*

▶ *Follow the instructions in sequence to set up the Michigan unemployment tax. The agency is TAX again. Accept all default general ledger accounts and click Next. The rate should be 2.7% for all time periods. Click Next in the Company tax rates for 2019 window and then click Finish in the Taxable compensation window. When the Schedule Payments window opens again, click OK. When you are asked about city taxes and the Michigan Obligation Assessment, click*

OK and then indicate there are no such taxes by deleting each item when the appropriate window opens. Click OK and click No if asked whether the employee is subject to the taxes you just deleted. Also do not set up vacation pay if you are asked if you want vacation/sick/personal pay set up.

▶ Before proceeding to set up the second employee, click Regina Rexrode in the Employees tab and make sure you correctly entered the information for each tab.

▶ Set up the payroll information for the second employee following the same procedures as you did for Regina Rexrode, except that you will not have to set up the MI Unemployment tax again.

SECOND EMPLOYEE–SALARIED

Personal tab
- First Name: Tina
- Last Name: Hanson
- Social Security Number: 642-55-1342
- Gender: Female
- Date of Birth: 02/18/1978

Address & Contact tab
- Address: 185 Post Ave.
- City: Zeeland
- State: MI
- Zip: 49464
- Main Phone: 616-555-0052

Additional Info tab
- Account No.:   HAN

Employment Info tab
- Hire Date: 01/19/2009

Payroll Info tab
- Pay Frequency: Biweekly
- Earnings:
  - Item Name (drop-down arrow):   Salary
  - Hourly/Annual Rate (regular):   65,000.00

Taxes button
- Federal and State of
  Michigan allowances: 1
- Federal Filing Status: Single

▶ Click OK to close the New Employee window. Open the Employees window again for Tina Hanson and make sure the information is correct in all windows.

Enter Chart of Accounts (General Ledger) Beginning Balances

Entering Beginning Balances In Accounts

When a manual system is converted to *QuickBooks,* it is almost always done at the beginning of a fiscal year, which means that only balance sheet accounts are carried forward from the manual system into the beginning trial balance. There is a distinction between the two types of balance sheet accounts:

- Accounts with subsidiary records, where the dates of the transactions that affected the account balance are important, must be set up through the subsidiary accounting modules. The three accounts for Super Office Furniture Plus with subsidiary accounts are accounts receivable, accounts payable, and inventory. The beginning balance for the inventory account was set up earlier. Accounts receivable is set up next using transactions prior to the beginning balance sheet date.
- Accounts with no subsidiary balances, which are all other balance sheet accounts with beginning balances, are set up with a general journal entry.

Setting Up Accounts Receivable

Only Thurgood Insurance Company has a beginning balance, which is $14,840. It resulted from a sale in December of 2018, but you will enter it on 12/31/2018. The invoice number is 1206.

▶ *Click the Home Page icon → Create Invoices icon to open the Create Invoices window.*

▶ *Type Thurgood Insurance Company (don't select THURGOOD) in the Customer: Job window. Click [Tab] to open the following window.*

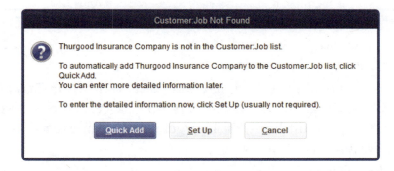

▶ *Click the Quick Add button.*

▶ *Add the Date (use 12/31/2018) and Invoice # (1206) in the appropriate boxes at the top of the window → click [Tab].* **Note:** *If accounts receivable—a/c #11000—is not selected in the Account box, do that now.*

▶ *Type Historical Transaction in the Item Code box → click [Tab] to open the window shown below.*

▶ *Click Yes to include Historical Transaction in the Item List and to open the window shown below.*

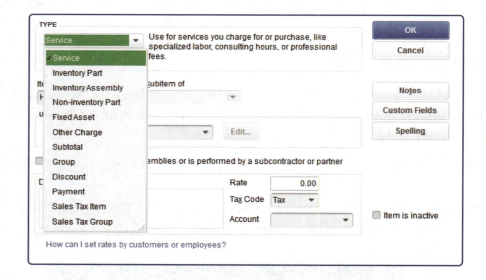

▶ *Enter Service for Type and 46000 in the Account box → click OK to return to the Create Invoices window.*

▶ *Type 14840 in the Amount box and select "Non" (Non-Taxable Sales) in the Tax box. Click Save & Close. Do not be concerned that the transaction is old or the credit limit for Thurgood has been exceeded. Click Yes if you receive any such prompts.*

▶ *If the Information Changed window opens, click No.*

Setting Up Accounts Payable

▶ *Set up accounts payable the same way as you just did for accounts receivable (use QuickAdd again), but use the Enter Bills window. The Oak Factory is the only vendor with a beginning balance. Invoice No. R2263 (Ref. No. box) for $2,000.00 is dated 12/29/2018. Use 12/31/2018 as the date, however, as you did for the opening account receivable balance. Select 51800 as the Account and enter $2,000.00 in the Amount box within the Expenses tab. Click Save & Close.*

Setting Up the Remaining Balance Sheet Account Balances

▶ *Click the Reports icon → Accountant & Taxes → Trial Balance. Change the To and From dates to 01/01/2019 → Refresh to open the trial balance. It should include the following trial balance.*

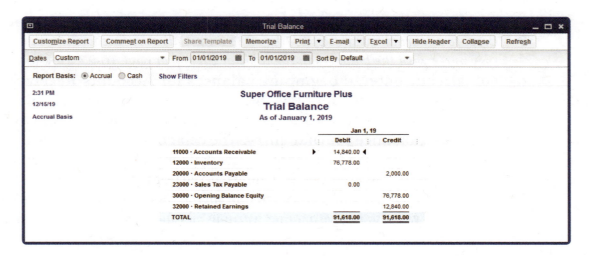

QuickBooks automatically records the effects of the inventory beginning balance to the account called Opening Balance Equity. Notice that the two other opening balance entries—for accounts payable and accounts receivable—were posted to Retained Earnings. You will reclassify things appropriately when you enter all other beginning balances shortly. The account balances that you want as of 01/01/2019 are shown at the top of page 10-27.

▶ *Use a journal entry to reclassify the amount in the Opening Balance Equity account to Retained Earnings ($76,778.00) as of 01/01/2019. Click OK when you receive a warning about posting to Retained Earnings.* Normally it is inappropriate to record transactions directly to retained earnings, but in this case it is acceptable.

▶ *Prepare, but do not print, a trial balance at 01/01/2019. The trial balance should now include four accounts with balances as follows:*

Accounts Receivable	14,840	Debit
Inventory	76,778	Debit
Accounts Payable	2,000	Credit
Retained Earnings	89,618	Credit

▶ *If the balances are different from those, make corrections until they are the same. Otherwise, proceed with the next step.*

The table at the top of page 10-27 shows what the beginning trial balance should eventually look like for the company at 01/01/19. Your next task will be to prepare a general journal entry to record the beginning balances for all accounts other than accounts receivable, accounts payable, and inventory. Recall that these three accounts already have beginning balances as a result of transactions you recorded earlier.

▶ *Prepare a journal entry as of 01/01/19 to adjust all other accounts so that they agree with the balances in the table at the top of page 10-27.* **Because you already entered beginning balances for Accounts Receivable, Inventory, and Accounts Payable through earlier work in this chapter, you do not need to include these three accounts in your beginning balance journal entry.** *The debit or credit to retained earnings will be used to balance the general journal entry. It is again appropriate to adjust retained earnings as part of the journal entry. If you receive a message about not posting to the payroll liabilities account, click Yes to continue. If you also receive another warning message about posting to Retained Earnings, click OK.*

Print a Beginning Trial Balance

▶ *Prepare a trial balance at 01/01/2019 and view it on your screen.* The trial balance should now include all accounts with a balance that are the same as the ones at the top of page 10-27. *If the balances are different from those, make corrections until they are the same.*

▶ *Print the trial balance.*

This completes Setup. Go to the following section where you will record a sample set of transactions for Super Office Furniture Plus.

Process Transactions and Do Other Activities for the New Company

In this section, you will process transactions and do other activities for Super Office Furniture Plus for the first week of January 2019 using *QuickBooks*.

In processing these transactions or doing other activities, you should first find the appropriate reference pages and then use the Reference book to guide you.

Transaction #1

▶ *Record the following transaction using QuickBooks.*

On January 2, 2019, the company sold merchandise on account to Leverett Enterprises. Other details of the credit sale follow.

- **Customer Job:** LEVERETT
- **Date:** January 2, 2019
- **Invoice No.:** 1207
- **Customer PO No.:** 89458
- **Items sold:**

Qty.	Item Code	Unit Price
10	Desk Chair	$175.00
15	Desk Lamp	42.00

- **General Ledger Account Information:** All default account numbers are correct
- **Invoice Total (check figure):** $2,522.80, including tax

Transaction #2

▶ *Record the following transaction using QuickBooks.*

On January 3, 2019, the company received a check from Thurgood Insurance Company (the one you set up via QuickAdd, not the THURGOOD customer) in full payment of an accounts receivable for Invoice No. 1206. Other details of the collection follow.

- **Amount:** $14,840.00 (no discount; terms were Net 30)
- **Customer's Check No.:** 10053
- **General Ledger Account Information:** All default accounts numbers are correct

Notes: (1) Do not forget to make a bank deposit for this receipt.
(2) Make sure to select the Quick Add version of Thurgood, not the regular one.

Transaction #3

▶ *Process the following purchase order using QuickBooks.*

On January 5, 2019, the company ordered inventory from The Oak Factory. The inventory was not received on January 5. Other details of the purchase order follow.

- Vendor: OAK
- Date: January 5, 2019
- PO No.:   1407
- Inventory items ordered:

Item	Qty.	Unit Price
Basic Desk	40	$236.00
Desk Chair	30	122.00

- Purchase order total
  (check figure): $13,100.00

Transaction #4

▶ *Record the following receipt of goods in QuickBooks.*

On January 9, 2019, Super Office Furniture Plus received the inventory merchandise from The Oak Factory for Purchase Order No. 1407 along with Invoice No. 56781. Other details of the purchase follow.

- Select PO (PO No.):   1407
- Vendor: OAK
- Date goods were received: January 9, 2019
- Ref. No. (Invoice No.):   56781
- Goods shipped/invoiced: All goods ordered were received no trade discount, freight, miscellaneous charges, or taxes
- Payment terms: Net 30 (change default but do not save the setting for future bills when prompted)
- General ledger
  account information: All default information is correct
- Invoice total (check figure): $13,100.00

Transaction #5

▶ *Record the following cash disbursement transaction.*

On January 11, 2019, the company issued a check to The Oak Factory in full payment of Invoice No. 56781. Other details of the cash disbursement follow.

- **Vendor:** OAK
- **Check No.:** 124307 (click the Assign check number radio button in the Pay Bills window if it is not already selected)
- **Check amount:** $13,100.00 (no discount; terms were Net 30)
- **Invoice paid:** 56781
- **Account:** 10100 Checking Account
  (Edit if necessary. Account 10100 must be selected in the Account box)
- **General ledger Account Information:** All default information is correct

Transaction #6

▶ *Record the biweekly payroll for Super Office Furniture Plus's two employees in QuickBooks. If you receive a Workers Comp Not Set Up window, click the Turn Off Workers Comp button.* Information necessary to process the biweekly payroll follows.

- **Pay End Date:** Jan 15, 2019
- **Check Date:** Jan 15, 2019
- **First Check #:** 124308

Hourly employee's info
- **Employee ID:** REX
- **Employee Name:** Regina Rexrode
- **Regular hours:** 88
- **Overtime hours:** 13

Other Activity

▶ *Print a hard copy of the general ledger trial for Super Office Furniture Plus as of January 31, 2019.* Notice that the January 2019 transactions you recorded are reflected in this trial balance.

Compare the balances on your printed trial balance with those in the table at the bottom of page 10-27. If there are any differences, make necessary corrections as you have learned in previous chapters. When your balances agree with those at the bottom of page 10-27, print a corrected trial balance and hand it in to your instructor.

Chapter Summary

After completing Chapter 10, you have now learned:

- ✔ how to set up a new company in the software.
- ✔ how to add records and account balances for vendors, customers, inventory, and the chart of accounts.
- ✔ how to add employee records.

You should now save your work by making a periodic backup of Super Office Furniture Plus using the instructions in the E-materials you downloaded from the Armond Dalton Resources website. Then proceed to the Chapter 10 homework assigned by your instructor either online or in the Student Problems & Cases book (consult your instructor).

Check Figures

Super Office Furniture Plus
Beginning Trial Balance as of 01/01/2019

Number	Account Name	Balance Dr.	(Cr.)
10100	Checking Account	$ 16,875.42	
10200	Savings Account	7,852.50	
11000	Accounts Receivable	14,840.00	
12000	Inventory	76,778.00	
14000	Prepaid Expenses	6,200.00	
15000	Equipment	36,980.00	
17000	Accumulated Depreciation		$ 14,409.48
20000	Accounts Payable		2,000.00
24000	Payroll Liabilities		224.98
25000	Income Taxes Payable		1,248.76
27000	Notes Payable		33,000.00
30100	Capital Stock		50,000.00
32000	Retained Earnings		58,642.70
		$159,525.92	$159,525.92

Super Office Furniture Plus
Trial Balance as of 01/31/2019

Number	Account Name	Balance Dr.	(Cr.)
10100	Checking Account	$ 15,531.86	
10200	Savings Account	7,852.50	
11000	Accounts Receivable	2,522.80	
12000	Inventory	88,238.00	
14000	Prepaid Expenses	6,200.00	
15000	Equipment	36,980.00	
17000	Accumulated Depreciation		$ 14,409.48
20000	Accounts Payable		2,000.00
23000	Sales Tax Payable		142.80
24000	Payroll Liabilities		1,584.97
25000	Income Taxes Payable		1,248.76
27000	Notes Payable		33,000.00
30100	Capital Stock		50,000.00
32000	Retained Earnings		58,642.70
36000	Sales		2,380.00
51800	Cost of Goods Sold	1,640.00	
66000	Payroll Expenses	4,443.55	
		$163,408.71	$163,408.71

This page is intentionally blank.

Notes

Notes

Notes

Notes

Notes

Notes